Synthetic Membranes
and
Membrane Separation Processes

Takeshi Matsuura, M.S., Ph.D.

British Gas, Consumers Gas,
NSERC Industrial Research Chair
in Membrane Science and Technology,
and
Professor
Industrial Membrane Research Institute
Department of Chemical Engineering
University of Ottawa
Ontario, Canada

CRC Press
Taylor & Francis Group
Boca Raton London New York

CRC Press is an imprint of the
Taylor & Francis Group, an **informa** business

CRC Press
Taylor & Francis Group
6000 Broken Sound Parkway NW; Suite 300
Boca Raton, FL 33487-2742

© 1993 by Taylor & Francis Group, LLC
CRC Press is an imprint of Taylor & Francis Group, an Informa business

First issued in paperback 2019

No claim to original U.S.Government works

ISBN 13: 978-0-367-44961-2 (pbk)
ISBN 13: 978-0-8493-4202-8 (hbk)

Visit the Taylor & Francis Web site at
http://www.taylorandfrancis.com

and the CRC Press Web site at
http://www.crcpress.com

Library of Congress catalog number: 93-15515

Library of Congress Cataloging-in-Publication Data

Catalog record is available from the Library of Congress

Preface

The area of applications of membrane separation processes has been growing steadily since cellulose acetate membranes were developed by Loeb and Sourirajan more than 30 years ago. When I joined the laboratory for membrane research at the National Research Council of Canada in 1969, reverse osmosis and ultrafiltration were the only major membrane applications for wastewater treatment and drinking water production. The emergence of industrial separation processes based on membrane gas separation, pervaporation, and vapor permeation, in the 1970s and 1980s, has widened the scope of membrane applications considerably. It is expected that the membrane processes will penetrate into all aspects of separation processes involved in chemical, petroleum, petrochemical, pharmaceutical, medical, food processing, and environmental industries, during the last decade of this century.

I have been giving an elective course on "membrane separation processes" since 1986, at the Department of Chemical Engineering of the University of Ottawa, to the fourth-year and graduate students. This book was written in response to the request from the students for a course textbook. The content of the lectures has been changing continuously since the course was started, and it will be changing in the future as well. This book should therefore be regarded as the 1992 version of my lectures.

Many books have already been written on the synthetic membranes and membrane separation processes. They include:

· *Reverse Osmosis*, S. Sourirajan, Academic Press, New York, 1970.
· *Synthetic Polymeric Membranes*, R.E. Kesting, McGraw-Hill, New York, 1971.
· *Membranes in Separations, Techniques of Chemistry*, Vol.VII, S.-T. Hwang and K. Kammermeyer, Wiley-Interscience, New York, 1975.
· *Hyperfiltration and Ultrafiltration in Plate-and-Frame Systems*, R.F. Madsen, Elsevier, Amsterdam, 1977.
· *Synthetic Polymeric Membranes*, R.E. Kesting, John Wiley & Sons, New York, 1985.
· *Reverse Osmosis/Ultrafiltration Process Principles*, S. Sourirajan and T. Matsuura, National Research Council of Canada, Ottawa, 1985.

- *Ultrafiltration Handbook*, M. Cheryan, Technomic, Lancaster, 1986.

- *Membrane Systems: Analysis and Design, Applications in Biotechnology, Biomedicine and Polymer Science*, W.R. Vieth, Hanser Publishers, New York, 1988.

- *Membrane Processes*, R. Rautenbach and R. Albrecht, John Wiley & Sons, New York, 1989.

- *Diffusion In and Through Polymers*, W.R. Vieth, Carl Hanser, Munich, 1991.

- *Basic Principles of Membrane Technology*, M. Mulder, Kluwer Academic Publishers, Dordrecht, the Netherlands, 1991.

- *Transport Mechanisms in Membrane Separation Processes*, J.G.A. Bitter, Plenum Press, New York, 1991.

Furthermore, there are about 30 edited books and conference proceedings.

Reverse Osmosis/Ultrafiltration Process Principles, coauthored with S. Sourirajan in 1985, and *Fundamentals of Synthetic Membranes*, authored by myself in Japanese in 1980, became the basis of this book. A large number of pages were given to the solution-diffusion model, to provide the students with a more balanced view of the membrane transport in its historical development. Equal importance was also given to the pore model, since the effect of the membrane morphology should not be overlooked when a transport model is established. Interaction forces working between permeant molecules and molecules that constitute the membrane material are another component that affects the membrane performance. Readers who are interested in a more thorough discussion of the pore model should refer to *Reverse Osmosis/Ultrafiltration Process Principles*. Other topics discussed in this book are: membrane material, membrane preparation, membrane morphology, module calculation, concentration polarization, and the membrane reactor. An outline of the applications of membrane separation processes, including reverse osmosis, ultrafiltration, microfiltration, and membrane gas and vapor separation and pervaporation, was also included. However, since the number of pages allowed for the book was limited, detailed descriptions for each application had to be abandoned.

In the theoretical part of the book an attempt was made to explain the fundamental concept underlying the membrane separation process, as thoroughly as possible. Discussions were made quantitatively, since the author believes strongly that progress in any engineering discipline can be achieved only when the subject is understood in quantitative terms.

It should also be noted that this book was not intended to be used as a handbook, but was written to help students develop their own theoretical approaches to the subject. Some assignments given to the students during the course have been selected and included as examples in the end of each chapter. The author wishes that those examples will help readers have a clearer understanding of the discussions.

Recent progress in the area of inorganic membranes, membrane characterization and membrane surface modifications have not been discussed in this book, although the author realizes that they are important topics. Membrane separation processes other than those with pressure as a "driving force" were not included, since the author's experience in the those fields was considered insufficient. Books are available that would be more appropriate to the readers who are interested in the latter membrane processes.

The author is deeply indebted to many of his colleagues and students for the completion of this book. He is especially thankful to Professor W. Vieth of Rutgers University for his advice and encouragement in writing this book. Drs. B.D. Taylor and C.E. Capes of the Institute of the Environmental Chemistry allowed me to use facilities at the National Research Council of Canada. Dr. Z. Duvnjak of the Department of Chemical Engineering of the University of Ottawa released me from department teaching and administrative duties until I could finish writing the manuscript. The author is also thankful to Mr. C. Tam and Mr. R. Tyagi for their helpful discussions that contributed to the improvement of the manuscript, to Ms. S. Page for her corrections of his English, and to Mr. S. Deng and Mr. Y. Fang for drawing figures. Mr. Y. Fang also assisted me in editing the computer files of the manuscript. Furthermore, it should be acknowledged that Mr. A. Tabemohhadi generously agreed with the inclusion of his computer program in this book.

Takeshi Matsuura

Dedicated to Dr. S. Sourirajan, who introduced me to the fascinating field of membrane research, and to Hizung Kim Matsuura, who has provided me with her unceasing support

Table of Contents

viii

1

Membranes for Separation Processes

Membranes for industrial separation processes can be classified into the following groups according to the driving force that causes the flow of the permeant through the membrane.

1. Pressure difference across the membrane is the driving force.

    · Reverse osmosis

    · Ultrafiltration

    · Microfiltration

    · Membrane gas and vapor separation

    · Pervaporation

2. Temperature difference across the membrane is the driving force.

    · Membrane distillation

3. Concentration difference across the membrane is the driving force.

    · Dialysis

    · Membrane extraction

4. Electric potential difference across the membrane is the driving force.

    · Electrodialysis

Reverse osmosis is a process to separate solute and solvent components in the solution. Although the solvent is usually water, it is not necessarily restricted to water. The pore radius of the membrane is less than 1 nanometer (nm) (1 nm = 10^{-9} m). While solvent water molecules, whose radius is about one tenth of 1 nm, can pass through the membrane freely, electrolyte solutes, such as sodium chloride and organic solutes that contain more than one hydrophilic functional group in the molecule (sucrose, for example), cannot pass through the membrane. These solutes are either rejected from the membrane

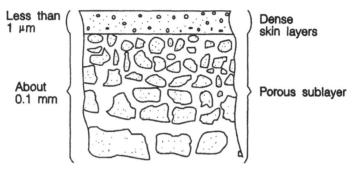

Figure 1.1. Schematic representation of the cross-section of an asymmetric membrane.

surface, or they are more strongly attracted to the solvent water phase than to the membrane surface. The preferential sorption of water molecules at the solvent-membrane interface, which is caused by the interaction force working between the membrane-solvent-solute, is therefore responsible for the separation. Polymeric materials such as cellulose acetate and aromatic polyamide are typically used for the preparation of reverse osmosis membranes. Figure 1.1 illustrates schematically the cross-sectional structure of such membranes. A thin, dense layer responsible for the separation, and therefore often called an active surface layer, is supported by a porous sublayer that provides the membrane with sufficient mechanical strength. The entire thickness of the membrane is about 0.1 mm, while the thickness of the active surface layer is only 30 to 100 nm.

When a membrane is placed between pure water and an aqueous sodium chloride solution, water flows from the chamber filled with pure water to that filled with the sodium chloride solution, whereas sodium chloride does not flow (Figure 1.2a). As water flows into the sodium chloride solution chamber, the water level of the solution increases until the flow of pure water stops (Figure 1.2b) at the steady state. The difference between the water level of the sodium chloride solution and that of pure water at the steady state, when converted to hydrostatic pressure, is called osmotic pressure. When a pressure higher than the osmotic pressure is applied to the sodium chloride solution, the flow of pure water is reversed; the flow from the sodium chloride solution to the pure water begins to occur. There is no flow of sodium chloride through the membrane. As a result, pure water can be obtained from the sodium chloride solution. The above separation process is called reverse osmosis.

The most successful application of the reverse osmosis process is in the production of drinking water from seawater. This process is known as seawater desalination and is currently producing millions of gallons of potable water daily in the Middle East. Fishing boats, ocean liners, and submarines also carry

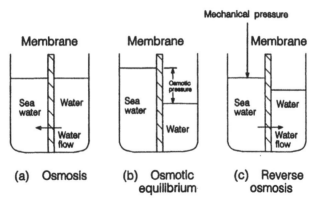

Figure 1.2. Principle of reverse osmosis.

reverse osmosis units to obtain potable water from the sea. In brackish water where the content of sodium chloride is much less than in seawater, lower osmotic pressures should be overcome for desalination. The reverse osmosis process is also being used to produce ultrapure water for the manufacture of semiconductors.

Ultrafiltration is a process based on the same principle as that of reverse osmosis. The main difference between reverse osmosis and ultrafiltration is that ultrafiltration membranes have larger pore sizes than reverse osmosis membranes, ranging from 1 to 100 nm. Ultrafiltration membranes are used for the separation and concentration of macromolecules and colloidal particles. Osmotic pressures of macromolecules are much smaller than those of small solute molecules, and therefore operating pressures applied in the ultrafiltration process are usually much lower than those applied in the reverse osmosis process. Membranes having pore sizes between those for reverse osmosis and ultrafiltration membranes are sometimes called nanofiltration membranes. The size of the solute molecules that are separated from water, and the range of operating pressures, are also between those for reverse osmosis and ultrafiltration. Ultrafiltration membranes are prepared from polymeric materials such as polysulfone, polyethersulfone, polyacrylonitrile, and cellulosic polymers. Inorganic materials such as alumina can also be used for ultrafiltration membranes.

Typical applications of ultrafiltration processes are the treatment of electroplating rinse water, the treatment of cheese whey, and the treatment of wastewater from the pulp and paper industry.

The pore sizes of microfiltration membranes are even larger than those of ultrafiltration membranes and range from 0.1 μm (100 nm) to several μm. The sizes of the particles separated by microfiltration membranes are therefore even larger than those separated by ultrafiltration membranes. However, the separation mechanism is not a simple sieve mechanism whereby the particles whose sizes are smaller than the pore size flow freely through the pore while the particles that are larger than the pore size are stopped completely. In many cases the particles to be separated are adsorbed onto the surface of the pore, resulting in a significant reduction in the pore size. Particles can also be deposited on top of the membrane, forming a cake-like secondary filter layer. Therefore, the sizes of the particles that can be separated by microfiltration membranes are often much smaller than those of the pores in the "uncontaminated" membranes. Several methods are used to prepare microfiltration membranes. One of the methods is to sinter small particles made of metals, ceramics, and plastics. The spaces formed between the particles become the pores through which materials can be transported. A second method is to stretch a polymeric film. When a polyethylene film is stretched, part of the film becomes opaque. Pores are found in this part by electron micrographic observation. Another method is to irradiate a plastic film, such as a polycarbonate film, with an electron beam. Pores are formed when sections hit by the electron beam are chemically etched with a strong alkaline solution. The phase-inversion technique, in which the polymeric solution is cast onto a film and then solidified by immersing the film in a nonsolvent gelation bath, is also applied to prepare microfiltration membranes.

Microfiltration membranes are used for the removal of microorganisms from the fermentation product. For example, various antibiotics are produced by the function of microorganisms. Microfiltration membranes are used to separate the microorganisms from the product antibiotics. Microfiltration membranes are also used to remove yeast from alcoholic beverages. For example, in the process of producing draught beer, yeast is removed by membrane filtration. Recently, a cartridge for cleaning tap water was developed. Microfiltration hollow fibers made of polyethylene are combined in the cartridge with calcium carbonate and activated carbon columns, as illustrated in Figure 1.3. Small organic molecules, such as halogenated hydrocarbons, are removed by adsorption to activated carbon. Mineral and CO_2 contents in water are increased while passing through a calcium carbonate layer. Finally, microorganisms, molds, and other turbid materials are removed by filtration with microfiltration hollow fibers.

There are several types of membranes for gas and vapor separation. Palladium is known to absorb a large amount of hydrogen gas. One mol of palladium can absorb 6 mol of hydrogen. When the palladium alloy is stretched into a film, and a hydrogen pressure gradient across the membrane is applied, the transport of hydrogen starts to occur by this mechanism, illustrated in Figure 1.4. Hydrogen molecules are decomposed into atoms and adsorbed on the high-pressure side of the membrane. Then the electrons that surround the proton are transferred to the free electron band of palladium. The naked proton permeates the membrane from the high-pressure to the low-pressure side of the membrane

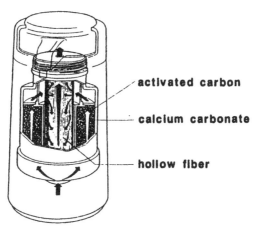

Figure 1.3. Waterpurifier using microfiltration hollow fibers.

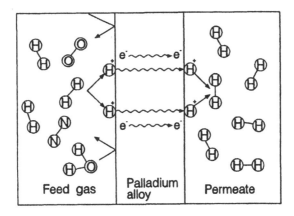

Figure 1.4. Hydrogen permeation through palladium membrane.

and receives an electron at the low-pressure side of the palladium membrane, forming a hydrogen atom. Two hydrogen atoms are combined to form a hydrogen molecule, which leaves the membrane at the low-pressure side. Since the above transport mechanism works only for hydrogen gas, it can be used for the separation of hydrogen from other gases. The production of ultrapure hydrogen by palladium membranes is currently used in semiconductor industries. The separation and recovery of hydrogen from industrial exhaust gases that are rich in hydrogen also are important processes.

The properties of polymeric materials are similar to those of rubber, at temperatures above the glass transition temperature. Polymers in this state are called rubbery polymers. Usually, gas molecules permeate through the rubbery polymer very quickly because the binding force between molecular segments of the polymer is not strong, and segments can move relatively easily to open a channel through which even large gas molecules can pass. Therefore, the more the permeant molecule is sorbed to the membrane material, the faster the gas molecule can be transported through the polymeric membrane. Thus, volatile organic compounds, to which the polymeric material exhibits a strong affinity, permeate through a rubbery polymeric membrane much faster than do oxygen and nitrogen, even though the latter molecules are smaller. Membranes prepared from rubbery polymers can therefore be used to remove volatile organic compounds from air. A typical example of rubbery polymers is polydimethylsiloxane.

When the temperature is below the glass transition temperature, the polymer is in a glassy state. There are regions where polymer segments are rigidly assembled in a crystalline form and also regions where they are loosely assembled in an amorphous form. The segmental motion of the polymeric molecule is, however, not as free as that of the rubbery polymer, even in the amorphous region, since the motion is restricted by the surrounding crystalline region. The size of the permeant molecule is a factor governing the permeation rate in the glassy polymer. Since hydrogen molecules have the smallest size, membranes prepared from glassy polymers can be used effectively for hydrogen enrichment. Carbon dioxide permeates through these membranes much faster than does methane, partly because a carbon dioxide molecule is slightly smaller than a methane molecule. Typical examples of these polymers are cellulose acetate, polysulfone, polyethersulfone, and aromatic polyimide. The affinity of the polymeric material to the permeating molecule is another contributing factor affecting the permeation rate. The permeation rate of carbon dioxide is much greater than that of methane, partially due to the stronger affinity of the polymeric material to carbon dioxide.

There are membranes that have ionic charges on the polymeric segments. Suppose the membrane is negatively charged, as illustrated in Figure 1.5; cations in the solution enter the membrane preferentially. When the valence of the cation is as high as three (for example Al^{+++}), the cation becomes immobile in the membrane, because of a strong coulombic attractive force between positive and negative charges. When the cationic valence is either two or one, the cation can move through the membrane, under an electrical potential difference between two sides of the membrane. Anions in the solution cannot enter the membrane, due to the coulombic repulsive force, and anions cannot pass through the membrane. Thus, a negatively charged membrane is permeable only to cations and therefore is called a cationic membrane. Similarly, when the membrane is positively charged, it is permeable to anions and is called an anionic membrane. Suppose anionic and cationic membranes are placed alternately, as illustrated in Figure 1.6, and an electrical potential is applied from a cathode and an anode stationed at both ends of the membrane assembly. When an aqueous

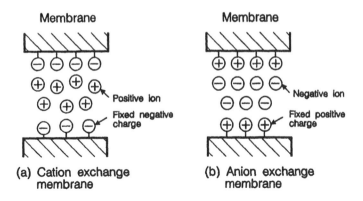

Figure 1.5. Schematic representation of ion exchange membrane.

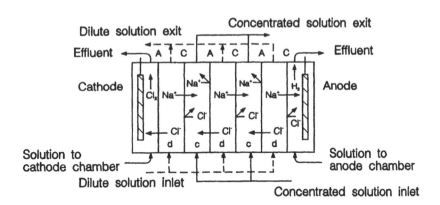

A: anion exchange membrane
C: cation exchange membrane
d: dilute solution chamber
c: concentrated solution chamber

Figure 1.6. Principle of electrodialysis.

sodium chloride solution is fed from the bottom into the dilute solution chamber, sodium ions are driven by the electrical potential to the anode and therefore start to move to the right. The sodium ions can permeate through the cationic membrane, but are rejected when they meet the anionic membrane. Chloride ions, on the other hand, are driven to the left, permeate through the anionic membrane, and are rejected at the cationic membrane. As a result, both sodium and chloride ions leave a chamber (dilution chamber) and are concentrated in the neighboring chambers (concentration chambers). Thus, the electrodialysis gives rise to dilution chambers where sodium chloride is diluted and concentration chambers where sodium chloride is concentrated, each placed alternately. Electrodialysis can thus be used either to concentrate the salt or to produce potable water from the seawater. Electrodialysis can also be applied to reduce the salt content of soy sauce. When comparing mother's milk with cow's milk, the ash content is three times higher, and the protein content is six times higher, in cow's milk. Therefore, when cheese whey, the leftover after cheese protein is removed from cow's milk, and fresh cow's milk are combined, the protein content becomes almost equal to that of mother's milk. Furthermore, by removing ash through electrodialysis, the ash content also becomes very similar to that of mother's milk. Artificial mother's milk is thus produced. Ion exchange membranes are also used for the electrolysis of sodium chloride solution, to manufacture sodium hydroxide and chlorine. A cationic membrane made of perfluorocarbon polymer is placed at the center of the electrolysis chamber (Figure 1.7). This polymeric material is known to be resistant to solutions of high alkaline concentration. When a sodium chloride solution is loaded to the left side of the cationic membrane, the sodium ion is attracted toward the anode and moves to the right side of the membrane. On the surface of the anode, water is decomposed into a proton and a hydroxyl ion. The proton is immediately reduced into a hydrogen atom by receiving one electron from the surface of the anode. Two hydrogen atoms are combined to form a hydrogen molecule, and leave the anode chamber. Sodium hydroxide solution is thus produced in the anode chamber. The chlorine ion moves toward the cathode. When it reaches the cathode, the chlorine ion donates an electron to the cathode and becomes a chlorine atom. Two chlorine atoms are combined to form a chlorine molecule, before leaving the cathode chamber.

    Suppose there are two kinds of solutes, one represented by a closed circle and the other represented by an open circle, in Figure 1.8. Suppose also that the solute represented by a closed circle cannot pass through a membrane, while the solute represented by an open circle can. When a solution that contains both solutes is fed to the left side of the membrane from the bottom of the chamber (see Figure 1.8) while the solvent (usually water) is fed to the other side of the membrane from the top of the chamber, only the solute represented by an open

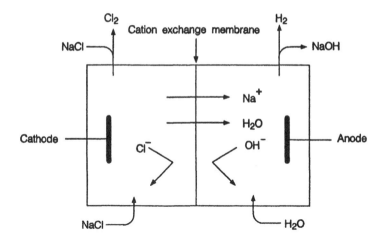

Figure 1.7. Principle of electrolysis of sodium chloride solution.

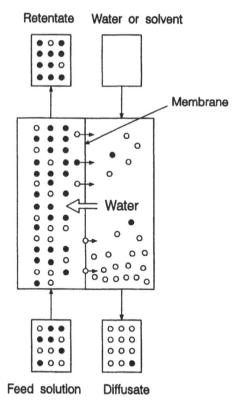

Figure 1.8. Schematic representation of dialysis.

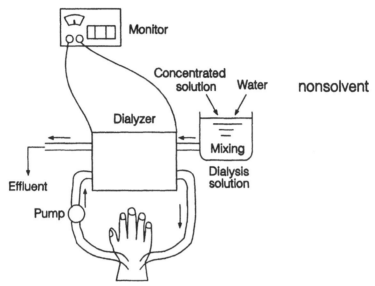

Figure 1.9. Schematic representation of hemodialysis.

circle can be transferred from the left side to the right side of the membrane. The solute represented by the closed circle is left behind. Thus, the solute mixture can be separated. The separation process based on the above principle is called dialysis. This process is being used in the artificial kidney, to remove urea, ureic acid, and creatinine from blood. Figure 1.9 shows schematically a dialysis system for blood treatment. A solution with an osmotic pressure nearly equal to that of blood is supplied into the dialyzer to remove toxic substances from the blood. A monitor is attached to control the dialyzing conditions. Toxic substances of relatively low molecular weights can be removed effectively by the above hemodialysis system. However, some toxic substances with high molecular weights, such as β_2-microglobulin, cannot be removed. Hemofiltration, in which ultrafiltration membranes are employed, can remove toxic substances with molecular weights ranging from 500 to tens of thousands.

2

Membrane Material

2.1 Introduction

Materials used for the membrane cover a wide range, from organic polymeric materials to inorganic materials. Normally they are solid. Membranes are prepared from these materials and used for various separation processes, such as membrane gas separation, pervaporation, reverse osmosis, ultrafiltration, and microfiltration, depending on the pore size at the surface of the membrane. In this chapter an attempt is made to answer the question of why a limited number of materials were chosen for a certain membrane separation process, and to give a scientific guideline for the choice and the design of the membrane material. Materials have chemical and physical structures. While chemical structure means the essential structure of the molecule that constitutes the material and can be described by the chemical formula, physical structure is defined as the way in which the molecules are assembled in the material. There are possibilities of many different physical structures for a given chemical structure. Therefore, chemical structure and physical structure are totally unrelated. However, it is believed by many people that there is an "ultimate physical structure" corresponding to each chemical structure, particularly when the molecules are most densely assembled. The properties of a material with such an ultimate structure are called "intrinsic properties." It is also believed by some people that materials used for the membrane gas separation and reverse osmosis have such an ultimate structure, and that the permeation properties exhibited by those materials are intrinsic properties. The presence of such an ultimate physical structure has not been proven, particularly for polymeric materials. Suppose the above concept is valid; the membrane separation performance for gas separation and reverse osmosis should be uniquely related to the chemical structure. In other words, the membrane gas separation and reverse osmosis separation should be predictable from the chemical structure of the material alone.

As there are chemical structures and physical structures of the material, so there are chemical properties and physical properties. Chemical properties

11

describe how the chemical structure of the material changes under certain circumstances. For example, cellulose acetate is susceptible to hydrolysis at high or low pH. Aromatic polyamide is susceptible to chlorination in the presence of chlorine in water, etc. On the other hand, physical properties include physical parameters such as density, melting point, glass transition temperature, Young's modulus, compressibility, etc. Physical properties also include the degree of change in the physical structure of a material in a given environment. For example, the degree of swelling of a polymeric material in an organic solvent and the change in crystallinity at high or low temperatures are physical properties. When a solid material is in contact with some liquid or gas, the interfacial properties (such as interfacial tension and adsorption) are also considered physical properties. Chemical and physical properties are closely related to chemical and physical structures.

Both chemical and physical properties have to be examined in order for some material to be chosen for membrane preparation. For example, resistance to the change of the material, either in chemical or in physical structure, is an important criteria for the choice of the membrane material. However, since membrane separation includes interfaces between the phase that involves the permeant mixture and the membrane phase, the interfacial properties are the most important physical properties of the membrane. Therefore, only the latter aspect will be taken into consideration in this chapter. The solubility parameter approach described in the first part of this chapter is to evaluate the interfacial interaction force working between molecules, which are in contact with each other at an interface, from the chemical structure of the molecules involved. The liquid chromatography method described in the second part of this chapter is one of the experimental methods to study interfacial properties.

2.2   Solubility Parameter

The solubility parameter is a parameter to express the nature and magnitude of the interaction force working between molecules. When applied to the membrane, the solubility parameter can give a measure to the interaction force working between the molecules that constitute the membrane material, and also the interaction force between the latter molecule and the permeant molecule. They are intrinsic to the chemical structure. In other words, the interaction force measured by the solubility parameter is uniquely determined when the molecular formula of the molecules involved in the interaction are given.

Dissolution of polymeric materials is accompanied by a free energy change, ΔG (J/mol), which can be written as

$$\Delta G = \Delta H - T\Delta S \qquad (2.1)$$

where ΔH (J/mol), T (K), and ΔS (J/mol·K) are the heat of mixing, absolute temperature, and entropy of mixing, respectively. Since the dissolution of polymeric materials involves large changes in entropy, whether ΔG is plus or minus is determined primarily by the magnitude of ΔH. Several methods have been proposed to evaluate ΔH. Among others the following equation proposed by Hildebrand [1] is by far the most popular:

$$\Delta H_M = V_M \left[(\Delta E_1/V_1)^{1/2} - (\Delta E_2/V_2)^{1/2} \right]^2 \phi_1 \phi_2 \qquad (2.2)$$

where ΔH_M is the total heat of mixing (J/mol), V_M is the total molar volume of the mixture (m^3/mol), ΔE is the heat of vaporization (J/mol), V is the molar volume (m^3/mol), ϕ is volume fraction, and subscripts 1 and 2 represent the components 1 and 2 of the solution mixture. ΔE is also called the cohesive energy. A physical interpretation of ΔE is the degree of attraction between molecules in a liquid. $\Delta E/V$ in Equation (2.2) is equal to the density of the heat of vaporization and is called "internal pressure" or "cohesive energy density" of the substance. Rearrangement of the above equation yields

$$\Delta H_M / V_M \phi_1 \phi_2 = \left[(\Delta E_1/V_1)^{1/2} - (\Delta E_2/V_2)^{1/2} \right]^2 \qquad (2.3)$$

It is obvious from Equation 2.3 that the heat of mixing, ΔH_M, is always positive and becomes smaller as cohesive energy densities of component 1 and component 2 become closer. Substituting the above ΔH_M for ΔH of Equation 2.1, ΔG decreases as cohesive energy densities of component 1 and component 2 become closer to each other. This favors the dissolution of polymer 1 in solvent 2. The quantity $\sqrt{\Delta E/V}$ is defined as the "solubility parameter" and can be obtained both for the polymer repeat unit and for the solvent [1,2]. A polymer becomes more soluble in a solvent when their respective solubility parameters are close to each other. On the other hand, a polymer becomes less soluble in a solvent when the difference in their solubility parameters increases.

The heat of vaporization can be divided into three components, with each component representing a molecular interaction force of different kind; i.e.,

$$\Delta E/V = \Delta E_d/V + \Delta E_p/V + \Delta E_h/V \qquad (2.4)$$

where ΔE_d is the London dispersion force, ΔE_p is the dipole force, and ΔE_h is the hydrogen bonding force component. In terms of solubility parameters,

$$\delta_{sp}^2 = \delta_d^2 + \delta_p^2 + \delta_h^2 \qquad (2.5)$$

where δ_d, δ_p, and δ_h are given as

$$\delta_d = (E_d/V)^{1/2} \qquad (2.6)$$

$$\delta_p = (E_p/V)^{1/2} \qquad (2.7)$$

$$\delta_h = (E_h/V)^{1/2} \qquad (2.8)$$

and are the dispersion force, dipole, and hydrogen bonding component of the solubility parameter, respectively [3]. Furthermore, δ_{sp}, δ_d, δ_p, and δ_h can be calculated by applying additivity rules to the structural components of the repeat unit of the macromolecule and to those of the solvent molecule, by the following equations [4]:

$$\delta_{sp} = \sqrt{\Sigma E_{coh}/V} \qquad (2.9)$$

$$\delta_d = \Sigma F_{di}/V \qquad (2.10)$$

$$\delta_p = \sqrt{\Sigma F_{pi}^2/V} \qquad (2.11)$$

$$\delta_h = \sqrt{\Sigma E_{hi}/V} \qquad (2.12)$$

Numerical values assigned to each structural component of organic compounds are listed in Tables 1 and 2. Using these numerical values, the overall solubility parameter δ_{sp}, and its components δ_d and δ_h, can be calculated when the molecular structure of the repeat unit of a macromolecule is known. Solubility parameters calculated using the numerical values listed in Tables 1 and 2 are shown in Table 3 for various polymer repeat units [5]. Pairs of solubility parameters (δ_d and δ_h) are plotted in Figure 2.1, with respect to polymeric materials for which the preparation of reverse osmosis membranes was attempted. δ_d and δ_h were chosen to represent hydrophobic and hydrophilic properties of the polymeric material. In the figure, closed circles indicate polymers from which reverse osmosis membranes with sufficiently high sodium chloride rejection and permeation rates could be produced, while open circles indicate polymers from which reverse osmosis membranes with sufficiently high performance data have not been produced so far. Clearly, all closed circles are located in a limited range of δ_d, δ_h combinations. These results indicate that the solubility parameter of the polymeric material can be employed to choose polymeric materials as candidates for the preparation of reverse osmosis membranes. Solubility parameters have also been used for the study of pervaporation [6,7] and vapor permeation [8].

Table 1. Group Contributions to $E_{coh,i}$ and V_i [5]

Structural group	$E_{coh,i}$ (cal/mol)	V_i (cm³/mol)	Structural group	$E_{coh,i}$ (cal/mol)	V_i (cm³/mol)
-CH₃	1,125	33.5	-NF₂	1,830	33.1
-CH₂-	1,180	16.1	-NF-	1,210	24.5
>CH-	820	-1.0	-CONH₂	10,000	17.5
-C- (four bonds)	350	-19.2	-CONH-	8,000	9.5
H₂C=	1,030	28.5	-CON<	7,050	-7.7
-CH=	1,030	13.5	HCON<	6,600	11.3
>C=	1,030	-5.5	HCONH-	10,500	27.0
HC≡	920	27.4	-NHCOO-	6,300	18.5
-C≡	1,690	6.5	-NHCONH-	12,000	—
Phenyl	7,630	71.4	-CONHNHCO-	11,200	19.0
Phenylene (o,m,p)	7,630	52.4	-NHCON<	10,000	—
Phenyl (trisubstituted)	7,630	33.4	>NCON<	5,000	-14.5
Phenyl (tetrasubstituted)	7,630	14.4	NH₂COO-	8,840	—
Phenyl (pentasubstituted)	7,630	-4.6	-NCO	6,800	35.0
Phenyl (hexasubstituted)	7,630	-23.6	-ONH₂	4,550	20.0
Conjugation in ring for each double bond	400	-2.2	>C=NOH	6,000	11.3
Halogen attached to carbon atom with double bond	0.8×$E_{coh,i}$ for halogen	4.0	-CH=NOH	6,000	24.0
			-NO₂ (aliphatic)	7,000	24.0
			-NO₂ (aromatic)	3,670	32.0
-F	1,000	18.0	-NO₃	5,000	33.5
-F (disubstituted)	850	20.0	-NO₂ (nitrite)	2,800	33.5

Table 1. (continued)

Structural group	$E_{coh,i}$ (cal/mol)	V_i (cm³/mol)	Structural group	$E_{coh,i}$ (cal/mol)	V_i (cm³/mol)
-F (trisubstituted)	550	22.0	-NHNO₂	9,500	28.7
-CF₂- (for perfluorocompds)	1,020	23.0	-NNO-	6,500	10
-CF₃- (for perfluorocompds)	1,020	57.5	-SH	3,450	28.0
-Cl	2,760	24.0	-S-	3,380	12
-Cl (disubstituted)	2,300	26.0	-S₂-	5,700	23.0
-Cl (trisubstituted)	1,800	27.3	-S₃-	3,200	47.2
-Br	3,700	30.0	-SO₂-	9,350	23.6
-Br (disubstituted)	2,950	31.0	>SO	9,350	—
-Br (trisubstituted)	2,550	32.4	SO₃	4,500	27.6
-I	4,550	31.5	SO₄	6,800	31.6
-I (disubstituted)	4,000	33.5	-SO₂Cl	8,850	43.5
-I (trisubstituted)	3,900	37.0	-SCN	4,800	37.0
-CN	6,100	24.0	-NCS	6,000	40.0
-OH	7,120	10.0	P	2,250	-1.0
-OH (disubstituted or on adjacent C atoms)	5,220	13.0	PO₃	3,400	22.7
-O-	800	3.8	PO₄	5,000	28.0
-CHO (aldehyde)	5,100	22.3	PO₃(OH)	7,600	32.2
-CO-	4,150	10.8	Si	810	0
-COOH	6,600	28.5	SiO₄	5,200	20.0
-CO₂-	4,300	18.0	B	3,300	-2.0
			BO₃	0	20.4

Structural group	$E_{coh,i}$ (cal/mol)	V_i (cm³/mol)	Structural group	$E_{coh,i}$ (cal/mol)	V_i (cm³/mol)
-CO₃- (carbonate)	4,200	22.0	Al	3,300	-2.0
-C₂O₃- (anhydride)	7,300	30.0	Ga	3,300	-2.0
HCOO- (formate)	4,300	32.5	In	3,300	-2.0
-CO₂CO₂- (oxalate)	6,400	37.3	Tl	3,300	-2.0
-HCO₃	3,000	18.0	Ge	1,230	-1.5
-COF	3,200	29.0	Sn	2,700	1.5
-COCl	4,200	38.1	Pb	4,100	2.5
-COBr	5,770	41.6	As	3,100	7.0
-COI	7,000	48.7	Sb	3,900	8.9
-NH₂	3,000	19.2	Bi	5,100	9.5
-NH-	2,000	4.5	Se	4,100	16.0
-N<	1,000	-9.0	Te	4,800	17.4
-N=	2,800	5.0	Zn	3,460	2.5
-NHNH₂	5,250	—	Cd	4,250	6.5
-NNH₂	4,000	16	Hg	5,450	7.5
-NHNH-	4,000	16			
-N₂ (diazo)	2,000	23			
-N=N-	1,000	—			
>C=N-N=C<	4,800	0			
-N=C=N-	2,740	—			
-NC	4,500	23.1			

Table 2. Solubility Parameter Component Group Contributions [5]

Structural group	$F_{d,i}$ cal$^{1/2}$ cm$^{3/2}$/mol	$F_{p,i}$ cal$^{1/2}$ cm$^{3/2}$/mol	$E_{h,i}$ cal/mol	$V_{g,i}$ cm^3/mol
-CH$_3$	205	0	0	23.9
-CH$_2$-	132	0	0	15.9
>CH-	39	0	0	9.5
-$\overset{\vert}{\underset{\vert}{\text{C}}}$-	-34	0	0	4.6
=CH$_2$	196	0	0	—
=CH-	98	0	0	13.1
=C<	34	0	0	—
Cyclohexyl	792	0	0	90.7
Phenyl	699	54	0	72.7
Phenylene (o,m,p)	621	54	0	65.5
-F	108	—	—	10.9
-Cl	220	269	96	19.9
-Br	269	—	—	—
-CN	210	538	597	19.5
-OH	103	244	4,777	9.7
-O-	49	196	717	10.0
-CHO	230	392	1,075	—
-CO-	142	376	478	13.4

Structural group	$F_{d,i}$ cal$^{1/2}$ cm$^{3/2}$/mol	$F_{p,i}$ cal$^{1/2}$ cm$^{3/2}$/mol	$E_{h,i}$ cal/mol	$V_{g,i}$ cm^3/mol
-COOH	259	205	2,388	23.1
-COO-	191	239	1,672	23.0
				18.25 (acrylic)
-COOH	259	—	—	—
-NH$_2$	137	—	2,006	—
-NH-	78	103	740	12.5
-N<	10	391	1,194	6.7
O H $\overset{\|\|}{=}\ \overset{\|}{-}$ -C-N- (aliphatic)	220	—	4,657	24.9
O H $\overset{\|\|}{=}\ \overset{\|}{-}$ -C-N- (aromatic)	220	479	7,762	24.9
O H H O $\overset{\|\|}{=}\ \overset{\|}{-}\ \overset{\|}{-}\ \overset{\|\|}{=}$ -C-N-N-C- (aromatic)	440	—	10,629	49.8
-NO$_2$	244	523	358	—
-S-	215	—	—	17.8
-SO$_2$-	289	—	3,224	31.8
=PO$_4$-	362	924	3,105	—

Table 2. (continued)

Structural group	$F_{d,i}$ cal$^{1/2}$ cm$^{3/2}$/mol	$F_{p,i}$ cal$^{1/2}$ cm$^{3/2}$/mol	$E_{h,i}$ cal/mol	$V_{g,i}$ cm^3/mol
Ring	93	—	—	—
One phase of symmetry	—	0.5 ×	—	—
Two phases of symmetry	—	0.25 ×	—	—
More phases of symmetry	—	0 ×	0 ×	—

Table 3. Solubility Parameters of Polymers

Polymer No.	Polymer	Structure of polymer repeat unit	δ_{sp} cal$^{1/2}$ cm$^{-3/2}$	δ_h cal$^{1/2}$ cm$^{-3/2}$	δ_d cal$^{1/2}$ cm$^{-3/2}$
1	Cellulose acetate-398	$(CH_2)_3(CH)_{20}(O)_8(OH)_{2\cdot19}(OCCH_3)_{9\cdot81}$	12.7	6.33	7.60
2	Cellulose acetate-376	$-(OH)_{3\cdot05}(OCCH_3)_{8\cdot95}$	13.1	6.72	7.59
3	Cellulose acetate-383	$-(OH)_{2\cdot78}(OCCH_3)_{9\cdot22}$	13.0	6.59	7.59
4	Cellulose triacetate	$-(OH)(OCCH_3)_{11}$	12.0	5.81	7.61
5	Cellulose acetate propionate-504	$-(OH)_4(OCC_2H_5)_8$	12.9	6.62	7.68
6	Cellulose acetate propionate-151	$-(OH)_{0\cdot8}(OCCH_3)_{8\cdot25}(OCC_2H_5)_{2\cdot95}$	11.6	5.57	7.65

Table 3. Solubility Parameters of Polymers (continued)

Polymer No.	Polymer	Structure of polymer repeat unit	δ_{sp} cal$^{1/2}$ cm$^{-3/2}$	δ_h cal$^{1/2}$ cm$^{-3/2}$	δ_d cal$^{1/2}$ cm$^{-3/2}$
7	Cellulose acetate propionate-482	$—(OH)_2(OCC_2H_5)_{10}$	11.8	5.72	7.72
8	Cellulose acetate propionate-063	$—(OH)_{2\cdot62}(OCCH_3)_{8\cdot33}(OCC_2H_5)_{1\cdot05}$	12.8	6.46	7.61
9	Cellulose acetate butyrate-553	$—(OH)_3(OCC_3H_7)_9$	11.8	5.83	7.78
10	Cellulose acetate butyrate-272	$—(OH)_{1\cdot78}(OCCH_3)_{5\cdot78}(OCC_3H_9)_{4\cdot94}$	10.7	5.68	7.71
11	Cellulose acetate butyrate-171	$—(OH)_{1\cdot00}(OCCH_3)_{8\cdot15}(OCC_3H_7)_{2\cdot85}$	11.5	5.52	7.68
12	Cellulose acetate butyrate-500	$—(OH)_{0\cdot58}(OCCH_3)_{1\cdot95}(OCC_3H_7)_{9\cdot47}$	10.9	4.85	7.80
13	Ethyl cellulose G	$—(OH)_{3\cdot06}(OC_2H_5)_{8\cdot94}$	11.5	5.76	7.23

#	Name	Structure			
14	Ethyl cellulose T	—(OH)$_{1-64}$(OC$_2$H$_5$)$_{10-36}$	10.5	4.93	7.36
15	Cellulose	—(OH)$_{12}$	24.1	11.85	7.36
16	Cellolose trihydrogen phthalate		14.1	5.71	8.92
17	Ethyl cellulose phthalate		12.0	4.62	8.45
18	Copolyamide 31B		15.9	9.27	8.30

Table 3. Solubility Parameters of Polymers (continued)

Polymer No.	Polymer	Structure of polymer repeat unit	δ_{sp} cal$^{1/2}$ cm$^{-3/2}$	δ_h cal$^{1/2}$ cm$^{-3/2}$	δ_d cal$^{1/2}$ cm$^{-3/2}$
19	Copolyamidohydrazide 81B (PPPH1115)	(polymer structure)	16.0	9.35	9.28
20	Copolyhydrazide 117B	(polymer structure)	16.3	9.60	9.20
21	Copolyamidohydrazide 107B (PPPH8273)	(polymer structure)	16.1	9.44	9.25

No.	Polymer	Structure			
22	Polysemicarbazide 94B	(structure)	16.5	8.22	8.99
23	Polyethylenimine-Toluene 2,4-Diisocyanate NS-100	a	12.2	6.25	7.87
24	Sulfonated polyfuran NS-200	(structure)	16.2	6.47	8.42
25	Polypiperazineamide t-2,5-DMPipF	(structure)	12.1	4.50	8.15
26	Polypiperazineamide t-2,5-DMPip-TEZ	(structure)	13.7	—	—

Table 3. Solubility Parameters of Polymers (continued)

Polymer No.	Polymer	Structure of polymer repeat unit	δ_{sp} cal$^{1/2}$ cm$^{-3/2}$	δ_h cal$^{1/2}$ cm$^{-3/2}$	δ_d cal$^{1/2}$ cm$^{-3/2}$
27	Polybenzimidazolone PBIL		17.1	7.84	9.32
28	Carboxylated aromatic polyamide		16.5	9.43	9.37
29	Nylon-6	$-NH(CH_2)_5\ C-$	12.4	6.69	8.45
30	Polyimide		19.0	8.23	9.75

31	Polyalanine	$-NH\,CH(CH_3)\,C(=O)-$	15.4	8.94	7.97
32	Bisphenol A polysulfone (Udel)		12.6	3.66	8.97
33	Polyacrylic acid	$-CH_2CH(COOH)-$	14.0	7.02	8.89
34	Polyvinyl alcohol	$-CH_2CH(OH)-$	19.1	11.68	7.82
35	Polyvinyl formal	$-CH_2-CH-CH_2-CH-$ with $O-CH_2-O$	11.2	4.07	6.61
36	Polyvinyl acetal	$-CH_2-CH-CH_2-CH-$ with $O-CHCH_3-O$	10.3	3.71	7.48

Table 3. Solubility Parameters of Polymers (continued)

Polymer No.	Polymer	Structure of polymer repeat unit	δ_{sp} cal$^{1/2}$ cm$^{-3/2}$	δ_h cal$^{1/2}$ cm$^{-3/2}$	δ_d cal$^{1/2}$ cm$^{-3/2}$
37	Polyvinyl butyral	$-CH_2-CH-CH_2-CH-$ (ring with O, O, CHC$_3$H$_7$)	9.8	3.25	7.67
38	Polyvinyl hydrogen phthalate	$-CH_2-CH(OC(=O)-C_6H_4-COOH)-$	13.4	5.54	9.39
39	Polyethylene	$-CH_2CH_2-$	8.6	0	8.32
40	Polypropylene	$-CH_2CH(CH_3)-$	8.0	0	7.65
41	Polystyrene	$-CH_2CH(C_6H_5)-$	10.6	0	8.88
42	Polyvinyl chloride	$-CH_2CH(Cl)-$	11.0	1.45	8.65
43	Polymethyl methacrylate	$-CH_2C(CH_3)(COCH_3)-$ (with =O)	9.9	4.40	8.08

No.	Polymer	Structure			
44	Polydiallyl phthalate	$-CH-CH_2-O-C(=O)-C(=O)-O-CH_2-CH-$ (phthalate ring) with $-CH_2-$ / $-CH_2-$	12.2	4.15	8.30
45	Polyethylene glycol	$-CH_2CH_2O-$	9.4	4.14	7.50
46	Polyprophlene glycol	$-CH_2CH(CH_3)O-$	8.7	3.48	7.18
47	Sulfonated 2,6-dimethyl polyphenylene oxide	aromatic CH_3, SO_3H units 0.2 / 0.8	12.6	4.36	8.10
48	Sulfonated polysulfone	aromatic structure with SO_2, O, SO_3H, CH_3-C-CH_3	14.1	5.62	8.87

Table 3. Solubility Parameters of Polymers (continued)

Polymer No.	Polymer	Structure of polymer repeat unit	δ_{sp} cal$^{1/2}$ cm$^{-3/2}$	δ_h cal$^{1/2}$ cm$^{-3/2}$	δ_d cal$^{1/2}$ cm$^{-3/2}$
49	Polyacrylonitrile	$-CH_2CH(CN)-$	14.4	3.65	8.51
50	PA-300	b	15.0	8.98	8.33

[a] Structure from Rozelle, L.T., J.E. Cadotted, K.E. Cobian, and C.V. Kopp, Jr., in *Reverse Osmosis and Synthetic Membranes,* S. Sourirajan, Ed., National Research Council of Canada, Ottawa, 1977, p. 249.

[b] Structure from Channabasappa, K.C. and J.J. Strobel, Status of sea water reverse osmosis membrane process technology, Proc. 5th Int. Symp. on Fresh Water from the Sea, Vol. IV, A. Delyannis and E. Delyannis, Eds., Athens, 1976, p. 267.

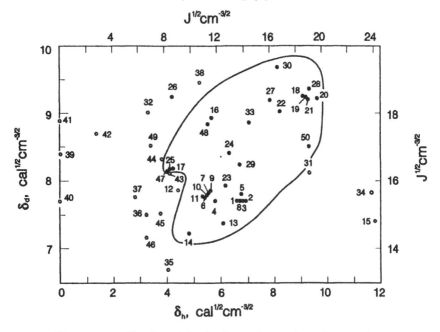

Figure 2.1. The δ_d and δ_h plot for various polymeric materials.

2.3   Liquid Chromatography

Water is pumped through a column packed with polymer powder, and a solution
sample is injected at the column inlet. Solute eluted from the column is sensed
by a detector connected to the column outlet (Figure 2.2). The residence time
of the solute (usually called retention time in chromatography theory) depends
on the partition of the solute between the solvent and the powder packed in the
column. A strong affinity of the solute for the solvent is indicated by a short
retention time. Conversely, a strong affinity to the packed powder results in a
long retention time. Chromatography is thus an effective means to measure the
partition of solute between solvent and the polymer particle. The partition measured according to the above method is considered to approximate the partition
at the surface of the membrane made of the polymeric material packed into the
chromatography column. It is only an approximation, since the difference in
morphology of the polymer powder and the polymeric membrane is expected
to affect the partition of the solute. Since the partition of the solute depends
on the interaction forces at the solvent/polymer interface, chromatography is a
system in which interaction forces working at the solvent/membrane interfaces
are simulated.

A fundamental question then arises as to the physical meaning of the partition of the solute between the solvent and polymer. Strictly speaking, the concept

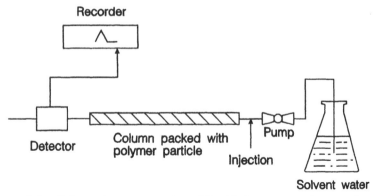

Figure 2.2. Schematic representation of liquid chromatography experiment.

of partition is not clear. The solvent forms a homogeneous liquid phase, but the polymeric material is microscopically heterogeneous. Should the polymer be regarded, for practical purposes, as a homogeneous phase and should the partition between two homogeneous phases be considered? Or should the polymer be regarded as a microscopically porous material and should the adsorption of solute from the solvent phase to the wall of the pore be considered? Each model may be justified according to the situation concerned. An attempt is made in this chapter to present a new approach [9].

Table 4 shows some retention volume (retention time × volumetric flow rate of solvent in the column) data for various solutes in a column packed with cellulose acetate powder through which water flows as the solvent. It is interesting to note that the retention volumes of raffinose and other organic and inorganic solutes are smaller than that of heavy water. Since the affinity between heavy water and the packed polymer is nearly equal to that between ordinary water and the packed polymer, despite a small difference due to the isotope effect of heavy water, the above experimental results indicate that several solutes in the column move faster than the solvent water. If only the partition equilibrium between the solvent water phase and the polymer phase is taken into consideration, the above experimental data cannot be explained, since the speed of a solute in the column should always be slower than that of the solvent water due to the partition into the polymer particle. The above experimental results can, however, be understood by assuming the presence of two kinds of water. One is ordinary bulk water, is freely mobile, and forms the mobile phase in the chromatography column. The other is interfacial-bound water that is bound at the surface of the polymer and forms the stationary phase in the chromatography column. This water has a structure similar to ice and does not dissolve carbohydrates and inorganic electrolytes as much as ordinary water does.

Liquid chromatography data can be quantitatively explained by chromatography theory. The retention volume of a solute, V_R (m^3), can be written as

$$V_R = V_m + KV_s \tag{2.13}$$

where V_m and V_s are the volumes (m^3) of the mobile and stationary phases, respectively, and K is the partition coefficient. Strictly speaking, the retention volume observed experimentally, V_R', should include the dead space of the instrument, V_d. Therefore,

$$V_R' = V_R + V_d = V_m + V_d + KV_s \tag{2.14}$$

Table 4. Retention Volume Data [9]

Solute	Retention volume, V_R', cm^3, of column(polymer)		
	CA-383-40	CA-398-3	CTA
Raffinose	2.379	3.375	1.806
D-Sorbitol	2.565	3.375	1.806
Magnesium sulfate	2.379	3.443	1.856
Sucrose	2.430	3.443	1.856
Sodium chloride	2.666	3.578	1.890
Glycerol	2.801	3.679	1.924
Sodium thiocyanate	3.004	3.814	1.890
Heavy water	3.105	4.050	2.126
Phenol	—	44.9	—

Furthermore, the following assumptions can be made with respect to K values:

1. $K = 1$ for heavy water. This assumption can be justified when considering that water molecules will be partitioned equally between the mobile and the stationary phases, since both phases consist of water. Furthermore, it can be justified that heavy water and ordinary water behave equally when they are partitioned between both phases.

2. $K = 0$ for the solute that exhibits the smallest retention volume (in Table 4 it is raffinose). Such a solute is not partitioned into the stationary phase and moves in the chromatography column together with the stream of the mobile phase.

These two assumptions can be written as

$$[V'_R]_{water} = V_m + V_d + V_s \qquad (2.15)$$

$$[V'_R]_{min} = V_m + V_d \qquad (2.16)$$

Subtracting Equation 2.16 from Equation 2.15 yields

$$V_s = [V'_R]_{water} - [V'_R]_{min} \qquad (2.17)$$

and

$$K = \frac{V'_R - [V'_R]_{min}}{[V'_R]_{water} - [V'_R]_{min}} \qquad (2.18)$$

Equations 2.17 and 2.18 allow for the calculation of V_s and K from experimental values for V'_R, $[V'_R]_{min}$ and $[V'_R]_{water}$.

The amount of the stationary-phase water relative to that of the mobile-phase water can be given by

$$\frac{[V'_R]_{water} - [V'_R]_{min}}{[V'_R]_{min} - V_d} = \frac{V_s}{V_m} \qquad (2.19)$$

According to Equation 2.19, V_s/V_m can be obtained by determining $[V'_R]_{water}$, $[V'_R]_{min}$, and V_d experimentally. It should be noted that V_m includes water that fills both the interparticular space and the pores of the polymeric particle that are sufficiently large to retain water unaffected by the influence of the interface (see Figure 2.3).

Experimental data for some solutes are shown in Table 4, with respect to cellulose acetate materials of different acetyl contents. The table shows that the solute of minimum retention volume is raffinose, for these particular columns. The table also shows that the results are not those of size exclusion chromatography, since phenol, the size of which is greater than water, exhibits a retention volume much larger than water does. Table 5 shows V_s and K values calculated on the basis of data shown in Table 4. For solutes between raffinose and heavy water, $K < 1$, indicating that those solutes are dissolved in the interfacial water (that constitutes the stationary phase) to a lesser extent than in ordinary water.

As mentioned earlier, V_s is considered to be the volume of the stationary-phase water that is bound strongly to the surface of the polymer. This interfacial water is supposed to have properties different from those of ordinary water. For example, the former water is less capable of dissolving inorganic electrolytes and organic solutes with more than one polar functional group such as sucrose and glycerol. The thickness of the interfacial water in a liquid chromatography

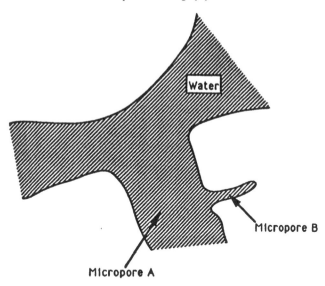

Figure 2.3. Schematic representation of the microscopic structure of a polymer particle.

column can be determined by knowing V_s and the surface area of the packed particles. V_s is given in Table 5.

The surface area can be determined by monolayer adsorption using p-nitrophenol as adsorbent. Figure 2.4 shows the adsorption isotherm of p-nitrophenol for cellulose acetate material. In the range of low p-nitrophenol concentration, the isotherm is an S shape, indicating that the OH group of the molecule is attached to the cellulose acetate surface. The aromatic ring and the nitro group are apart from the polymer surface. In the high-concentration range of p-nitrophenol, on the other hand, a plateau is found that corresponds to the monolayer coverage of the polymer surface by p-nitrophenol [10]. Since the amount of p-nitrophenol adsorbed to a unit weight of cellulose acetate at the monolayer coverage is known from the figure, and the area occupied by a single p-nitrophenol is also known to be 25×10^{-20} m^2, the surface area per unit weight of cellulose acetate can be calculated. The total weight of polymer powder in the chromatography column is known. Hence, it is possible to calculate the total area of the cellulose acetate powder in the column. The thickness of the interfacial water layer can then be determined. The results are summarized in Table 5. It should be noted that a surface area of more than 200×10^{-20} m^2 was obtained by the above method. It should also be noted that the surface area so obtained corresponds to that of highly swollen polymers [11].

Conclusions drawn from the above experimental results can be summarized as follows:

Table 5. Data on Interfacial Water Layer Thickness
and Distribution Coefficient [9]

	Column = polymer		
	CA-383-40	CA-398-3	CTA
V_s (cm^3)	0.726	0.675	0.320
Specific surface area (m^2/g)	237.8	244.3	218.4
Weight of polymer in column (g)	1.445	2.901	2.143
Total surface area (m^2)	343.6	708.7	468.0
Interfacial water layer thickness ($\times 10^{10}$ m), t	21.2	9.5	6.8
Solute		K	
Raffinose	0	0	0
D-Sorbitol	0.256	0	0
Magnesium sulfate	0	0.101	0.156
Sucrose	0.070	0.101	0.156
Sodium chloride	0.395	0.301	0.263
Glycerol	0.581	0.450	0.369
Sodium thiocyanate	0.861	0.650	0.253
Heavy water	1	1	1

1. The partition of solute between water and polymer can be interpreted as the partition between the ordinary bulk water and the interfacial-bound water. The interface is considered to include all surfaces where water and polymer are in contact.

2. The thickness of interfacial water is $10 - 20 \times 10^{-10}$ m.

3. The thickness and the property of the interfacial water depend on the nature of the polymer.

The values for V_s/V_m were obtained, for different polymers, from Equation 2.19 and are correlated to the overall solubility parameter of polymer, δ_{sp}, in Figure 2.5. The following conclusions can be drawn from this figure:

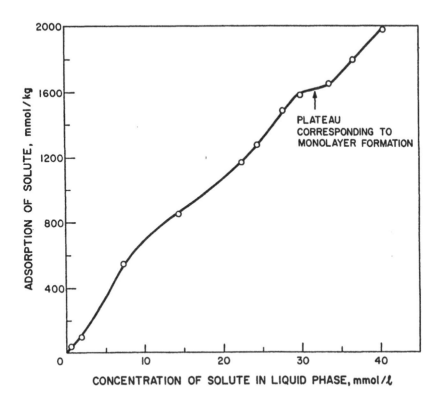

Figure 2.4. Adsorption isotherm of *p*-nitrophenol. (Reproduced from [9] with permission.)

1. The solubility parameters of polymers from which reverse osmosis membranes of excellent performance have been prepared are between 24 and 35 $J^{1/2}cm^{-3/2}$.

2. The above polymers exhibited V_s/V_m values higher than the V_s/V_m values of polymers from which reverse osmosis membranes of reasonably high performance could not be prepared.

The ratio V_s/V_m shows a maximum at a solubility parameter corresponding to that of cellulose acetate-398 (cellulose acetate with acetyl content of 39.8%) and decreases when the solubility parameter either increases or decreases from that particular value. These results indicate that an increase in the hydrophilicity

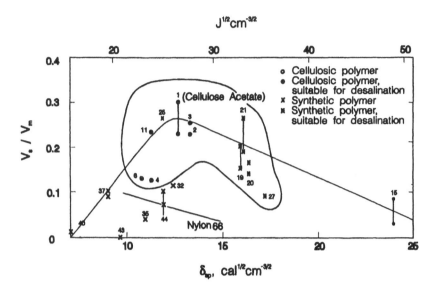

Figure 2.5. V_s/V_m vs. δ_{sp}.

of the polymer, which tends to increase the solubility parameter, does not necessarily lead to an increase in the amount of the interfacial water. The structure
of water in the vicinity of the surface of a very hydrophilic polymer is similar
to that of the ordinary water, and consequently, the formation of the interfacial water with properties different from the ordinary water is not necessarily
favored. On the other hand, when the hydrophobicity of the polymer is very
high, which is usually associated with low solubility parameter values, the surface of the polymer is not effectively wetted by water, and water cannot fill the
microscopic pores, leading to a decrease in the amount of the interfacial water.
Therefore, the amount of the interfacial water becomes greater in the intermediate range of solubility parameters where the hydrophilicity and hydrophobicity
are appropriately balanced. The desalination process by reverse osmosis is a
process to remove the interfacial pure water that fills the microscopic pores
of the polymeric membrane. Therefore, it seems natural that those polymeric
materials that contain the largest amounts of interfacial water exhibit the best
performance as reverse osmosis membranes.

Figure 2.6 illustrates a correlation between the permeability coefficient of
water vapor, P_{water}, through various polymeric films [12], and the solubility

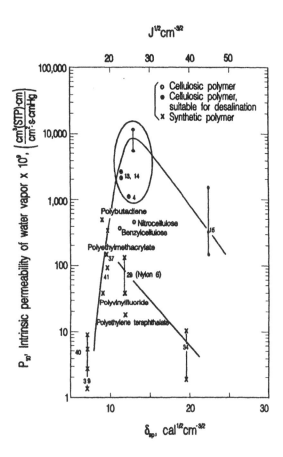

Figure 2.6. Permeability coefficient of water vapor vs. solubility parameter.

parameters of those polymers. This figure also shows that some cellulosic materials exhibit a maximum in permeability coefficient, the value of which is far greater than those of other polymeric materials with similar solubility parameter values. The patterns of the curves in Figures 2.5 and 2.6 are strikingly similar, indicating that the presence of the special interfacial water influences the transport of water vapor through polymeric membranes. Another correlation between the permeability coefficient of oxygen and nitrogen and the solubility parameter of polymers is given in Figure 2.7 [13].

The chromatography method has been used for the study of membrane material-permeant interactions for various systems [14]–[27].

Example 1

The structure of the repeat unit of cellulose acetate can be written as

$$(CH_2)_4(CH)_{20}(O)_8\ (OH)_{2.19}(OCCH_3)_{9.81}$$
(including four D-glucopyranose rings).

Calculate δ_d, δ_p, δ_h, and δ_{sp}, using numerical values given in Table 1.

- $\sum V_{g,i} = 4 \times 15.9) + (20 \times 9.5) + (8 \times 10.0) + (2.19 \times 9.7) + (9.81 \times 23.0) + (9.81 \times 23.9) = 814.93$

- $\sum F_{d,i} = (4 \times 132) + (20 \times 39) + (8 \times 49) + (2.19 \times 103) + (9.81 \times 191) + (9.81 \times 205) + (4 \times 93) = 6182.4$

- $\sum F_{p,i}^2 = (4 \times 0)^2 + (20 \times 0)^2 + (8 \times 196)^2 + (2.19 \times 244)^2 + (9.81 \times 239)^2 + (9.81 \times 0)^2 = 8{,}241{,}267$

- $\sum E_{h,i} = (4 \times 0) + (20 \times 0) + (8 \times 717) + (2.19 \times 4777) + (9.81 \times 1672) + (9.81 \times 0) = 32{,}600$

- $\sum V_i = (4 \times 16.1) + (20 \times -1.0) + (8 \times 3.8) + (2.19 \times 10.0) + (9.81 \times 18.0) + (9.81 \times 33.5) = 601.92$

- $\sum E_{coh,i} = (4 \times 1180) + (20 \times 820) + (8 \times 800) + (2.19 \times 7120) + (9.81 \times 4300) + (9.81 \times 1125) = 96{,}332.1$

- $\delta_d = 6182.4/814.93 = 7.57 \text{cal}^{1/2}\ \text{cm}^{-3/2} = 15.54 J^{1/2}\ \text{cm}^{-3/2}$

- $\delta_p = \sqrt{8{,}241{,}267}/814.93 = 3.52\ \text{cal}^{1/2}\ \text{cm}^{-3/2} = 7.20\ J^{1/2}\ \text{cm}^{-3/2}$

- $\delta_h = \sqrt{32{,}600/814.93} = 6.32\ \text{cal}^{1/2}\ \text{cm}^{-3/2} = 12.95\ J^{1/2}\ \text{cm}^{-3/2}$

- $\delta_{sp} = \sqrt{96{,}322.1/601.92} = 12.65\ \text{cal}^{1/2}\ \text{cm}^{-3/2} = 25.88\ J^{1/2}\ \text{cm}^{-3/2}$

Example 2

Calculate the solubility parameters of (1) aromatic polyamide, (2) aromatic polyhydrazide, and (3) aromatic polyamidehydrazide.

The structure of the repeat unit of the polymer is

1. $NH\phi_m NHCO\phi_r CO$ [including 2(aromatic CONH) and 2ϕ]

2. $NHNHCO\phi_m CONHNHCO\phi_r CO$ [including 2(aromatic CONHNHCO) and 2ϕ]

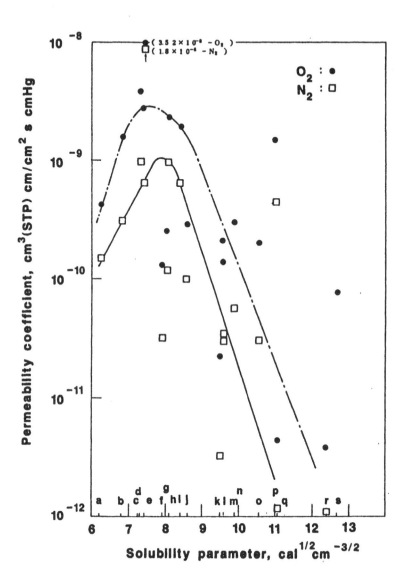

Figure 2.7.   Permeability coefficient of gas vs. solubility parameter.

3. $NH\phi_rCONHNHCO\phi_rCO$

where ϕ_m and ϕ_r indicate phenylene groups with a substitution at the meta position and a random substitution either at the meta or at the para position.

1.

· $\sum V_{g,i} = (2 \times 65.5) + (2 \times 24.9) = 180.8$

· $\sum F_{d,i} = (2 \times 621) + (2 \times 220) = 1682$

· $\sum F_{p,i}^2 = (2 \times 54)^2 + (2 \times 479)^2 = 929{,}428$

· $\sum E_{h,i} = (2 \times 0) + (2 \times 7762) = 15{,}524$

· $\sum V_i = (2 \times 52.4) + (2 \times 9.5) = 123.8$

· $\sum E_{coh,i} = (2 \times 7630) + (2 \times 8000) = 31{,}260$

· $\delta_d = 1682/180.8 = 9.30 \text{ cal}^{1/2} \text{ cm}^{-3/2} = 19.03 \text{ J}^{1/2} \text{ cm}^{-3/2}$

· $\delta_p = \sqrt{929{,}428/180.8} = 5.33 \text{ cal}^{1/2} \text{ cm}^{-3/2} = 10.90 \text{ J}^{1/2} \text{ cm}^{-3/2}$

· $\delta_h = \sqrt{15{,}524/180.8} = 9.27 \text{ cal}^{1/2} \text{ cm}^{-3/2} = 18.96 \text{ J}^{1/2} \text{ cm}^{-3/2}$

· $\delta_{sp} = \sqrt{31{,}260/123.8} = 15.89 \text{ cal}^{1/2} \text{ cm}^{-3/2} = 32.51 \text{ J}^{1/2} \text{ cm}^{-3/2}$

2.

· $\sum V_{g,i} = (2 \times 65.5) + (2 \times 49.8) = 230.6$

· $\sum F_{d,i} = (2 \times 621) + (2 \times 440) = 2122$

· $\sum F_{p,i}^2 = (2 \times 54)^2 + (2 \times - - - -)^2 = - - - -$

· $\sum E_{h,i} = (2 \times 0) + (2 \times 10{,}629) = 21{,}258$

· $\sum V_i = (2 \times 52.4) + (2 \times 19.0) = 142.8$

· $\sum E_{coh,i} = (2 \times 7630) + (2 \times 11{,}200) = 37{,}660$

· $\delta_d = 2122/230.6 = 9.20 \text{ cal}^{1/2} \text{ cm}^{-3/2} = 18.82 \text{ J}^{1/2} \text{ cm}^{-3/2}$

· $\delta_p = - - - - -$

· $\delta_h = \sqrt{21{,}258/230.6} = 9.60 \text{ cal}^{1/2} \text{ cm}^{-3/2} = 19.65 \text{ J}^{1/2} \text{ cm}^{-3/2}$

· $\delta_{sp} = \sqrt{37{,}660/142.8} = 16.24 \text{ cal}^{1/2} \text{ cm}^{-3/2} = 33.25 \text{ J}^{1/2} \text{ cm}^{-3/2}$

3.

· $\delta_d = (9.30 + 9.20)/2 = 9.25 \text{ cal}^{1/2} \text{ cm}^{-3/2}$

· $\delta_h = (9.27 + 9.60)/2 = 9.44 \text{ cal}^{1/2} \text{ cm}^{-3/2}$

· $\delta_{sp} = (15.89 + 16.24)/2 = 16.07 \text{ cal}^{1/2} \text{ cm}^{-3/2}$

Example 3

The structure of the repeat unit of polyethylene terephthalate is:

$$\phi_p COCH_2CH_2OC \text{ [including } 1(\phi), 2(COO) \text{ and } 2(CH_2)].$$

Calculate the solubility parameters (δ_d and δ_h) of the polymer. Is this polymer suitable for reverse osmosis membranes for desalination purposes?

- $\sum V_{g,i} = (1 \times 65.5) + (2 \times 23.0) + (2 \times 15.9) = 143.3$
- $\sum F_{d,i} = (1 \times 621) + (2 \times 191) + (2 \times 132) = 1267$
- $\sum E_{h,i} = (1 \times 0) + (2 \times 1672) + (2 \times 0) = 3344$
- $\delta_d = 1267/143.3 = 8.84 \text{ cal}^{1/2} \text{ cm}^{-3/2} = 18.09 \text{ J}^{1/2} \text{ cm}^{-3/2}$
- $\delta_h = \sqrt{3344/143.3} = 4.83 \text{ cal}^{1/2} \text{ cm}^{-3/2} = 9.89 \text{ J}^{1/2} \text{ cm}^{-3/2}$

The position of the δ_d, δ_h plot on Figure 2.1 is in the range of polymer materials suitable to produce desalination membranes. However, the position is close to the boundary.

Example 4

Calculate the solubility parameters δ_d and δ_h of polyvinyl alcohol and polyvinyl acetal polymers. The structures of the repeat units of the polymers are

polyvinyl alcohol $-CH_2-CH-$
                         $|$
                        OH

[including $1(CH_2)$, $1(CH)$ and $1(OH)$]

polyvinyl acetal $-CH_2-CH-CH_2-CH-$
                         $|$ $|$
                         O O
                          $\diagdown$ $\diagup$
                            CH
                            $|$
                            CH_3

[including $1(CH_3)$, $2(CH_2)$, $3(CH)$ and $2(O)$].

Are these polymers suitable to produce membranes for desalination purposes? Is a mixture of 20% polyvinyl alcohol and 80% polyvinyl acetal suitable to produce desalination membranes?

Polyvinyl Alcohol

- $\sum V_{g,i} = 15.9 + 9.5 + 9.7 = 35.1$

- $\sum E_{d,i} = 132 + 39 + 103 = 274$
- $\sum E_{h,i} = 0 + 0 + 4777 = 4777$
- $\delta_d = 274/35.1 = 7.83$ cal$^{1/2}$ cm$^{-3/2}$ = 16.0 J$^{1/2}$ cm$^{-3/2}$
- $\delta_h = \sqrt{4777/35.1} = 11.68$ cal$^{1/2}$ cm$^{-3/2}$ = 23.9 J$^{1/2}$ cm$^{-3/2}$

Polyvinyl Acetal

- $\sum V_{g,i} = (1 \times 23.9) + (2 \times 15.9) + (3 \times 9.5) + (2 \times 10.0) = 104.2$
- $\sum F_{d,i} = (1 \times 205) + (2 \times 132) + (3 \times 39) + (2 \times 49) + (93) = 777$
- $\sum E_{h,i} = (1 \times 0) + (2 \times 0) + (3 \times 0) + (2 \times 717) = 1434$
- $\delta_d = 777/104.2 = 7.45$ cal$^{1/2}$ cm$^{-3/2}$ = 15.26 J$^{1/2}$ cm$^{-3/2}$
- $\delta_h = \sqrt{1434/104.2} = 3.71$ cal$^{1/2}$ cm$^{-3/2}$ = 7.59 J$^{1/2}$ cm$^{-3/2}$

Both polymers are outside the range of the solubility parameters that enable the manufacture of satisfactory desalination membranes. When 20% of polyvinyl alcohol is mixed with 80% of polyvinyl acetal,

- $\delta_d = (0.2 \times 7.83) + (0.8 \times 7.45) = 7.53$ cal$^{1/2}$ cm$^{-3/2}$
- $\delta_h = (0.2 \times 11.68) + (0.8 \times 3.71) = 5.31$ cal$^{1/2}$ cm$^{-3/2}$

The mixture seems to be suitable to produce reverse osmosis membranes for seawater desalination.

Nomenclature for Chapter 2

ΔE = heat of vaporization, J/mol

E_{coh} = structural component for the overall solubility parameter, J/mol

E_{hi} = structural component for the hydrogen bonding component of the solubility parameter, J/mol

F_{di} = structural component for the dispersion force component of the solubility parameter, $J^{1/2}m^{3/2}/mol$

F_{pi} = structural component for the dipole component of the solubility parameter, $J^{1/2}m^{3/2}/mol$

ΔG = free energy of mixing, J/mol

ΔH = heat of mixing, J/mol

ΔH_M = total heat of mixing, J/mol

K = partition coefficient,—

ΔS = entropy of mixing, J/mol K

T = absolute temperature, K

V = molar volume, m^3/mol

V_d = dead space of the instrument, m^3

V_m = volume of mobile phase, m^3

V_M = total molar volume of the mixture, m^3/mol

V_R = retention volume, m^3

V'_R = retention volume observed experimentally, m^3

$[V'_R]_{water}$ = retention volume of water represented by that of heavy water, m^3

$[V'_R]_{min}$ = smallest retention volume, m^3

V_s = volume of stationary phase, m^3

δ = solubility parameter, $J^{1/2}/m^{3/2}$

ϕ = volume fraction

Subscripts

1 = component 1

2 = component 2

3

Membrane Preparation

3.1   Introduction

When a material is given, membranes should be prepared from the given material. The membranes so prepared should then be tested by separation experiments. Two kinds of data are obtained from such experiments: one is the degree of separation of fluid (either liquid or gas) mixtures, and the other is the total permeation rate. There are several different ways to measure the degree of separation, but all are concerned with the concentration of a targeted component of the mixture in the permeate relative to that in the feed. Often, there are several targeted components. The total permeation rate is the sum of the permeation rates of the individual components. While the separation depends on the pore size and the pore size distribution of the surface layer of the membrane that is in contact with the feed fluid, the permeation rate depends on the permeation rate, which is normally desirable, the thickness should be as small as possible. Thus, it is necessary to provide a membrane with an asymmetric structure that consists of at least two layers of different pore sizes. One is a thin surface layer with pore sizes small enough to make the separation possible, and the other is a layer with much larger pore sizes, called a porous layer. The latter layer is much thicker than the surface layer in order to provide a membrane with sufficient mechanical strength. The permeation rate is, however, governed primarily by the surface layer, since the pore size of the porous layer is very large.

There are different methods to prepare membranes. The principle underlying the membrane preparation is, however, always the same; i.e., it is to control the pore size and the pore size distribution at the surface layer and to decrease the thickness of the surface layer. Several methods of membrane preparation are described in the following section.

3.2   Methods of Membrane Preparation

3.2.1   Preparation of Asymmetric Membranes — Phase-Inversion Technique

Cellulose acetate membranes developed by Loeb and Sourirajan for the purpose of seawater desalination continue to be useful in various membrane applications, despite the development of new membrane materials and new membrane preparation techniques. Because of its historical importance, the casting method of the first successful reverse osmosis membrane is described below in detail.

A polymer solution of the following composition is prepared (numbers are given as weight percent.):

- · Cellulose acetate (E-398-3, acetyl content 39.8%): 22.2
- · Acetone: 66.7
- · Water: 10.0
- · Magnesium perchlorate: 1.1

The film is cast in a cold box at 0 to $-10°C$ on a glass plate with 0.025-cm side runners to give this thickness to the as-cast film. A thick glass rod resting on the side runners is passed across the top of the glass plate in a cold box. Acetone is allowed to evaporate partially, in the cold box, from the surface for a period of 3 to 4 min. This step is called the evaporation step. The glass plate containing the partially hardened film is then immersed into ice cold water in a tray where it is left for at least 1 h. While standing in ice cold water, the film detaches from the surface of the glass plate. This step is called the gelation step. This film in the as-cast condition is too porous to be used as a reverse osmosis membrane. Hence, it is subjected to posttreatments, including heat and pressure treatment. The heat treatment consists of holding the as-cast membrane floating freely between glass plates in water while increasing the water temperature gradually from the laboratory temperature to about 80°C. After holding the above temperature for 10 min, the water is cooled rapidly. The pressure treatment consists of holding the heat-treated membrane in the permeation cell at a pressure about 20% higher than the operating pressure.

As can be seen from the above description, there are many variables involved in the phase-inversion technique. Among others the composition of the polymer solution, the solvent evaporation temperature and evaporation period, the nature and the temperature of the gelation media, and the heat treatment temperature are the primary factors affecting the reverse osmosis performance of the membrane. When polymers other than cellulose acetate are used, solvents and nonsolvent additives appropriate to prepare membranes from the particular polymer must be found. Depending on the combination of variables, membranes of different polymeric materials with different pore sizes can be prepared.

3.2.2   Formation of Dry Asymmetric Cellulose Acetate Membranes for Gas Separation

The wet cellulose acetate membranes prepared in the foregoing step for reverse osmosis should be dried to be used for gas separation. The water in the membrane cannot be evaporated in air; the asymmetric porous structure of the membrane is destroyed during the evaporation step. Instead, the following multiple-stage solvent exchange method can be applied.

In this technique the water in the membrane is first replaced by a water-miscible solvent, called the first solvent, which is a nonsolvent for the membrane polymeric material. Then, the first solvent is replaced by a second solvent that is volatile. The second solvent is subsequently air evaporated to obtain the dry membrane. A number of different solvents, and a combination thereof, can be used both for the first and the second solvents. First solvents include ethyl alcohol, isopropyl alcohol, tertiary butyl alcohol, ethylene glycol, diethylene glycol, triethylene glycol, and ethylene glycol monoethyl ether. The second solvents include pentane, hexane, cyclohexane, benzene, toluene, carbon disulfide, and isopropyl ether. The replacement of water in the membrane with the first solvent is done by successive immersion in first solvent-water mixtures that are progressively more concentrated in the first solvent. For example, in four-stage replacement, 25, 50, 75, and 100 vol% aqueous solutions of the first solvent are used. When one of the glycols is used as a first solvent, an intermediate solvent should be used between the first and second solvents, since glycols are not miscible with the second solvents. Intermediate solvents are ethyl alcohol and *n*-butyl alcohol. The usage of different solvents, and the combination thereof, results in membranes of different average pore size and pore size distributions. Upon completion of the solvent exchange, membranes are transferred to a desiccator where solvents are evaporated slowly.

3.2.3   Preparation of Ultrathin Membranes

Recognizing that the key to prepare a reverse osmosis membrane of high flux is its asymmetric structure, a thin selective layer deposited on top of a porous sublayer, an ultrathin top layer was made by spreading a dilute polymer solution on water. The resultant ultrathin membrane had thicknesses ranging from 140 to 500 nm [28].

A polymer is first dissolved in an organic solvent to give the solution a consistency about equal to a varnish or light syrup. The polymer solution is then poured onto the surface of water. The solution spreads spontaneously over the water surface. The solvent is evaporated completely, leaving the polymer residue in the form of a membrane. The membrane thickness can be controlled by adjusting the solution concentration and viscosity and sometimes by manually stretching the spread solution. Esters and ketones such as isobutyl acetate and cyclohexane were found to be excellent solvents for cellulosic polymers. In some cases when polymers are not soluble in the solvent at room temperature, dissolution of the polymer can be achieved by warming the solvent to 50 to 60°C.

The ultrathin membranes prepared by the above method are transferred to the top surface of a porous substrate membrane before being mounted in the permeation cell. Normally, the air-dried side of the membrane is brought into contact with the feed solution. When the other side (the side that faced water during solvent evaporation) is brought into contact with the feed solution, there is no significant change in the reverse osmosis performance of the membrane. Therefore, the ultrathin membrane has no asymmetric structure as far as membrane permeation is concerned.

3.2.4   Formation of a Thin Layer on a Porous Support Membrane

Since the transfer of an ultrathin membrane to a porous support layer was difficult, an attempt was made by Riley et al. to cast an ultrathin layer directly onto the support membrane [29]. The composite membrane consists of the following three layers:

1. A porous sublayer made of cellulose nitrate-cellulose acetate material
2. An intermediate layer of polyacrylic acid acting as a masking layer of the porous substrate membrane
3. A selective skin layer made of cellulose triacetate material

According to the method of Riley et al., the porous support membrane is prepared by casting a solution of the following compositions (wt%) onto a glass plate:

- Cellulose nitrate (DuPont DHA35E) 7.8
- Cellulose acetate (Eastman E383-40) 1.3
- Acetone 53.7
- Absolute ethanol 19.9
- N-butyl alcohol 13.3
- Glycerol 1.9
- Surfactant (Rohm and Haas Triton X-100) 0.5

The membrane is allowed to gel under controlled air flow, temperature, and humidity conditions. The membrane is asymmetric, with a very thin skin layer on the air-dried surface, which ranges from about 5 to 15% in porosity, supported by a graduated porous substrate of interconnected pores. The thickness of the support membrane is 100 to 125 μm.

The intermediate masking layer is applied by an airless spray of a 2.5-wt% solution of polyacrylic acid (Rohm and Haas Co., Acrysol A-1, molecular weight 50,000) in equal volumes of ethanol and water to the finely porous glossy surface of the porous substrate membrane. The support membrane is preheated and maintained at 85°C for spraying. The thickness of the intermediate layer is 3 μm.

The selective layer is formed by contacting the surface of the masking layer with a 0.5-wt% solution of cellulose triacetate (Eastman E432-130B) in chloroform and withdrawing the surface vertically. The withdrawal rate is 1.9 cm/s. During the withdrawal a stream of dry air is passed over the membrane surface.

3.2.5 In Situ Polymerization

A thin polymer film can be formed in situ on the surface of a porous substrate membrane by an interfacial polycondensation process. A classical example of this method was described by Rozelle et al. in detail for the formation of the North Star NS-100 membrane [30]. According to their description, the polysulfone support films are placed, shiny surface upwards, into a 0.67% aqueous polyethylenimine (PEI) solution in an aluminum tray. After 1 min, the PEI solution is poured off, and the tray held in a vertical position for 1 min to allow the excess solution to drain from the surface of the film. Then the wet surface is contacted with a 0.5% solution of toluene 2,4-diisocyanate (TDI) for 1 min at room temperature. After draining the excess TDI solution, the tray is placed horizontally at 115°C for 10 min. After the heat curing, the composite membrane is easily peeled off from the aluminum surface.

The NS-100 membrane so prepared is believed to consist of three layers depicted by Figure 3.1.

1. Polysulfone porous sublayer of about 1.7 mil thickness, having water flux of 500 to 5000 gfd with no separation capability for salt solution

2. Polyethylenimine layer exhibiting the water flux of 50 gfd with 70% salt separation

3. Polyethylenimine-2,4-diisocyanate layer considered to be responsible for the selectivity of the composite membrane with a structure depicted in Figure 3.2.

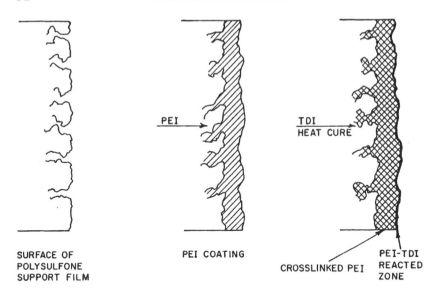

PEI
TDI
HEAT CURE
SURFACE OF
POLYSULFONE
SUPPORT FILM
PEI COATING
CROSSLINKED PEI
PEI-TDI
REACTED
ZONE

Figure 3.1. Structure of NS-100 membrane. (Reproduced from [30] with permission.)

CH$_2$CH$_2$ GROUPS REPRESENTED BY ⟛

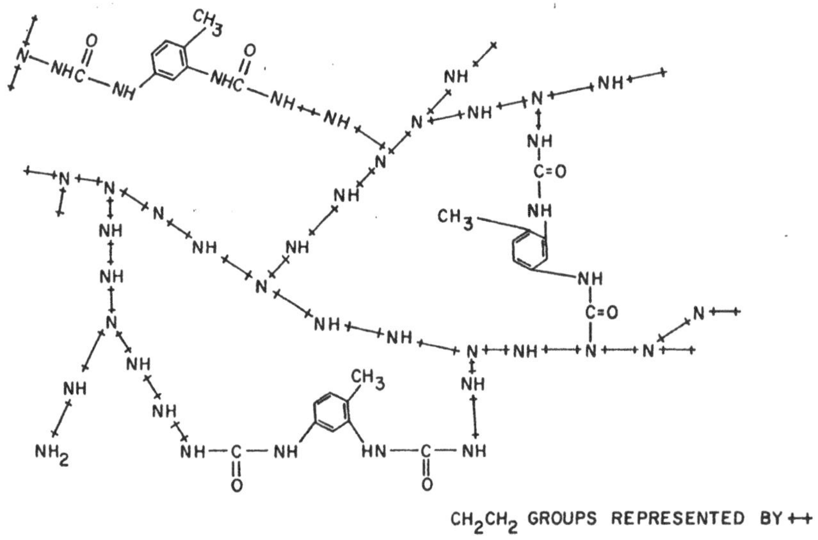

Figure 3.2. Structure of polyethylenimine-toluene 2,4-diisocyanate layer. (Reproduced from [30] with permission.)

3.2.6 Coating of a Thin Silicone Rubber Layer on the Surface of a Porous Polyethersulfone Substrate Membrane

According to Chen et al. [31], the porous substrate membrane is prepared in the following way. A solution of polyethersulfone (Victrex 200P) polymer in dimethyl sulfoxide is prepared (21 to 26 wt% polymer) and cast onto polyethylene backing cloth, to the thickness of 254 μm. Then, the cast membrane is gelled in ice cold water for 1 h.

The coating of silicone rubber on the porous sublayer is done in the following way. The wet polyethersulfone membrane prepared above is dried in a desiccator for about a week. The membrane is then kept in an oven with forced air circulation at 60°C for 1 h. The dry polyethersulfone membrane so prepared is mounted at the bottom of a cylindrical test cell that can hold 50 ml of silicone in a hexane solution in the feed side chamber. The latter solution is prepared by dissolving 5 g of polydimethylsiloxane, 0.25 g of tetraethyl orthosilicate, and 0.25 g of dibutyltin dilaurate in 100 ml of hexane. The solution is left in the test cell for 1 h. The hexane solution is then poured out of the test cell, and the cell is kept upside down for 30 min so that any residual solution is drained and a thin silicone layer is formed after hexane evaporation. The silicone layer is subjected to cross-linking by keeping the cell in an oven at 60°C for 1 h. The coating can be repeated to form a multilayer. The silicone-coated membrane, together with the test cell, can be used for the permeation experiment immediately.

3.2.7 Plasma Coating

When a vacuum is maintained inside a tubular reactor and a high frequency field is applied outside, a glow discharge is generated inside the reactor. A plasma that consists of various ions, radicals, electrons, and molecules is formed in the glow discharge. Those species originate from the inserted or residual gas in the reactor and repeat decomposition and recombination while emitting a glow. When a substrate membrane is placed into the plasma, the surface of the membrane is subject to various changes corresponding to the property of plasma. In particular, (1) when the inserted or residual gas is a nonpolymerizing gas such as nitrogen or air, the substrate surface is modified and/or chemically active sites are introduced to the surface; and (2) when the gas in the reactor is an organic compound, an irregular polymerization occurs on the substrate surface. This is called plasma polymerization.

Figure 3.3 illustrates the tubular reactor for the plasma polymerization. The reactor diameter is 75 mm and its length is 300 mm. A porous polysulfone substrate membrane (Brunswick SDM 90-25, average pore diameter 0.1 μm) disk of 70 mm in diameter is placed in the reactor. After evacuating the reactor, a mixed

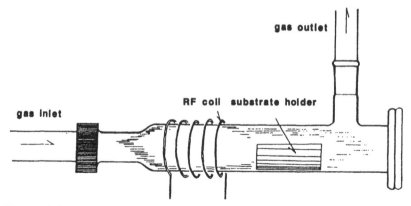

Figure 3.3. Tubular reactor for plasma polymerization. (Reproduced with permission of the author.)

gas stream of hexafluoroethane and allylamine (hexafluoroethane/allylamine ratio, 0 to 2.0) is introduced to the reactor at the allylamine flow rate of 2.0 cm^3 STP/min under the pressure of 10 to 20 Pa. Then a low-temperature plasma is excited by a 13.56-MHz power source at 25 to 30 W. The reaction time is 4 h, and the plasma layer thickness is 10 μm.

Figure 3.4 shows a reactor used for the plasma graft-polymerization. The reactor diameter is 30 mm, and the length is 240 mm. A square polypropylene substrate membrane (Celanese, Celgard 2400, 200 × 2000 Å and 2500, 400 × 4000 Å, 6 × 6 cm) is placed in the reactor and is subjected to plasma treatment for 30 s in the presence of residual air under the pressure of 30 Pa. A low-temperature plasma is excited by a 13.56-MHz power source at less than 20 W to prevent the polypropylene substrate membrane from heat damage. Then an aqueous solution of methacrylic acid-2-hydroxyethyl and acrylic acid mixture (monomer concentration 5 wt%) is transferred from the monomer solution reservoir to the reactor. The reactor is immersed into a temperature-controlled bath where polymerization is carried out for 2 h at 50 to 70°C [32].

3.2.8   Dynamic Formation of a Thin Layer

A porous polyvinyl chloride-acrylonitrile membrane with a nominal pore size of 0.4 μm cast on a nylon fabric is available commercially as Acropor AN from Gelman Co. This porous filter material was wrapped around a porous stainless steel tube (pore size 5 μm) of 1.5875 cm (5/8 in.) outer diameter. A 0.05-mol/l aqueous sodium chloride solution containing 10^{-4} mol/l hydrous Zr(IV) oxide was circulated in the annulus between the outside of the tube and a pressure jacket. When the feed velocity was 10.67 m/s (35 ft/s), the sodium chloride separation was 64% and the solution flux 480 gfd at 6.55 MPa (950 psig) [33].

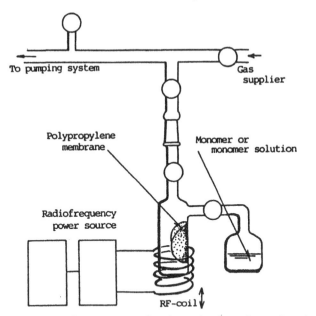

To pumping system

Gas
supplier

Polypropylene
membrane

Monomer or
monomer solution

Radiofrequency
power source

RF-coil

Figure 3.4.   Reactor for plasma graft-polymerization. (Reproduced with permission of the author.)

A substantial increase in the sodium chloride separation was observed when the above Zr(IV) oxide-coated membrane was exposed to a 0.05-mol/l sodium chloride solution containing 25 ppm of polyacrylic acid (PAA, Rohm and Haas Co. Acrysol A-3). The pH of the solution was adjusted to 7 by adding sodium hydroxide.

3.2.9   Preparation of Inorganic Membranes

Figure 3.5 illustrates the alumina membrane preparation procedures. An aqueous aluminum-s-butoxide solution is heated at a temperature above 80°C to precipitate γ-AlOOH (boehmite), and then 0.03 to 0.15 mol/l of acid (HCl, HNO$_3$, HClO$_4$) is added at a temperature above 90°C to produce stable γ-AlOOH sol by peptization. The size of the γ-AlOOH sol particle is determined by the temperature, the type of acid, the concentration of acid, and the precipitation period. The control of the size of the particle is important, since it will govern the size of the pore of the membrane. In order to produce a composite γ-Al$_2$O$_3$ membrane, one side of a porous support membrane with a pore size from 0.12 to 0.8 μ is brought into contact with the γ-AlOOH sol. When the pore of the support membrane is sufficiently small, water in the sol enters the pores of the support membrane by capillary action; the boehmite particle is left at the pore entrance, and its concentration increases. At a certain concen-

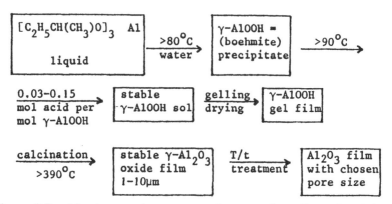

Figure 3.5. Alumina membrane preparation procedure. (Reproduced from [34] with permission.)

tration the boehmite sol turns into gel. The type and the amount of acid used to peptize the sol plays an important role in the gel layer formation. The gel layer thickness increases with an increase in the period of contact between the porous support membrane and the sol solution. The gel film so prepared is then calcinated at a temperature above 390°C to form a stable γ-Al_2O_3 film with a thickness controlled between 1 and 10 μm. The γ-Al_2O_3 film is further heat treated in an oven at an elevated temperature. The temperature and the period of the heat treatment determines the pore size of the γ-alumina membrane. The observed pore size is shown in Table 1 for different temperatures and periods of the heat treatment. As the temperature increases from 400 to 800°C, the pore diameter increases from 2.5 to 4.8 nm, while the porosity changes slightly. At 900°C the pore size becomes 5.4 nm, but the porosity decreases, indicating the formation and the sintering of α-alumina. The pore size distributions observed by the gas adsorption and the mercury intrusion methods are given in Figure 3.6 for different heat treatment temperatures [34].

3.2.10   Hollow Fiber Formation

A simple experimental device for the hollow fiber spinning is given in Figures 3.7 and 3.8 [35]. The spinning system consists of a dope tank (1), a filter (2), a spinneret (3), gelation baths (5, 6, 7), a nitrogen cylinder (8), and valves. The spinneret has a tube-in-orifice structure as shown in Figure 3.8. A polymer solution (typical polyethersulfone polymer [23 wt%], N,N-dimethylacetamide [65 wt%] and polyvinylpyrrolidone [12 wt%]) under nitrogen pressure (typical 35 to 200 kPa) passes through the filter, goes into the spinneret, and is extruded from the annular space of the spinneret (see Figure 3.8). The internal coagulant, water, in a funnel (4) comes out from the central tube of the spinneret as a water

Table 1. Microstructural Characteristics of Alumina Films
as a Function of Sintering Temperature and Sintering Time [34]

Temperature (°C)	Time (h)	Phase	BET-surface (m^2/g)	Pore size (nm)	Porosity (%)
200	34	γ-AlOOH	315(2)[a]	2.5	41
400[A]	34	γ-Al_2O_3	301(7)	2.7	53
	170	γ-Al_2O_3	276(4)	2.9	53
	850	γ-Al_2O_3	249(2)	3.1	53
500	34	γ-Al_2O_3	240(1)	3.2	54
700	5	γ-Al_2O_3	207(2)	3.2	51
	120	γ-Al_2O_3	159(2)	3.8	51
	930	γ-Al_2O_3	149(2)	4.3	51
800[B]	34	γ-Al_2O_3	154(2)	4.8	55
900	34	θ-Al_2O_3	99(2)	5.4	48
1000[C]	34	α-Al_2O_3	15(3)	78	41
550[b]	34	γ-Al_2O_3	147(4)	6.1	59

[a] The standard deviation is given in parentheses.
[b] Prepared from a sol obtained by an autoclave treatment at 200°C.

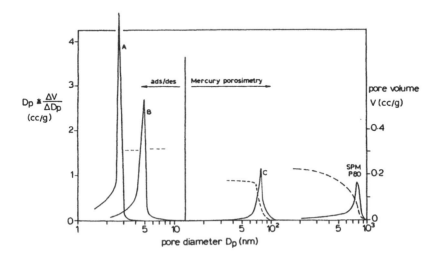

Figure 3.6. Pore size distribution of inorganic membranes. (Reproduced from [34] with permission.)

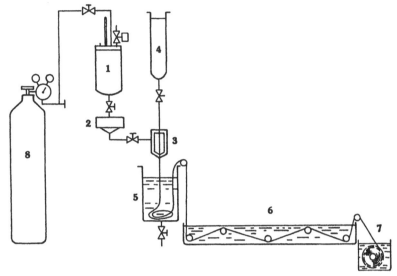

Figure 3.7.   Schematic diagram of hollow fiber spinning system. (Reproduced from [35] with permission.)

jet under gravitational force at the same time. Typical flow rates of the internal coagulant and the polymer solution are 1 m/s and 3.5 cm/s, respectively. A nascent hollow fiber thread is partially coagulated by the internal coagulant while polymer solution is travelling over the distance from the spinneret outlet to the gelation bath (5), which is called an air gap. A typical air gap is 60 cm. The coagulation is completed in the coagulation bath by an external coagulant. A fiber with sufficient mechanical integrity passes over the rollers in a bath (6), followed by a bath (7) where the fiber is collected.

3.3   Membrane Permeation Experiments

3.3.1   Reverse Osmosis and Ultrafiltration

The detail of the permeation cell for reverse osmosis experiments and the flow system are shown schematically in Figure 3.9. The permeation cell is made of stainless steel 310 and consists of two detachable parts. The upper part is a high-pressure chamber. A wet membrane is mounted on a stainless steel porous plate embedded in the lower part of the cell such that the active surface layer of the asymmetric membrane faces the feed solution under high pressure. A wet Whatman filter paper is placed between the membrane and the porous plate to protect the membrane from abrasion. The feed solution is supplied to the feed chamber of the permeation cell by a pressure pump, while the permeate

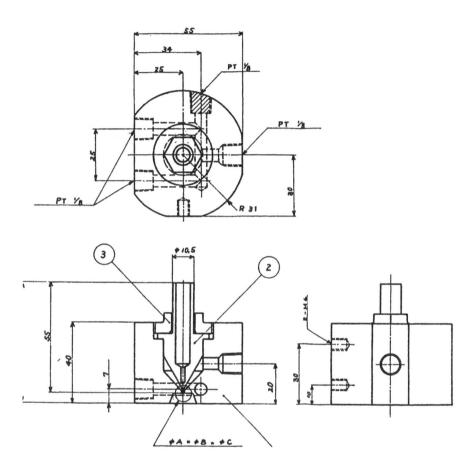

Figure 3.8. Hollow fiber spinneret (with permission of Daicel Chemical Industries Ltd.)

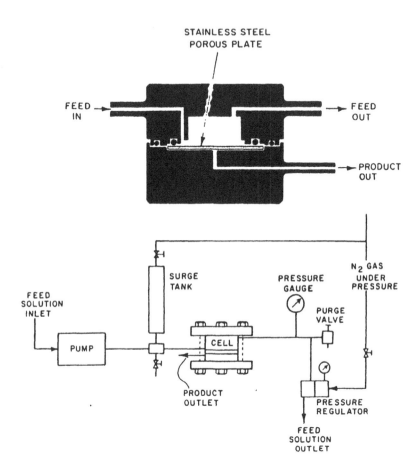

Figure 3.9. Reverse osmosis cell and test system. (Reproduced from [5] with permission.)

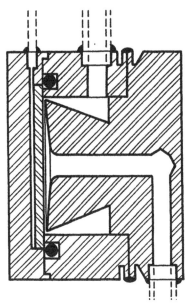

Figure 3.10. Ultrafiltration cell design. (Reproduced from [5] with permission.)

(product) solution is discharged from the permeate (product) outlet, which is open to the atmosphere. The pressure of the feed solution is measured by a pressure gauge, and the solution is released to an atmospheric pressure through a pressure regulator. Nitrogen gas under pressure from a gas cylinder is used to load the dome of the pressure regulator and to provide a gas cushion in the surge tank.

The design of the permeation cell used for ultrafiltration experiments is shown in Figure 3.10. The permeation cell has a thin-channel flow design above the membrane surface, which allows relatively high fluid velocity parallel to the membrane surface. As shown in the figure, the feed fluid enters the cell through the center opening, flows radially through the thin channel, and leaves the cell through the side opening.

The design of the static cell is illustrated in Figure 3.11. This cell is used for the batchwise operation when the quantity of the feed solution is limited. The cell is a stainless steel pressure vessel consisting of two detachable parts. The membrane is mounted on a stainless steel porous plate embedded in the lower part of the cell through which the permeate liquid is discharged into the atmosphere. The upper part of the cell holds the feed fluid under pressure. Compressed nitrogen is used to apply pressure on the feed fluid. The feed solution is kept well stirred during the permeation experiment, by means of a magnetic stirrer fitted in the cell about 0.3 cm above the membrane surface.

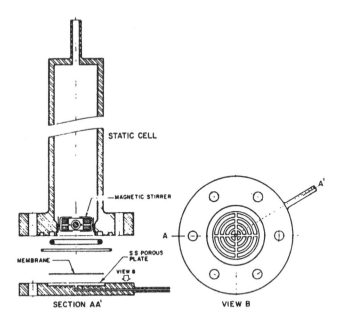

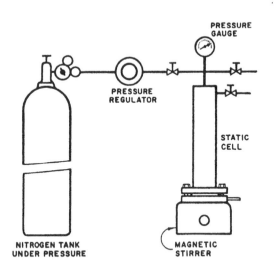

Figure 3.11.   Design of the static cell. (Reproduced from [5] with permission.)

3.3.2 Membrane Gas Separation

The static cell used for reverse osmosis and ultrafiltration experiments can be used to test the separation of gas mixtures. Air in the feed chamber of the test cell and the feed gas line is removed by flushing them with the feed gas stream. The feed gas is then supplied to the feed gas chamber under pressure. The gas permeation velocity is measured by a bubble flow meter connected to the permeate side of the test cell. The permeate sample is also subjected to analysis by gas chromatography. This simple device is useful when an asymmetric membrane is tested and when the permeation rate is high.

3.3.3 Pervaporation

The schematic diagram of the pervaporation apparatus is shown in Figure 3.12. The same permeation cell as the reverse osmosis static cell can be used for the experiment. The feed liquid mixture in the permeation cell (2) is either open to the atmosphere or under the pressure applied from a nitrogen cylinder (1). Vacuum is applied on the permeate side of the membrane by a vacuum pump (6). The permeate vapor is condensed and collected in a cold trap (5) cooled with liquid nitrogen. After a steady state is reached, the line is switched to the second cold trap. The permeation rate is determined by weighing the sample collected in the cold trap during a prescribed period. The sample is also subjected to analysis.

3.3.4 Vapor Permeation

The flow diagram of the experimental setup is shown schematically in Figure 3.13. A mixture of organic vapor and nitrogen is produced by bubbling nitrogen gas from a porous sintered stainless steel ball immersed in a chosen organic liquid at a temperature slightly lower than the permeation cell temperature. The permeation cell is housed in an isothermal chamber. The permeate side of the membrane is connected to two cold traps followed by a vacuum pump. The permeation of the organic vapor is induced by maintaining its partial pressure on the permeate side lower than the feed side. The membrane-permeated organic vapor is condensed and collected initially in one of the cold traps, and then the cold trap is switched to the other after the steady state is reached. The permeation rate is determined gravimetrically by weighing the sample collected for a predetermined period.

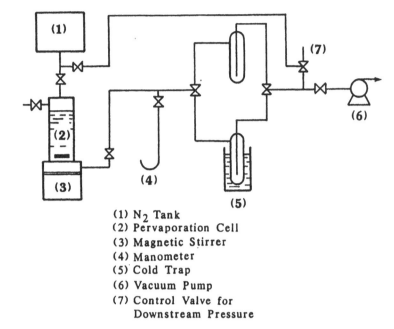

(1) N$_2$ Tank
(2) Pervaporation Cell
(3) Magnetic Stirrer
(4) Manometer
(5) Cold Trap
(6) Vacuum Pump
(7) Control Valve for
    Downstream Pressure

Figure 3.12. Schematic flow diagram of pervaporation experimental system. (Reproduced from [239] with permission.)

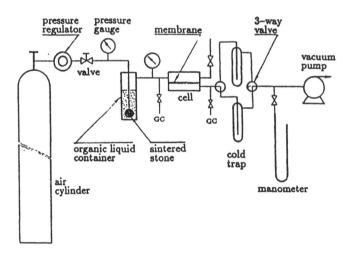

Figure 3.13. Schematic flow diagram of vapor permeation experimental system.

3.4   Triangular Phase Diagram

As mentioned earlier, the phase-inversion technique by the Loeb-Sourirajan method is by far the most popular method to prepare polymeric membranes with asymmetric structure, although there are many other methods of membrane preparation. Hence, it is meaningful to discuss the formation mechanism of such an asymmetric structure during the phase-inversion process. In this process the polymer solution that is homogeneous when cast on a glass plate should cross a phase boundary and split into two phases during the solvent evaporation step or during the gelation step. The whole process can be described by a change of the composition in the cast membrane, along a line on a ternary nonsolvent (N)-solvent (S)-polymer (P) composition diagram, because gelation media are nonsolvent to the membrane polymer. A phase boundary line is also drawn on the ternary composition diagram.

Since the separation of a given composition into two phases of different compositions that are in equilibrium is a thermodynamic process, some thermodynamic consideration is necessary to draw a phase boundary line. On the other hand, the composition change on the ternary diagram is affected by the speed of solvent evaporation or by the speed of solvent-nonsolvent exchange during the gelation step. Hence, kinetic consideration is necessary to draw a line of composition change on the ternary diagram. Therefore, theoretical treatment of the membrane formation by the phase-inversion technique consists of two aspects: one the thermodynamic aspect, and the other the kinetic aspect. An attempt is made in this chapter to discuss these two aspects in detail.

It should be noted that this approach deals with the composition of the nonsolvent, solvent, and polymer in a bulk phase. This approach does not provide any information on the way by which the nonsolvent, solvent, and polymer of a given composition are dispersed microscopically. Hence, this approach is not necessarily useful to discuss the formation of the microstructure of the membrane.

The thermodynamics of the polymer solution have been discussed thoroughly by Tompa [36] in his book of 1956, based on the Flory-Huggins expression for the free energy of mixing.

Let the free energy of mixing, ΔG_m, for a binary mixture of solvent and polymer be given as a function of the mole fraction of polymer X_p, by a curve like that in Figure 3.14. Consider 1 mol of a mixture whose composition is given by the point P and whose free energy can be given by the point Q. Suppose this mixture is separated into two solutions whose compositions correspond to points P' and P''; the amounts of the two phases are in the ratio PP':PP''. The free energy of each phase per mole is given by Q' and Q'', respectively, and the total free energy of the two phases is given by Q^+ for one mole. Q^+ is higher than Q; therefore there is an increase in free energy by separating a mixture P

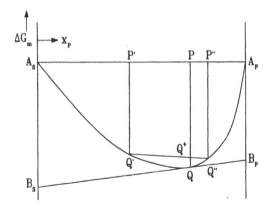

Figure 3.14.   Free energy of mixing vs. mole fraction of polymer. I.

into two phases with compositions in the neighborhood of P. In other words, a homogeneous solution with a composition P is thermodynamically stable. The above argument is valid for any composition as long as the free energy curve is concave.

Since

$$\Delta G_m = x_s \Delta \mu_s + x_p \Delta \mu_p \tag{3.1}$$

$$\left(\frac{\partial \Delta G_m}{\partial x_p} \right)_{P,T} = \Delta \mu_p - \Delta \mu_s \tag{3.2}$$

Solving the above two equations for $\Delta \mu_s$ and $\Delta \mu_p$,

$$\Delta \mu_s = \Delta G_m - x_p \left(\frac{\partial \Delta G_m}{\partial x_p} \right)_{P,T} \tag{3.3}$$

and

$$\Delta \mu_p = \Delta G_m + x_s \left(\frac{\partial \Delta G_m}{\partial x_p} \right)_{P,T} \tag{3.4}$$

Equations 3.3 and 3.4 mean that intercepts of a tangent of the ΔG_m (J/mol) curve at $x_p = 0$ and $x_p = 1$ give $\Delta \mu_s$ (J/mol) and $\Delta \mu_p$ (J/mol), respectively. As illustrated in Figure 3.14; $\Delta \mu_s$ is given by the line $A_s B_s$, and $\Delta \mu_p$ by the line $A_p B_p$, respectively.

When the free energy of mixing is given by a curve as illustrated in Figure 3.15, the mixtures whose compositions are represented by points P' and P" are considered stable against the separation into two phases, by the same argument as in Figure 3.14. Consider the composition represented by a point P; the

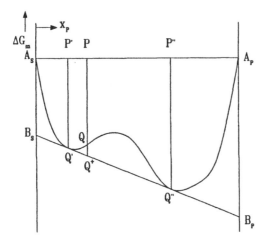

Figure 3.15. Free energy of mixing vs. mole fraction of polymer. II.

corresponding free energy, Q, is on a part of the curve that is concave above, and therefore, a mixture with this composition is still resistant to separation into two phases of neighboring compositions. However, this mixture can be separated into two phases whose compositions are P′ and P″, since Q is higher than Q^+. Of course, the mixture of composition P can be separated into any two phases if the total free energy of the two phases comes under the free energy represented by Q. It is obvious from Figure 3.15, however, that Q^+ is the minimum of the total free energies of two phases for any possible phase separation. The mixture whose composition is represented by P is therefore against the separation into two phases with neighboring compositions, but tends to separate into two phases whose compositions are P′ and P″. The mixture is metastable. When the point P is further shifted towards the right and goes beyond the inflection point of the free energy curve, the mixture is no longer against the separation into two phases of any neighboring compositions and separates spontaneously into two phases of compositions P′ and P″. The mixture is unstable.

Since the intercepts of the tangent at $x_p = 0$ and $x_p = 1$ are the same for the two coexisting phases whose compositions are represented by P′ and P″,

$$\Delta\mu'_s = \Delta\mu''_s \tag{3.5}$$

and

$$\Delta\mu'_p = \Delta\mu''_p \tag{3.6}$$

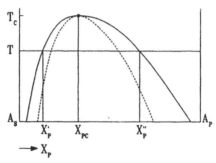

Figure 3.16. Phase diagram of a binary mixture.

The boundary between metastable and unstable regions is given by the second derivative of ΔG_m equal to zero. Then, differentiating Equation 3.3 once more,

$$\frac{\partial \Delta \mu_s}{\partial x_p} = \frac{\partial \Delta \mu_p}{\partial x_p} \tag{3.7}$$

In conjunction with the Gibbs-Duhem relation,

$$x_s d\mu_s + x_p d\mu_p = 0 \tag{3.8}$$

$$\frac{\partial \Delta \mu_s}{\partial x_p} = \frac{\partial \Delta \mu_p}{\partial x_p} = 0 \tag{3.9}$$

When the temperature is increased, the two minima appearing on the free energy curve as Q' and Q'' approach each other until they, and the two inflection points appearing between them, coincide. At this point the third derivative of ΔG_m becomes zero. Again in conjunction with the Gibbs-Duhem relation,

$$\frac{\partial^2 \Delta \mu_s}{\partial x_p^2} = \frac{\partial^2 \Delta \mu_p}{\partial x_p^2} = 0 \tag{3.10}$$

Such a point is called a plait point, and Equations 3.9 and 3.10 have to be satisfied simultaneously. At a given temperature two mole fractions x_p' and x_p'' that can satisfy Equation 3.5 and Equation 3.6 are searched for. Then a similar search for mole fractions is carried out for another temperature, and so on. When the mole fractions so obtained are plotted vs. temperature, a solid line (illustrated in Figure 3.16 as a phase diagram of binary systems) can be obtained. Equation 3.9 is solved at different temperatures for the mole fraction. When the mole fraction so obtained is plotted vs. temperature, a broken line (illustrated in Figure 3.16) can be obtained. The solid and the broken lines merge at a plait point, and the mole fraction that represents the composition of the plait point satisfies Equation 3.10. The meaning of the phase diagram of binary systems must be clear. For a given temperature T, the binary mixture is homogeneous and stable in the range from the A_s axis to the solid line (or $x_p = 0$ to $x_p = x_p'$). The mixture in the range from the solid line to the broken line is metastable. The mixture is against the separation into two phases of neighboring compositions, but tends to separate into two phases whose compositions are on the solid line (x_p' and x_p'' in Figure 3.16). When the mixture is in the range between two broken lines, the mixture is unstable and tends to separate spontaneously into two phases whose compositions are on the solid line. The mixture then goes into another metastable region between a broken and a solid line and becomes homogeneous and stable in the region from a solid line to the A_p axis (from $x_p = x_p''$ to $x_p = 1$). With an increase in temperature the range of the homogeneous and stable region increases, and at the temperature T_c, a plait point appears. At temperatures above T_c, the mixtures are homogeneous and stable from $x_p = 0$ to $x_p = 1$.

The same argument holds valid for a ternary system including nonsolvent(n), solvent(s), and polymer(p). Instead of Equations 3.5, 3.6, 3.9, and 3.10, which give the mole fractions corresponding to compositions of two coexisting phases, those of the boundary between metastable and unstable mixtures, and those of a plait point, the following equations can be used for a ternary system.

$$\Delta\mu_n' = \Delta\mu_n'' \tag{3.11}$$
$$\Delta\mu_s' = \Delta\mu_s''$$
$$\Delta\mu_p' = \Delta\mu_p''$$

$$G_{ss}G_{pp} = G_{sp}^2 \tag{3.12}$$

$$G_{sss} - 3gG_{ssp} + 3g^2G_{spp} - g^3G_{ppp} = 0 \tag{3.13}$$

where G_i, $G_{i,j}$, and $G_{i,j,k}$ are the appropriate partial derivatives of ΔG_m with respect to x_i, x_j, and x_k, and g is G_{ss}/G_{sp}. A typical example of the solutions of Equations 3.11 to 3.13 is illustrated in Figure 3.17. In analogy to the binary system illustrated in Figure 3.16 (the solution of the equation for two coexisting phases), Equation 3.11 produces a solid line on Figure 3.17. A pair of compositions is obtained as the compositions for the coexisting phases, and straight lines connecting these compositions are called tie lines. The solid line corresponding to the two coexisting phases is called a binodial curve. The solution of Equation 3.12 produces a broken line, as illustrated in Figure 3.17. This line is called a spinodial curve. The binodial and the spinodial curves touch each other at a plait point whose composition is obtained by solving Equa-

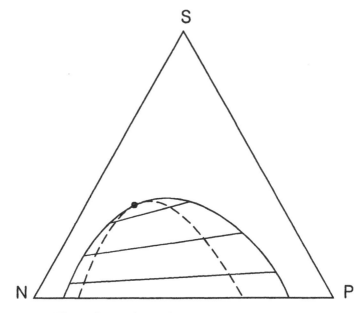

Figure 3.17. Phase diagram of a ternary system.

tion 3.13. The meanings of the binodial and the spinodial curves and the tie
lines illustrated in Figure 3.17 become clear in analogy to the binary system
whose phase separation diagram is depicted in Figure 3.16. Ternary mixtures
whose compositions are in the region surrounded by the NS and SP axes and
the solid line are homogeneous and stable. The mixtures whose compositions
are in the region surrounded by the solid and the broken lines are metastable.
The mixtures whose compositions are inside the broken line are unstable and
tend to separate spontaneously into two phases whose compositions are on the
solid line. The compositions of the two existing phases are at the ends of a tie
line that passes through a point representing the composition of the mixture.

Using the Flory-Huggins expression, the free energy of mixing for a ternary
system is given by

$$
\begin{aligned}
\Delta G_m/RT \; = \; & n_n \ln \phi_n + n_s \phi_s + n_p \phi_p \\
& + (\chi_{ns}\phi_n\phi_s + \chi_{sp}\phi_s\phi_p + \chi_{pn}\phi_p\phi_n) \\
& \cdot (m_n n_n + m_s n_s + m_p n_p)
\end{aligned}
\tag{3.14}
$$

where n_i is the number of moles of component i, ϕ_i is the volume fraction of
component $i(-)$, m_i is the ratio of molar volumes of component i and solvent (in
other words, $m_s = 1$), and χ_{ij} is the interaction constant between components i

and j. Rearranging Equation 3.14, the free energy can be written in a form

$$\Delta G_m/RT = \left[\sum \frac{\phi_i}{m_i}\ln\phi_i + \sum \chi_{ij}\phi_i\phi_j\right]\sum m_i n_i \qquad (3.15)$$

where the summation over i and j is to be taken over all different pairs of i and j. Since

$$\Delta\mu_n/RT = \frac{\partial\Delta G_m/RT}{\partial n_n} \qquad (3.16)$$

$$\Delta\mu_n/RT = \ln\phi_n + 1 - m_n\sum \frac{\phi_i}{m_i} + m_n\sum \chi_{ni}\phi_i - m_n\sum \chi_{ij}\phi_i\phi_j \qquad (3.17)$$

Further rearranging yields

$$\Delta\mu_n/RT = \ln\phi_n + (1 - m_n/m_s)\phi_s + (1 - m_n/m_p)\phi_p$$
$$+ m_n\left[\chi_n\left(\phi_s + \phi_p\right)^2 + \chi_s\phi_s^2\chi_p\phi_p^2\right] \qquad (3.18)$$

Similarly,

$$\Delta\mu_s/RT = \ln\phi_s + (1 - m_s/m_p)\phi_p + (1 - m_s/m_n)\phi_n$$
$$+ m_s\left[\chi_s(\phi_p + \phi_n)^2 + \chi_p\phi_p^2\chi_n\phi_n^2\right] \qquad (3.19)$$

$$\Delta\mu_p/RT = \ln\phi_p + (1 - m_p/m_n)\phi_n + (1 - m_p/m_s)\phi_s$$
$$+ m_p\left[\chi_p(\phi_n + \phi_s)^2 + \chi_n\phi_n^2\chi_s\phi_s^2\right] \qquad (3.20)$$

where χ_n, χ_s, and χ_p are given as

$$2\chi_n = \chi_{ns} + \chi_{np} - \chi_{sp} \qquad (3.21)$$
$$2\chi_s = \chi_{sp} + \chi_{sn} - \chi_{pn} \qquad (3.22)$$
$$2\chi_p = \chi_{pn} + \chi_{ps} - \chi_{ns} \qquad (3.23)$$

The equations for the spinodial and the plait point can be written as

$$\sum m_i\phi_i - 2\sum m_i m_j(\chi_i + \chi_j)\phi_i\phi_j + 4m_n m_s m_p\sum \chi_i\chi_j\phi_n\phi_s\phi_p = 0 \qquad (3.24)$$

and

$$\sum \frac{m_i^2\phi_i}{(1 - 2\chi_i m_i\phi_i)^3} = 0 \qquad (3.25)$$

For the coexisting two phases, Equation 3.11 and the following two equations must be satisfied.

$$\phi_n' + \phi_s' + \phi_p' = 1 \tag{3.26}$$

$$\phi_n'' + \phi_s'' + \phi_p'' = 1 \tag{3.27}$$

If considering Equations 3.11, 3.26, and 3.27, there are altogether five equations to be satisfied simultaneously, while there are six unknowns to be solved. These unknowns are ϕ_n', ϕ_s', ϕ_p', ϕ_n'', ϕ_s'', and ϕ_p''. Therefore, for a given ϕ_n', the other five unknowns can be obtained by solving the above five equations. The connection of two points representing the composition (ϕ_n', ϕ_s', ϕ_p') and the composition (ϕ_n'', ϕ_s'', ϕ_p'') on a triangular diagram produces a tie line, while a binodial line or a coexisting curve can be drawn by changing the value of ϕ_n'.

As for the spinodial equation, there are two equations to be solved simultaneously; i.e., Equation 3.24 and

$$\phi_n + \phi_s + \phi_p = 1 \tag{3.28}$$

while there are three unknowns: ϕ_n, ϕ_s, and ϕ_p. Therefore, a curve can be drawn on a triangular diagram by solving the above two equations for different values of ϕ_n. This curve is, of course, a spinodial curve. A point exists on this curve, corresponding to a composition that satisfies Equations 3.28, 3.24, and 3.25, simultaneously. This point is a plait point.

Zeman and Tkacik used interaction parameters, $\chi_{ns} = 1.0$, $\chi_{np} = 1.5$, and $\chi_{sp} = 0.5$, and the size parameters, $m_n = 1.0$, $m_s = 5.3$, and $m_p = 998$, for a ternary system water/n-methylpyrrolidone/polyethersulfone Victrex 4100 (200P) to calculate the binodial line, the tie line end points, the spinodial line, and the plait point [37]. The results are illustrated in Figure 3.18. Figure 3.19 illustrates an experimental cloud point curve and the tie lines. Phase boundary lines have been produced experimentally for various nonsolvent-solvent-polymer systems relevant to the preparation of separation membranes [38]–[42].

3.5   Change of the Solution Composition on the Triangular Diagram During the Gelation Process

The cast polymer solution is immersed into a gelation bath after the partial evaporation of the solvent. From this moment the exchange between the solvent (which is in the cast film) and the gelation medium (which is usually a nonsolvent for the membrane polymer) starts to occur. The solvent-nonsolvent exchange is controlled by the counterdiffusion of solvent and nonsolvent through

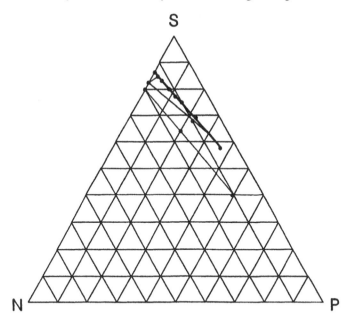

Figure 3.18. Phase diagram of the ternary system H_2O (N)-NMP (S)-PES (P). [Experimental cloud point curve (•) and tie lines (o).]

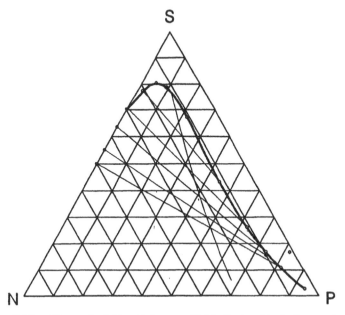

Figure 3.19. Theoretical binodial curve (thick line) with tie-line end points (o) and spinodial curve (thin line). [$\chi_{ns} = 1.0$, $\chi_{np} = 1.5$, $\chi_{sp} = 0.5$; critical point is given by (•).]

Polymer film

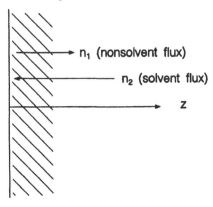

Figure 3.20. Schematic representation of nonsolvent-solvent exchange as pseudobinary diffusion.

the macromolecular network. Assuming the polymer concentration becomes very high at the very beginning of the gelation process, the net macroscopic movement of the polymer can be neglected. Based on such an assumption, Yilmaz and McHugh [43] discussed the system as a pseudobinary diffusion, to describe the kinetics involved in the gelation process [44].

In the following discussions density, ρ (kg/m^3), is defined as the weight of a nonsolvent-solvent mixture, free from polymer, in a unit volume of the film. Polymer is included in the latter unit volume. A quantity, ω, is defined as a mass fraction in the solvent-nonsolvent mixture, also on a polymer-free basis. In order to emphasize the pseudobinary approach, subscripts 1 and 2 are used instead of n(nonsolvent) and s(solvent). Then the mass flux of the nonsolvent becomes

$$n_1 = -\rho D \frac{\partial \omega_1}{\partial z} + \omega_1 (n_1 + n_2) \tag{3.29}$$

where n is the mass flux (kg/m^2 s), and z is the distance (m) from the film-gelation bath interface into the film vertical to the film surface (see Figure 3.20), and D is the diffusion coefficient of the nonsolvent.

The equations of continuity can be written as

$$\frac{\partial \rho}{\partial t} = -\frac{\partial (n_1 + n_2)}{\partial z} \tag{3.30}$$

$$\frac{\partial \rho_1}{\partial t} = -\frac{\partial n_1}{\partial z} \tag{3.31}$$

Assuming further that the directions of the nonsolvent and solvent flows are opposite and the ratio of the flux is constant regardless of position and time,

$$n_2 = -k'n_1 \tag{3.32}$$

Under the above assumption, Equation 3.29 can be written as follows:

$$n_1 = \frac{-\rho D}{1 + kw_1} \frac{\partial w_1}{\partial z} \tag{3.33}$$

where $k = k' - 1$. Substituting Equation 3.32 for n_2 of Equation 3.30,

$$\frac{\partial \rho}{\partial t} = -\frac{\partial (n_1 - k'n_1)}{\partial z} \tag{3.34}$$

$$= -(1 - k')\frac{\partial n_1}{\partial z} \tag{3.35}$$

$$= k\frac{\partial n_1}{\partial z} \tag{3.36}$$

Combining Equations 3.33 and 3.36 yields

$$\frac{\partial \rho}{\partial t} = k\frac{\partial}{\partial z}\left(\frac{-\rho D}{1 + kw_1}\frac{\partial w_1}{\partial z}\right) \tag{3.37}$$

$$= -\frac{\partial}{\partial z}\left(\frac{k\rho D}{1 + kw_1}\frac{\partial w_1}{\partial z}\right) \tag{3.38}$$

Furthermore, since

$$\rho_1 = \rho w_1 \tag{3.39}$$

$$\frac{\partial \rho_1}{\partial t} = \frac{\partial \rho}{\partial t}w_1 + \rho\frac{\partial w_1}{\partial t} \tag{3.40}$$

Substituting Equation 3.37 for $\frac{\partial \rho}{\partial t}$ in Equation 3.40,

$$= k\frac{\partial}{\partial z}\left(\frac{-\rho D}{1 + kw_1}\frac{\partial w_1}{\partial z}\right)w_1 + \rho\frac{\partial w_1}{\partial t} \tag{3.41}$$

From Equations 3.31 and 3.33,

$$\frac{\partial \rho_1}{\partial t} = -\frac{\partial}{\partial z}\left(\frac{-\rho D}{1 + kw_1}\frac{\partial w_1}{\partial z}\right) \tag{3.42}$$

Combining Equations 3.38, 3.40, and 3.42 and rearranging yields

$$\rho \frac{\partial w_1}{\partial t} = (1 + kw_1)\frac{\partial}{\partial z}\left(\frac{\rho D}{1 + kw_1}\frac{\partial w_1}{\partial z}\right) \tag{3.43}$$

$$= (1 + kw_1)\frac{\partial}{\partial z}\left(\rho D\frac{\partial w_1}{\partial z}\right)\frac{1}{1 + kw_1} \tag{3.44}$$

$$+ (1 + kw_1)\frac{-k}{(1 + kw_1)^2}\frac{\partial w_1}{\partial z}\left(\rho D\frac{\partial w_1}{\partial z}\right)$$

$$= \frac{\partial}{\partial z}\left(\rho D\frac{\partial w_1}{\partial z}\right) - \frac{k\rho D}{(1 + kw_1)}\left(\frac{\partial w_1}{\partial z}\right)^2 \tag{3.45}$$

For the given formula of k and D as functions of the solution composition, differential Equations 3.45 and 3.38 can be solved in terms of w_1 and ρ. For simplification, however, it can be justified to assume that k remains constant during the entire gelation process if the film thickness is large enough.

Combination of Equations 3.45 and 3.38 can yield a differential equation that relates ρ to w_1. Since

$$\frac{\partial \rho}{\partial t} = \frac{\partial \rho}{\partial w_1}\frac{\partial w_1}{\partial t} \tag{3.46}$$

Therefore,

$$\frac{\partial \rho}{\partial w_1} = \frac{\partial \rho}{\partial t}\bigg/\frac{\partial w_1}{\partial t} \tag{3.47}$$

$$= \frac{-\frac{\partial}{\partial z}\left(\frac{k\rho D}{1+kw_1}\frac{\partial w_1}{\partial z}\right)\rho}{\frac{\partial}{\partial z}\left(D\rho\frac{\partial w_1}{\partial z}\right) - \frac{k\rho D}{(1+kw_1)}\left(\frac{\partial w_1}{\partial z}\right)^2} \tag{3.48}$$

$$= \frac{\frac{-k\rho}{1+kw_1}\left[\frac{\partial}{\partial z}\left(D\rho\frac{\partial w_1}{\partial z}\right) - \frac{k\rho D}{(1+kw_1)}\left(\frac{\partial w_1}{\partial z}\right)^2\right]}{\frac{\partial}{\partial z}\left(D\rho\frac{\partial w_1}{\partial z}\right) - \frac{k\rho D}{(1+kw_1)}\left(\frac{\partial w_1}{\partial z}\right)^2} \tag{3.49}$$

$$= -\frac{k\rho}{1 + kw_1} \tag{3.50}$$

Rearranging,

$$\frac{\partial \rho}{\rho} = -\frac{\partial kw_1}{(1 + kw_1)} \tag{3.51}$$

Finally,

$$\ln \rho = -\ln(1 + kw_1) + C \tag{3.52}$$

where C is an integration constant.

Since $\rho = \rho_i$ when $w_1 = w_{1i}$, where the subscript i indicates the quantity at time zero,

$$\ln \rho_i = -\ln(1 + kw_{1i}) + C \qquad (3.53)$$

Combining Equations 3.52 and 3.53,

$$\ln \frac{\rho}{\rho_i} = -\ln \frac{1 + kw_1}{1 + kw_{1i}} \qquad (3.54)$$

Therefore,

$$\rho = \frac{(1 + kw_{1i})\rho_i}{1 + kw_1} \qquad (3.55)$$

The quantities ρ and w_1 are uniquely related by Equation 3.55, regardless of the position and time. Substituting the above equation for ρ of Equation 3.45,

$$\frac{1}{1 + kw_1}\left(\frac{\partial w_1}{\partial t}\right) = \frac{\partial}{\partial z}\left(\frac{D}{1 + kw_1}\frac{\partial w_1}{\partial z}\right) - \frac{kD}{(1 + kw_1)^2}\left(\frac{\partial w_1}{\partial z}\right)^2 \qquad (3.56)$$

Equation 3.56 is a partial differential equation by which w_1 can be solved as a function of (t, z) when D is given as a function of w_1. There is, however, no definite information on the functional form of the diffusivity, particularly corresponding to the process of the membrane formation. Therefore, a simple assumption is made that D decreases exponentially with an increase in the polymer concentration; i.e.,

$$D = D_0 \exp\left(\frac{a\rho}{\rho_i^*}\right) \qquad (3.57)$$

where $\rho_i^* = (1 + kw_{1i})\rho_i$, and a and D_0 are constants.

Some explanation is in order to validate Equation 3.57. Using Equation 3.55,

$$\frac{\rho}{\rho_i^*} = \frac{1}{1 + kw_1} \qquad (3.58)$$

Suppose $k > 0$ (and $k' > 1$), the ratio ρ/ρ_i^* decreases as more nonsolvent flows into the film, and w_1 increases during the gelation process. The outflux of solvent from the film is greater than the influx of nonsolvent, however, since $k' > 1$. As a result, as gelation progresses, the amount of the binary mixture in a unit volume decreases, which means a decrease in density ρ, while the amount of the polymer in a unit volume increases relative to the binary mixture. Consequently, the polymer concentration increases as the gelation progresses.

The decrease in the ratio ρ/ρ_i^* occurs, therefore, parallel to an increase in the polymer concentration. Equation 3.57 means that diffusivity D is proportional to the exponential of the ratio ρ/ρ_i^* and, therefore, decreases as the gelation progresses.

Combining Equations 3.56, 3.57, and 3.58, we obtain

$$\frac{1}{(1+k\omega_1)}\frac{\partial\omega_1}{\partial t}$$

$$= \frac{\partial}{\partial z}\left(\frac{D_0\exp\left[a/(1+k\omega_1)\right]}{1+k\omega_1}\frac{\partial\omega_1}{\partial z}\right)$$

$$- \frac{k\exp\left[a/(1+k\omega_1)\right]}{(1+k\omega_1)^2}\left(\frac{\partial\omega_1}{\partial z}\right)^2 \tag{3.59}$$

$$= D_0\left[\left(\frac{-k\exp\left[a/(1+k\omega_1)\right]}{(1+k\omega_1)^2}\right.\right.$$

$$\left. + a\exp\left(\frac{a}{1+k\omega_1}\right)\frac{-k}{(1+k\omega_1)^2}\frac{1}{(1+k\omega_1)}\right)$$

$$\left. \cdot\left(\frac{\partial\omega_1}{\partial z}\right)^2 + \frac{\exp\left[a/(1+k\omega_1)\right]}{1+k\omega_1}\frac{\partial^2\omega_1}{\partial z^2}\right]$$

$$- \frac{k\exp\left[a/(1+k\omega_1)\right]}{(1+k\omega_1)^2}\left(\frac{\partial\omega_1}{\partial z}\right)^2 \tag{3.60}$$

$$= D_0\left[\left(\frac{-2k\exp\left[a/(1+k\omega_1)\right]}{(1+k\omega_1)^2} - \frac{ak}{(1+k\omega_1)^2}\frac{\exp\left[a/(1+k\omega_1)\right]}{(1+k\omega_1)}\right)\right.$$

$$\left. \cdot\left(\frac{\partial\omega_1}{\partial z}\right)^2 + \frac{\exp\left[a/(1+k\omega_1)\right]}{1+k\omega_1}\frac{\partial^2\omega_1}{\partial z^2}\right] \tag{3.61}$$

$$= \left(\frac{-2kD}{(1+k\omega_1)^2} - \frac{akD}{(1+k\omega_1)^3}\right)\left(\frac{\partial\omega_1}{\partial z}\right)^2 + \frac{D}{(1+k\omega_1)}\frac{\partial^2\omega_1}{\partial z^2} \tag{3.62}$$

Therefore, the rate of nonsolvent composition change is

$$\frac{\partial\omega_1}{\partial t} = -D\left[\left(\frac{ak}{(1+k\omega_1)^2} + \frac{2k}{(1+k\omega_1)}\right)\left(\frac{\partial\omega_1}{\partial z}\right)^2 - \frac{\partial^2\omega_1}{\partial z^2}\right] \tag{3.63}$$

If the density of the nonsolvent-solvent mixture is ρ_i, and the mass fraction of the first component (nonsolvent) is ω_{1i} at time zero (the subscript i prepresents the initial condition), the boundary conditions to solve the differential

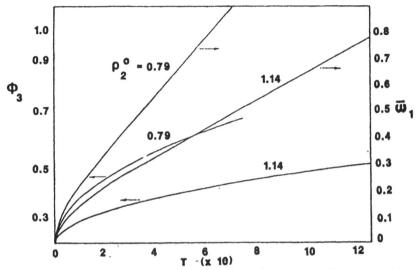

Figure 3.21. Schematic representation of density and mass fraction profiles in the film as a function of time. (Reproduced from [43] with permission.)

Equation 3.63 is

$$\rho = \rho_i \text{ at } t = 0 \text{ when } z > 0 \qquad (3.64)$$
$$\rho = \rho_i \text{ at } t > 0 \text{ when } z = \infty \qquad (3.65)$$

Equation 3.65 implies that we are assuming an infinite film thickness, and the density of the binary mixture at a film position that is far away from the gelation media/film interface maintains the initial density value regardless of the time. Similarly,

$$\omega_1 = \omega_{1i} \text{ at } t = 0 \text{ when } z > 0 \qquad (3.66)$$
$$\omega_1 = \omega_{1i} \text{ at } t > 0 \text{ when } z = \infty \qquad (3.67)$$

In most cases the initial mass fraction of nonsolvent in the film is equal to zero, and therefore, $\omega_{1i} = 0$. Figure 3.21 illustrates schematically the profiles of the density and mass fraction in the film, as a function of time. As for the boundary condition, it is assumed that the mass flux of the nonsolvent, n_1, is constant at the gelation media/film interface. Then using Equation 3.33,

$$\frac{-\rho D}{(1 + k\omega_1)} \left(\frac{\partial \omega_1}{\partial z} \right) = \text{constant at } t > 0 \text{ for } z = 0 \qquad (3.68)$$

Equation 3.63 can be solved with Equations 3.66 and 3.67 as the initial conditions and with Equation 3.68 as the boundary condition.

The discussion so far has been made on the basis of binary mixtures. We shall now convert the quantities involved in the binary system to those of the ternary nonsolvent/solvent/polymer system. The following equations can be used to obtain the volume fractions of the ternary system.

$$\phi_n = \frac{w_1 \rho}{\rho_1^0} \tag{3.69}$$

$$\phi_s = \frac{w_2 \rho}{\rho_2^0} \tag{3.70}$$

$$\phi_p = 1 - \phi_n - \phi_s \tag{3.71}$$

where ρ_1^0 and ρ_2^0 are the densities of the pure nonsolvent and solvent, respectively. Recall that the pseudobinary system is converted into the ternary system, and subscripts n, s, and p are used to indicate nonsolvent, solvent, and polymer.

Starting from Equation 3.69,

$$\phi_n = \frac{w_1 \rho}{\rho_1^0} \tag{3.72}$$

$$= \frac{\rho_i^* \frac{w_1}{\rho_1^0}}{(1 + kw_1)} \tag{3.73}$$

$$= \left(\frac{\rho_i^*}{\rho_1^0}\right) \bigg/ \left(\frac{1}{w_1} + k\right) \tag{3.74}$$

$$\phi_n / \phi_s = \frac{w_1}{\rho_1^0} \bigg/ \frac{w_2}{\rho_2^0} \tag{3.75}$$

$$= \frac{w_1}{\rho_1^0} \bigg/ \frac{(1 - w_1)}{\rho_2^0} \tag{3.76}$$

Therefore,

$$\phi_n(1 - w_1) / \rho_2^0 = \phi_s w_1 / \rho_1^0 \tag{3.77}$$

and

$$\left(\phi_n / \rho_2^0 + \phi_s / \rho_1^0\right) w_1 = \phi_n / \rho_2^0 \tag{3.78}$$

Then

$$w_1 = \frac{\phi_n / \rho_2^0}{\phi_n / \rho_2^0 + \phi_s / \rho_1^0} \tag{3.79}$$

Since

$$\phi_s = 1 - \phi_n - \phi_p \tag{3.80}$$

$$\omega_1 = \frac{\phi_n / \rho_2^0}{\phi_n / \rho_2^0 + (1 - \phi_n - \phi_p) / \rho_1^0} \tag{3.81}$$

Combining Equations 3.74 and 3.81,

$$\phi_n = \frac{\rho_i^* / \rho_1^0}{\frac{\phi_n / \rho_2^0 + (1 - \phi_n - \phi_p) / \rho_1^0}{\phi_n / \rho_2^0} + k} \tag{3.82}$$

$$= \frac{\phi_n / \rho_2^0 \cdot \rho_i^* / \rho_1^0}{\phi_n / \rho_2^0 + (1 - \phi_n - \phi_p) / \rho_1^0 + k\phi_n / \rho_2^0} \tag{3.83}$$

Then

$$\phi_n / \rho_2^0 + (1 - \phi_n - \phi_p) / \rho_1^0 + k\phi_n / \rho_2^0 = \rho_i^* / \rho_1^0 \rho_2^0 \tag{3.84}$$

$$\phi_n \rho_1^0 + (1 - \phi_n - \phi_p) \rho_2^0 + k\phi_n \rho_1^0 = \rho_i^* \tag{3.85}$$

$$\phi_n \left(\rho_1^0 - \rho_2^0 + k\rho_1^0 \right) = \left(\rho_i^* - \rho_2^0 \right) + \rho_2^0 \phi_p \tag{3.86}$$

$$\phi_n = \frac{\rho_2^0 - \rho_i^*}{\rho_2^0 - (1 + k) \rho_1^0} - \frac{\rho_2^0}{\rho_2^0 - (1 + k) \rho_1^0} \phi_p \tag{3.87}$$

and

$$\phi_n = \frac{\rho_2^0 - \rho_i^*}{\rho_2^0 - k' \rho_1^0} - \frac{\rho_2^0}{\rho_2^0 - k' \rho_1^0} \phi_p \tag{3.88}$$

where ρ_i^* was defined earlier as $(1 + k\omega_{1i})\rho_i$.

Equation 3.88 allows us to assess the route of the composition change on the triangular diagram during the gelation process. It should be noted that the equation was derived using the assumptions that k (or k') and the density of the solvent and nonsolvent are constant and that Equation 3.58 is valid. Equation 3.88 does not depend on the functional form of the diffusivity or on the boundary conditions adopted to solve differential equations.

Equation 3.88 further indicates that ϕ_n and ϕ_p are linearly related, which means that the concentration paths on the triangular diagram should be straight lines. This aspect of phase inversion was confirmed for the wet spinning of fibers, which shares many common features with the formation of membranes. Figures 3.22 and 3.23 illustrate such concentration paths for some sets of k and ρ_2^0. The gelation media is water, and therefore ρ_1^0 is 1.0×10^3 kg/m³. The

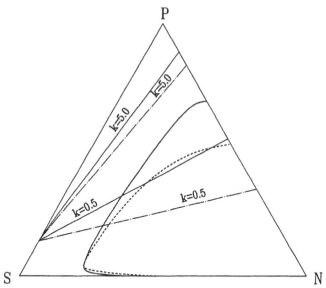

Figure 3.22. Mass transfer path on the triangular diagram. I. [$\rho_2^0 = 0.79$ (———); $\rho_2^0 = 1.14$ (— · —); the solid line and dotted line represent the binodial curve and the spinodial curve, respectively.]

binodial and spinodial curves are also written in the figures. (The positions of the N-S-P corners are different in Figures 3.22 and 3.23 from those in Figures 3.18 and 3.19. Figures 3.22 and 3.23 should be turned 60° clockwise to be superimposed to Figures 3.18 and 3.19.) Figure 3.22 shows the effect of k and ρ_2^0 on the concentration path. The phase separation diagram is for cellulose acetate. Obviously, k exhibits a larger effect than ρ_2^0, and when k is 5 (meaning the solvent outflux from the film is six times faster than the nonsolvent influx), the composition path does not cross the phase separation lines. On the other hand, when k is 0.5 (meaning the solvent outflux from the film is only 1.5 times faster than nonsolvent influx), the composition path crosses the phase separation lines. An increase in the solvent density also brings the composition path downwards on the triangular diagram. Figure 3.23 illustrates the effect of the initial polymer concentration in the cast film. The phase separation diagram on the figure corresponds to that of polyethersulfone. The figure indicates clearly that the composition path is shifted upwards with an increase in the initial polymer volume fraction. When both k and ϕ_{pi} are sufficiently high, the composition path can be completely out of the region of the phase separation, even though a very narrow region of the homogeneous solution mixture is left on the triangular diagram.

    The implication of these composition paths on the structure and performance of the membrane, after the gelation step, is very significant [45]–[47].

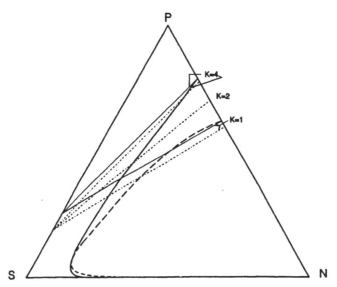

Figure 3.23. Mass transfer path on the triangular diagram. II. [$\phi_{pi} = 0.20$ (————); $\phi_{pi} = 0.13$ (– · –); the solid line and dotted line represent the binodial curve and the spinodial curve, respectively.]

As illustrated schematically in Figure 3.24, three different composition paths ultimately approach three different positions on the P-N axis, at the end of the gelation process. Path A starts from a higher initial polymer volume fraction on the P-S axis and does not cross the phase separation line. The membrane prepared under such conditions with a high-volume fraction of polymer on the P-S axis is homogeneous and dense throughout the entire cross-section of the membrane, since no phase separation occurs during the entire gelation process. Path B starts from a lower initial polymer volume fraction on the P-S axis, and the ratio of the rate of solvent outflux to that of the nonsolvent influx is relatively low. The membrane ultimately obtained at the P-N axis is asymmetric. The reason for the formation of the asymmetric structure is as follows. As illustrated in Figure 3.24, the film composition starts from the point α, passes through points β and γ, and finally reaches a point δ along the composition path. The meaning of the points α, β, and δ is obvious. Point γ is called the polymer densification point. On the left side of this point, the polymer-rich phase, the composition of which is indicated by a triangular symbol on the phase separation line, is liquid. On the other hand, on the right side of point γ, the polymer-rich phase is no longer liquid, but is solid.

As mentioned earlier, the composition path on Figure 3.24 should be the same regardless of the position (distance z from the film-gelation medium interface), but the speed of moving along the composition path depends on the

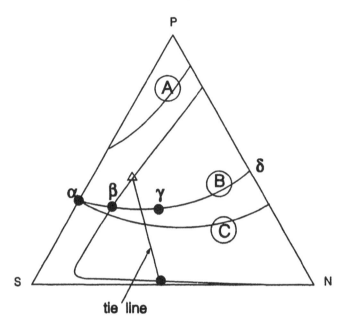

Figure 3.24. Schematic representation of mass transfer path on the triangular diagram.

position. The greater the distance z, the slower the movement. Considering the composition change near the film-gelation medium interface, the speed of the movement on the composition path is very fast. The point where the phase separation starts to begin, β, and the point where the polymer solidification starts to occur, γ, are attained very quickly. There is not enough time for the solution to undergo the phase separation, and therefore, polymer and water are interdispersed, resulting in a dense skin layer with a composition indicated by the point γ. The solvent-nonsolvent exchange proceeds further along the composition path, and the point δ is attained, while the dense structure of the surface layer is unperturbed. For a larger z, it takes a longer time to achieve the solidification point γ and there is enough time for a polymer-rich phase and a polymer-poor phase to be separated. If the amount of the polymer-rich phase is large relative to the amount of the polymer-poor phase, droplets of the latter phase are dispersed in the continuous polymer-rich phase and form pores in the continuous polymer matrix. As the value of z increases, the separation of the phases becomes clearer and the size of the pores increases. The pore size is fixed at the time of the polymer densification, and it remains the same while the solvent-nonsolvent exchange proceeds along the composition path. As a result an asymmetric structure of the membrane is formed across the cross-section of the membrane. In this structure there is a dense surface layer at the film/gelation media interface, followed by a porous sublayer with increasingly larger pore sizes as the distance from the interface increases. The membranes of this type demonstrate the best performance, giving high flux and high separation.

Path C in Figure 3.24 starts from the same polymer volume fraction on the P-S axis as path B; however, the ratio of the rate of solvent outflux to that of nonsolvent influx is smaller than that of path B. The volume fraction of polymer at the densification point is no longer high enough to form a continuous polymer-rich phase. Instead, a polymer-rich phase is dispersed in a nonsolvent-rich phase. The membranes of this type are too porous and mechanically too weak.

The above discussions indicate that not only the composition path and its position relative to the phase separation line, but also the speed of the composition change along the path, govern the structure of the membrane formed after the gelation process. In order to have some idea on this speed, the complete solution of the differential equation, Equation 3.63, with initial conditions, Equations 3.66 and 3.67, and the boundary condition, Equation 3.68, is necessary. Assuming a special case of a constant diffusivity, the pseudobinary equations are simplified considerably, and a complete analytical solution is possible. For $D = D_0$, Equation 3.63 can be transformed by the substitution of

$$\bar{f} = k\omega_1/(1 + k\omega_1) \tag{3.89}$$

to the following equation:

$$\frac{\partial \bar{f}}{\partial t} = D_0 \frac{\partial^2 \bar{f}}{\partial z^2} \tag{3.90}$$

The boundary condition, Equation 3.68, becomes

$$\frac{\partial \bar{f}}{\partial z} = -\frac{kN_1}{\rho_i^* D_0} \text{ at } z = 0 \tag{3.91}$$

These equations are made dimensionless by the following transformation:

$$\tau = \left(\frac{N_1}{\rho_1^0}\right)^2 \frac{t}{D_0} \tag{3.92}$$

and

$$x = z/L_0 \tag{3.93}$$

where L_0 is the initial film thickness.

Then Equation 3.90 becomes the following:

$$\frac{\partial f}{\partial \tau} = \alpha^2 \frac{\partial^2 f}{\partial x^2} \tag{3.94}$$

where

$$f(x, \tau) = \bar{f}(x, \tau) - f_0 \tag{3.95}$$

and

$$f_0 = \frac{k\omega_{1i}}{(1 + k\omega_{1i})} \tag{3.96}$$

The parameter α can be defined as

$$
\begin{aligned}
\alpha &= \frac{D_0 / L_0}{N_1 / \rho_1^0} \\
&= \left(\frac{\text{internal mass transfer resistance}}{\text{external mass transfer resistance}} \right)^{-1}
\end{aligned}
\tag{3.97}
$$

The dimensionless differential Equation 3.94 should be subject to the boundary condition,

$$-\frac{k\rho_1^0}{\rho_i^*} = \alpha \frac{\partial f}{\partial x} \quad \text{at } x = 0 \tag{3.98}$$

and initial conditions,

$$f(x, 0) = 0 \tag{3.99}$$
$$f(\infty, \tau) = 0 \tag{3.100}$$

The solution to the above set of equations is as follows:

$$f = \frac{k\rho_1^0}{\rho_i^*} \sqrt{4\tau} \left[\frac{1}{\sqrt{\pi}} \exp\left(\frac{-x^2}{4\alpha^2 \tau} \right) - \frac{x}{\sqrt{4\alpha^2 \tau}} \, \text{erfc}\left(\frac{x}{\sqrt{4\alpha^2 \tau}} \right) \right] \tag{3.101}$$

The composition at a given set of (z, t), or (x, τ) after the conversion of the variables, can be solved in the following way. First we obtain f from Equation 3.101. $\bar{f}$ is calculated by Equations 3.95 and 3.96. Then ω_1 is calculated by Equation 3.89. Using ω_1 so obtained, we can further calculate ρ by Equation 3.55. These ρ and ω_1 (and ω_2, which is equal to $1 - \omega_1$) values are used in Equations 3.69, 3.70, and 3.71 to calculate ϕ_n, ϕ_s, and ϕ_p.

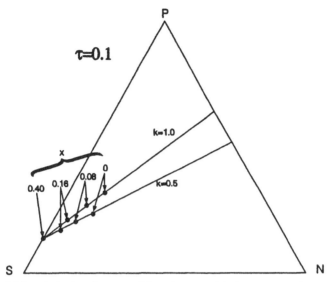

Figure 3.25. Calculated composition change with time on the triangular diagram I.

The above solutions are subject to some oversimplifications. Among others, the most obvious one is constancy in diffusivity. Nevertheless, the solution gives some insight into how the mass transfer proceeds during the gelation process. Some results of the numerical solutions are given, therefore, in a triangular diagram in Figures 3.25 and 3.26. Together with information on the position of polymer solidification or the point γ in Figure 3.24, the plot of the calculation results similar to those on Figures 3.25 and 3.26 allows us to know how far in the film the front of the solidification advances within a time frame of τ. The above value of x or the variable z, from which x stemmed, can give an approximate evaluation of the thickness of the dense surface layer. It has to be noted, however, that the evaluation of γ is not at all easy.

At the surface of the film, Equation 3.101 becomes

$$f_{x=0} = \frac{k\rho_b\sqrt{4\tau}}{\rho_i^*\sqrt{\pi}} \qquad (3.102)$$

where ρ_b is the density of gelation media (kg/m^3). More explicit equations were derived by Yilmaz and McHugh for the gradients of the weight fraction and volume fraction. They are

$$\frac{\partial \omega_1}{\partial x} = -\frac{\rho_b}{\rho_i^*}\frac{(1+k\omega_1)^2}{\alpha}\, erfc\left(\frac{x}{\sqrt{4\alpha^2\tau}}\right) \qquad (3.103)$$

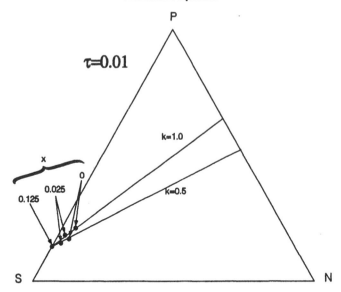

Figure 3.26. Calculated composition change with time on the triangular diagram II.

$$\frac{\partial \phi_p}{\partial x} = \frac{\rho_b}{\alpha} \left[\frac{(k+1)}{\rho_2^0} - \frac{1}{\rho_1^0} \right] erfc \left(\frac{x}{\sqrt{4\alpha^2 \tau}} \right) \qquad (3.104)$$

Going back to Equation 3.88, it was shown that k, ρ_2^0, and ϕ_{pi} affect the composition path. The ways to change the latter two parameters are obvious. According to Strathmann [46] and Frommer [48], there are several methods to change k, or the ratio of the rate of solvent outflux from the film to that of nonsolvent influx to the film. They are

1. By adding electrolytes or organic additives into the casting dope. This increases the rate of influx of the nonsolvent and decreases the k value.

2. By adding electrolytes or organic solutes into the gelation media. This decreases the activity of the gelation media, usually water, and consequently decreases the rate of influx of nonsolvent, resulting in an increase in k value.

3. By adding solvent into the gelation media. This has two opposing effects: i.e., a decrease in the rate of solvent outflux due to an increased activity of solvent in the gelation bath, and a decrease in the nonsolvent influx due to a decrease in the activity of nonsolvent in the gelation bath. While the former effect causes a decrease, the latter effect causes an increase in the value of k. Usually, however, the former effect is the greater, and k is decreased substantially. Often, the suppressed speed of nonsolvent influx allows sufficient time for the phase separation, and homogeneous dense membranes are not formed. .

4. By choosing a solvent with low affinity to nonsolvent (water). This decreases the rate of outflux of solvent and results in a decrease in k value.

More discussions on the solvent-nonsolvent exchange during the gelation process can be found in the literature [49], [50].

Example 1

The most commonly encountered ternary system containing polymers is that of two nonpolymeric liquids and a polymer species. As a simplest case we consider a ternary system with $m_n = m_s = 1$. Furthermore, it is assumed for simplification that the interaction parameter between the nonsolvent and polymer is equal to that between the nonsolvent and solvent, while that between solvent and polymer is zero. This means that the properties of the solvent molecule are the same as those of the polymer segment. We put, therefore, $\chi_{ns} = \chi_{np} = \chi$ and $\chi_{sp} = 0$. For such a simplified system Tompa has given the chemical potentials of three components as follows [36]:

$$\frac{\Delta\mu_n}{RT} = \ln\phi_n + \left(1 - \frac{1}{m_p}\right)\phi_p + \chi(1 - \phi_n)^2 \tag{3.105}$$

$$\frac{\Delta\mu_s}{RT} = \ln\phi_s + \left(1 - \frac{1}{m_p}\right)\phi_p + \chi\phi_n^2 \tag{3.106}$$

$$\frac{\Delta\mu_p}{RT} = \ln\phi_p - (m_p - 1)(1 - \phi_p) + m_p\chi\phi_n^2 \tag{3.107}$$

Draw the binodial curve and tie lines for the case when $m_p = 100$ and $\chi = 1.5$.

Distinguishing the phases by superscripts $'$ and $''$ and using Equations 3.105, 3.106, and 3.107 in Equation 3.11, we obtain

$$\ln\phi_n' + \left(1 - \frac{1}{100}\right)\phi_p' + 1.5\left(1 - \phi_n'\right)^2$$
$$= \ln\phi_n'' + \left(1 - \frac{1}{100}\right)\phi_p'' + 1.5\left(1 - \phi_n''\right)^2 \tag{3.108}$$

$$\ln\phi_s' + \left(1 - \frac{1}{100}\right)\phi_p' + 1.5\phi_n'^2$$
$$= \ln\phi_s'' + \left(1 - \frac{1}{100}\right)\phi_p'' + 1.5\phi_n''^2 \tag{3.109}$$

$$\ln\phi_p' - (100 - 1)\left(1 - \phi_p'\right) + 150\phi_n'^2$$
$$= \ln\phi_p'' - (100 - 1)\left(1 - \phi_p''\right) + 150\phi_n''^2 \tag{3.110}$$

Furthermore,

$$\phi'_n + \phi'_s + \phi'_p = 1 \qquad (3.111)$$
$$\phi''_n + \phi''_s + \phi''_p = 1 \qquad (3.112)$$

Rearranging Equation 3.109,

$$\ln \frac{\phi''_s}{\phi'_s} = -\left(1 - \frac{1}{100}\right)\left(\phi''_p - \phi'_p\right) - 1.5\left(\phi''^2_n - \phi'^2_n\right) \qquad (3.113)$$

Therefore,

$$-\ln \frac{\phi'_s}{\phi''_s} = 0.99\left(\phi'_p - \phi''_p\right) + 1.5\left(\phi'^2_n - \phi''^2_n\right) \qquad (3.114)$$

From Equation 3.110,

$$\frac{1}{100}\ln \phi'_p + \left(1 - \frac{1}{100}\right)\phi'_p + 1.5\phi'^2_n = \frac{1}{100}\ln \phi''_p + \left(1 - \frac{1}{100}\right)\phi''_p + 1.5\phi''^2_n \qquad (3.115)$$

Equation 3.115–Equation 3.109 yields

$$\frac{1}{100}\ln \phi'_p - \ln \phi'_s = \frac{1}{100}\ln \phi''_p - \ln \phi''_s \qquad (3.116)$$

Therefore,

$$-\ln \frac{\phi'_s}{\phi''_s} = -\frac{1}{100}\ln \frac{\phi'_p}{\phi''_p} \qquad (3.117)$$

Finally, Equation 3.108–Equation 3.109 yields

$$\ln \phi'_n - \ln \phi'_s + 1.5\left[\left(1 - \phi'_n\right)^2 - \phi'^2_n\right] = \ln \phi''_n - \ln \phi''_s + 1.5\left[\left(1 - \phi''_n\right)^2 - \phi''^2_n\right] \qquad (3.118)$$

Further rearranging,

$$\ln \frac{\phi'_s}{\phi''_s} = 3\left(\phi'_n - \phi''_n\right) - \ln \frac{\phi'_n}{\phi''_n} \qquad (3.119)$$

From Equations 3.114 and 3.119,

$$\phi'_p - \phi''_p = \frac{1}{0.99} \times \left[3\left(\phi'_n - \phi''_n\right) - \ln \frac{\phi'_n}{\phi''_n} - 1.5\left(\phi'^2_n - \phi''^2_n\right)\right] \qquad (3.120)$$

From Equations 3.117 and 3.119,

$$\ln \frac{\phi_p'}{\phi_p''} = -100 \times \left[3 \left(\phi_n' - \phi_n'' \right) - \ln \frac{\phi_n'}{\phi_n''} \right] \qquad (3.121)$$

From Equation 3.119,

$$\ln \frac{\phi_s'}{\phi_s''} = - \left[3 \left(\phi_n' - \phi_n'' \right) - \ln \frac{\phi_n'}{\phi_n''} \right] \qquad (3.122)$$

Using Equations 3.111, 3.112, 3.120, 3.121, and 3.122, ϕ_n'', ϕ_s', ϕ_s'', ϕ_p', and ϕ_p'' can be obtained for a given value of ϕ_n', in the following way.

For a given $\phi_n' = 0.5$, ϕ_n'' is assumed to be 0.3; then from Equation 3.120,

$$\phi_p' - \phi_p'' = \frac{1}{0.99} \times \left[3(0.5 - 0.3) - \ln \frac{0.5}{0.3} - 1.5(0.5^2 - 0.3^2) \right]$$
$$= 0.1523$$

and from Equation 3.121,

$$\frac{\phi_p'}{\phi_p''} = \exp(-100) \left[3(0.5 - 0.3) - \ln \frac{0.5}{0.3} \right]$$
$$= 0.0001337$$

Solving the above two equations simultaneously, $\phi_p' = 0.00002$, and $\phi_p'' = 0.1496$.

Using Equations 3.111 and 3.112, ϕ_s' and ϕ_s'' are calculated to be 0.49998 and 0.5504, respectively, and these results lead to $\ln \frac{\phi_s'}{\phi_s''} = \ln \frac{0.49998}{0.5504} = -0.0960$. On the other hand, from Equation 3.122,

$$\ln \frac{\phi_s'}{\phi_s''} = - \left[3(0.5 - 0.3) - \ln \frac{0.5}{0.3} \right]$$
$$= -0.0892 \qquad (3.123)$$

The difference between two $\ln \frac{\phi_s'}{\phi_s''}$ values is $\Delta = (-0.0892) - (-0.0960) = 0.0068$.

Similar calculations were carried out for the different ϕ_n'' values assumed. The results of the calculations are

ϕ_n'' value assumed	Δ
0.3	0.068
0.29	0.0004
0.28	−0.0013
0.25	−0.0113
0.2	−0.222

It is obvious that the best result was achieved for $\phi_n'' = 0.29$, and the following volume fractions were obtained corresponding to the above ϕ_n'' value.

$$\phi_n' = 0.5$$
$$\phi_s' = 0.49997$$
$$\phi_p' = 0.00003$$

$$\phi_n'' = 0.29$$
$$\phi_s'' = 0.5447$$
$$\phi_p'' = 0.1653$$

The above results show two compositions on a binodial line, which are connected by a tie line. In other words, these compositions are in equilibrium with each other when phase separation of the polymer solution takes place.

Example 2

Determine the composition path on a triangular diagram when the initial composition of a polymer solution is $\phi_{si} = 0.9$ and $\phi_{pi} = 0.1$, and the ratio of the solvent and nonsolvent flux is $k' = 1.5$. The solvent density ρ_2^0 is 0.79.

From Equation 3.58,

$$\rho_i^* = (1 + k\omega_{1i})\rho_i \qquad (3.124)$$

where $k = k' - 1 = 0.5$. The initial mass fraction of nonsolvent is zero, and therefore, $\omega_{1i} = 0$ (and $\omega_{2i} = 1.0$). From Equation 3.70,

$$\begin{aligned}
\rho_i &= \frac{\phi_{si}\rho_2^0}{\omega_{2i}} \\
&= \frac{(0.9)(0.79)}{1.0} \\
&= 0.711
\end{aligned}$$

Then

$$\rho_i^* = [1 + (0.5)(0)](0.711)$$
$$= 0.711$$

Inserting all numerical parameters in Equation 3.88,

$$\phi_n = \frac{0.79 - 0.711}{0.79 - (1.5)(1.0)} - \frac{0.79}{0.79 - (1.5)(1.0)}\phi_p$$
$$= -0.11127 + 1.1127\phi_p \qquad (3.125)$$

The composition of the polymer solution changes on a N-S-P triangular diagram, along the straight line represented by Equation 3.125. In particular, at the end of the gelation, when solvent is completely exchanged with nonsolvent water, $\phi_{se} = 0$ and $\phi_{ne} + \phi_{pe} = 1.0$. Then, from Equation 3.125,

$$1 - \phi_{pe} = -0.11127 + 1.1127\phi_{pe}$$

Solving the above equation for ϕ_{pe}, $\phi_{pe} = 0.526$ and $\phi_{ne} = 0.474$.

Example 3
How will the composition path be when $\rho_2^0 = 1.14$?

$$\rho_i^* = [1 + (0.5)(0)] \frac{(0.9)(1.14)}{1.0}$$
$$= 1.026$$

Therefore,

$$\phi_n = \frac{1.14 - 1.026}{1.14 - 1.5} - \frac{1.14}{1.14 - 1.5}\phi_p$$
$$= -0.3167 + 3.167\phi_p$$

At the end of the gelation step,

$$1 - \phi_{pe} = -0.3167 + 3.167\phi_{pe}$$

Solving the above equation for ϕ_{pe}, $\phi_{pe} = 0.315$, and therefore $\phi_{ne} = 0.685$.

Example 4
Determine the composition path when $k' = 6.0$ for $\rho_2^0 = 0.79$ and 1.14.

For $\rho_2^0 = 0.79$,

$$\phi_i^* = [1 + (5.0)(0)] \frac{(0.9)(0.79)}{1.0}$$
$$= 0.711$$

$$\phi_n = \frac{0.79 - 0.711}{0.79 - 6.0} - \frac{0.79}{0.79 - 6.0}\phi_p$$
$$= -0.01516 + 0.1516\phi_p$$

At the end of the gelation step,

$$1 - \phi_{pe} = -0.01516 + 0.1516\phi_{pe}$$

Solving the above equation for ϕ_{pe}, $\phi_{pe} = 0.882$, and therefore $\phi_{ne} = 0.118$.
For $\rho_2^0 = 1.14$,

$$\phi_i^* = [1 + (5.0)(0)] \frac{(0.9)(1.14)}{1.0}$$
$$= 1.026$$

$$\phi_n = \frac{1.14 - 1.026}{1.14 - 6.0} - \frac{1.14}{1.14 - 6.0}\phi_p$$
$$= -0.02346 + 0.2346\phi_p$$

at the end of the gelation step,

$$1 - \phi_{pe} = -0.02346 + 0.2346\phi_{pe}$$

Solving the above equation for ϕ_{pe}, $\phi_{pe} = 0.829$, and therefore $\phi_{ne} = 0.171$.

Example 5

Calculate the numerical value for ϕ_p at $x = 0.0379$ under the following conditions.

$$\phi_{pi} = 0.13$$
$$\rho_1^0 = 1.0 \times 10^3$$
$$\rho_2^0 = 0.79 \times 10^3$$
$$k = 1.0$$
$$\alpha = 1.0$$

$$\tau = 0.001$$

Considering ϕ_p is the function of x, from Equation 3.104,

$$
\begin{aligned}
d\phi_p &= -\frac{\rho_b}{\alpha}\left[\frac{(k+1)}{\rho_2^0} - \frac{1}{\rho_1^0}\right] erfc\left(\frac{x}{\sqrt{4\alpha^2\tau}}\right) dx \\
&= -\frac{\rho_b}{\alpha}\left[\frac{(k+1)}{\rho_2^0} - \frac{1}{\rho_1^0}\right] erfc\left(\frac{x}{\sqrt{4\alpha^2\tau}}\right) \sqrt{4\alpha^2\tau}\, d\left(\frac{x}{\sqrt{4\alpha^2\tau}}\right)
\end{aligned}
$$

Setting

$$X = \frac{x}{\sqrt{4\alpha^2\tau}} \tag{3.126}$$

$$d\phi_p = -\frac{\rho_b}{\alpha}\left[\frac{(k+1)}{\rho_2^0} - \frac{1}{\rho_1^0}\right]\sqrt{4\alpha^2\tau}\, erfc(X)dX \tag{3.127}$$

The polymer volume fraction, ϕ_p, should retain its initial value when the distance from the gelation media-polymer solution interface, z, is very large. Such a large value in z is arbitrarily represented by $X = 3.0$. Equation 3.127 is then integrated with a boundary condition $\phi_p = \phi_{pi}$ when $X = 3.0$. Then

$$\int_{\phi_p}^{\phi_{pi}} d\phi_p = -\frac{\rho_b}{\alpha}\left[\frac{(k+1)}{\rho_2^0} - \frac{1}{\rho_1^0}\right]\sqrt{4\alpha^2\tau}\int_X^{3.0} erfc(X)\, dX \tag{3.128}$$

Rearranging,

$$\int_{\phi_{pi}}^{\phi_p} d\phi_p = \frac{\rho_b}{\alpha}\left[\frac{(k+1)}{\rho_2^0} - \frac{1}{\rho_1^0}\right]\sqrt{4\alpha^2\tau}\int_X^{3.0} erfc(X)\, dX \tag{3.129}$$

In Equation 3.129, ρ_b is the density of the gelation media (kg/m^3) and, therefore, equal to the nonsolvent density ρ_1^0. $1/\rho_1^0 = 1.0 \times 10^{-3}$ (m^3/kg) and $1/\rho_2^0 = 1.266 \times 10^{-3}$ (m^3/kg). $X = (0.0379)/\sqrt{(4)(1.0^2)(0.001)} = 0.6$.

Inserting all the numerical parameters,

$$
\begin{aligned}
\int_{0.13}^{\phi_p} d\phi_p = {}&\frac{1.0 \times 10^3}{1.0}\Big[(1.0 + 1)\left(1.266 \times 10^{-3}\right) - \\
&\left(1.0 \times 10^{-3}\right)\Big]\sqrt{(4)(1.0^2)(0.001)}\int_{0.6}^{3.0} erfc(X)\, dX
\end{aligned}
$$

The above equation yields

$$\phi_p = 0.13 + 0.0969 \int_{0.6}^{3.0} erfc(X)\, dX$$

Since

$$\int_{0.6}^{3.0} erfc(X)\, dX = 0.1752$$

$$\phi_p = 0.13 + (0.0969)(0.1752)$$
$$= 0.147$$

Example 6
Calculate ϕ_p for $x = 0$.
   Since $x = 0$, $X = 0$ and

$$\phi_p = 0.13 + 0.0969 \int_{0.0}^{3.0} erfc(X)\, dX$$

$$\int_{0.0}^{3.0} erfc(X)\, dX = 0.5545$$

Therefore,

$$\phi_p = 0.13 + (0.0969)(0.5545)$$
$$= 0.1837$$

The polymer volume fraction of the gelation media-polymer mixture is much higher than the initial polymer volume fraction, which means a significant contraction of the polymer solution film in the gelation bath.

Nomenclature for Chapter 3

3.4 Triangular Phase Diagram

$$\Delta G_m \; = \; \text{free energy of mixing, J/mol}$$
$$\Delta G_m \; = \; \text{free energy of mixing, J/mol}$$
$$m_i \; = \; \text{ratio of molar volume of component } i \text{ and solvent}$$
$$n_i \; = \; \text{number of moles of component } i$$
$$\mathbf{R} \; = \; \text{gas constant, 8.314 J/mol K}$$
$$T \; = \; \text{absolute temperature, K}$$
$$x \; = \; \text{mole fraction, —}$$
$$\Delta\mu \; = \; \text{change in chemical potential accompanying mixing, J/mol}$$
$$\phi \; = \; \text{volume fraction, —}$$
$$\chi \; = \; \text{interaction constant, —}$$

Superscripts

$$' \; = \; \text{quantities concerning the phase }'$$
$$'' \; = \; \text{quantities concerning the phase }''$$

Subscripts

$$n \; = \; \text{nonsolvent}$$
$$s \; = \; \text{solvent}$$
$$p \; = \; \text{polymer}$$

3.5 Change of the Solution Composition on the Triangular Diagram During the Gelation Process

$$a \; = \; \text{constant defined by Equation 3.57}$$
$$D \; = \; \text{diffusion coefficient, m}^2\text{/s}$$
$$D_0 \; = \; \text{constant defined by Equation 3.57}$$
$$f \; = \; \text{quantity defined by Equation 3.95}$$
$$f_0 \; = \; \text{quantity defined by Equation 3.96}$$
$$\bar{f} \; = \; \text{quantity defined by Equation 3.89}$$
$$k \; = \; k' - 1$$
$$k' \; = \; \text{ratio of solvent and nonsolvent flux}$$
$$L_0 \; = \; \text{initial film thickness, m}$$
$$N_1 \; = \; n_1 \text{ at } z = 0, \text{ kg/m}^2 \text{ s}$$
$$n \; = \; \text{mass flux, kg/m}^2 \text{ s}$$
$$t \; = \; \text{time, s}$$
$$x \; = \; \text{quantity defined by Equation 3.93}$$

Membrane Preparation

z = distance, m

α = quantity defined as (internal mass transfer resistance external mass transfer resistance)$^{-1}$

ρ = density defined as the weight of nonsolvent-solvent mixture (free from polymer) in a unit volume of film, kg/m^3

ρ_b = density of the gelation media, kg/m^3

ρ_i^* = $(1 + k\omega_{1i})\rho_i$

τ = quantity defined by Equation 3.92

ϕ = volume fraction, —

ω = mass fraction in nonsolvent-solvent mixture on a polymer-free basis

Superscripts

0 = quantities of pure liquid

Subscripts

1 = nonsolvent in the nonsolvent-solvent mixture

2 = solvent in the nonsolvent-solvent mixture

e = quantity at the end of gelation

i = quantity at time = zero

n = nonsolvent

p = polymer

s = solvent

4

Microscopic Structure of the Membrane and the State of the Permeant

It has been shown in the previous chapter that membranes prepared by the phase inversion technique should be asymmetric, with a dense skin layer on the top side of the membrane that is supported by a relatively thick porous sublayer. This structure was confirmed by the electron microscopic observation. Moreover, some electron micrographs revealed the microscopic structure inside the dense skin layer. Since the membrane separation is governed by the skin layer, the control of its microscopic structure is considered the most crucial for the membrane design.

When the permeant molecules are retained in the membrane, they are confined in the space created between polymer molecules. The space should be penetrating from one side of the membrane to the other, since the permeant molecule can permeate through the membrane, under the influence of a driving force. The latter space can therefore be regarded as a "pore." The size of the pore in the dense layer of the membrane should be so small that the permeant molecule is under the influence of a strong interaction force exerted from the material that constitutes the "pore wall." Hence, the structure of such permeant molecules should be different from that of the molecules in the bulk phase. This structural change was confirmed by Differential Scanning Calorimetry (DSC) and Nuclear Magnetic Resonance (NMR) methods.

4.1 Microscopic Structure of the Membrane

It was shown in the previous chapter that the phase separation lines and the composition path on a triangular diagram allow quantitative discussions on the macroscopic structure of the membrane produced during the gelation process, such as membrane asymmetricity and the porosity of the porous sublayer. As will be shown later, when the cross-section of a cellulose acetate reverse osmosis membrane is observed under an electron microscope, the membrane is not homogeneous, but possesses remarkable asymmetricity. The presence of asymmetricity was anticipated long before the micrographic picture was taken. The following experimental observations supported the idea of a dense skin layer superimposed on a porous layer:

1. The membrane is capable of desalting electrolyte solutions only when one side of the membrane is in contact with a feed solution. This side, called side A, is faced toward air when the solvent is evaporated from the cast film of the polymer solution. When the other side, called side B, is in contact with the feed, the same membrane shows no desalting capacity.

2. When grease is coated on side A of the membrane, a large amount of salt solution is taken into the membrane by immersing the membrane into the solution. On the other hand, when grease is coated on the other side (side B), salt solution does not enter the membrane [51].

Microscopic technique is one method of physically proving the asymmetric structure of the membrane. Keilin is the first person who attempted to observe the cross-sectional structure of the membrane, by a dyeing method [52]. A technique similar to that applied for the investigation of cellulose rayon was adopted [53]. According to the method, a cellulose acetate membrane was converted into cellulose by hydrolysis in a 0.4 N sodium hydroxide solution. The membrane was then immersed successively into 2% Victoria Blue "B" and Calamine Brilliant Yellow solutions in a mixed solvent of water:dioxane (1:9). The membrane was incorporated into paraffin wax and sliced in a direction vertical to the membrane surface to the thickness of 4 μm. The thin slice so prepared was investigated under an optical microscope. A green-colored layer of 4-μm thickness was observed on the side that was faced toward air during the solvent evaporation step. This layer could be dyed effectively, since the macromolecular density was high. The thickness of the layer extended over about 9% of the entire membrane thickness. The rest of the cross-section was porous, and dyeing was not as effective as with the dense layer. There was a sharp boundary between the two layers. A cellulose acetate membrane was also dyed using 1.2% aqueous solution of Diphenyl Fast Red 5BL Supra I and 2.8% aqueous solution of Chlorantine Fast Green BLL, before slicing the membrane sample. The micrographic investigation revealed that the cellulose acetate membrane had the same structure as that of the cellulose membrane, thus confirming that hydrolysis did not alter the membrane structure.

Scanning electron microscopy (SEM) offers higher magnification than optical microscopy, extending the resolution down to the nanometer level. The first electron micrographic picture of a cellulose acetate membrane was taken

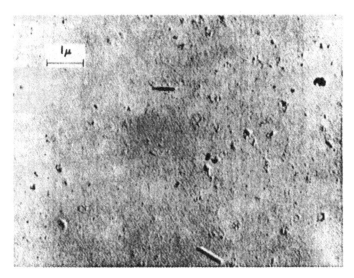

Figure 4.1.   Electron micrographic picture of a cellulose acetate membrane from the dense skin layer side. (Reproduced from [55] with permission.)

by Riley et al [54]. A cellulose acetate membrane was dried and shadowed in a vacuum deposition chamber with palladium from a 20° angle. A thin carbon layer (thickness about 20 nm) was deposited before polymer was removed by acetone solvent in an extraction vessel equipped with a reflux condenser, leaving a palladium carbon replica. Electron micrographic pictures of the replica so prepared were taken. Since direct printing was applied, the dark region looked bright in the picture. When the cellulose acetate membranes contained above 60% (by weight) water, the removal of water without changing the microscopic structure of the membrane seemed the factor governing the quality of the electron micrographic picture. Therefore, the membrane was dried using a freeze-drying method [55]. A small membrane sample was immersed into isopentane cooled to $-170°C$ with liquid nitrogen. After a few minutes the membrane was transferred to a freezer that was maintained at $-60°C$. The pressure inside the freezer was kept at 0.1 mmHg, and the membrane was dehydrated by sublimation. The membrane so dried could exhibit the same reverse osmosis performance as it did before drying when the membrane was rewetted with water. Figure 4.1 shows an electron micrographic picture of a sample prepared by the method described above. This picture was taken from above the dense surface of the membrane. Several virus-shaped particles (about 50 × 500 nm^2) are observed. The membrane surface is smooth and there are no pores larger than 10 nm in diameter. Figure 4.2 shows the picture of the membrane cross-section near the porous surface. The picture indicates that the pore size in this region is about 0.4 μm. Figure 4.3 shows the picture of the surface opposite

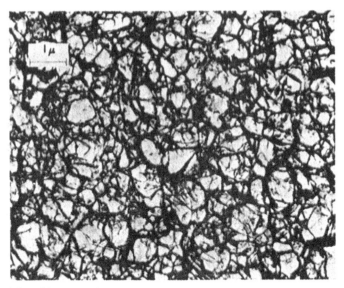

Figure 4.2. Electron micrographic picture of the cross-section of a cellulose acetate membrane. (Reproduced from [55] with permission.)

to the dense skin layer. The surface is not smooth, and many wrinkles are observed. This porous surface has, therefore, a structure totally different from the dense surface. Riley's electron micrographic observation can be summarized as follows: cellulose acetate reverse osmosis membranes are asymmetric. A dense layer is formed on the side of the membrane that faces air during the solvent evaporation period. Its thickness is 0.5 μm. The other side of the membrane is porous, and the pore size is 0.4 μm. According to the above observation, Riley et al. concluded that there are no detectable pores larger than 10 nm in the dense layer.

The pore size of interest in reverse osmosis is much smaller than the 10-nm resolution used by Riley et al. Schultz and Asunmaa took high-resolution electron microscopic pictures that revealed more detail of the membrane surface [56]. According to their method, an ultrathin membrane 60 nm in thickness was prepared from cellulose acetate (Eastman E-398-10) by casting an acetone solution of the polymer on a glass plate and evaporating the solvent completely. The film on the glass plate was shadowed with Pt-Pd (80:20) alloy by vacuum deposition from a 7° angle. A carbon layer 20 nm in thickness was deposited, and a cellulose acetate film removed, by dissolving the film successively into amyl acetate and acetone. The carbon replica so prepared was observed under an electron microscope (Hitachi HU-11), magnified 19,000 and 30,000 times. Printing was carried out in two steps, and thus a reversed print with a magnification of 165,000 was obtained. The dark object appears dark on the print.

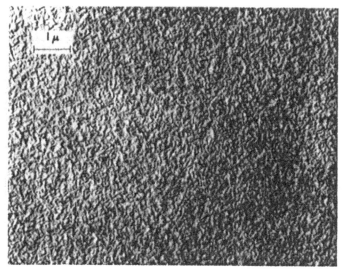

Figure 4.3. Electron micrographic picture of a cellulose acetate membrane from the side opposite to the dense skin layer. (Reproduced from [55] with permission.)

Figure 4.4 shows the picture of an ultrathin membrane. It can be seen that circular unit cells are compacted in an irregular fashion. The average diameter of the unit cell was found to be 18.8 ± 0.3 nm. Schultz and Asunmaa have also taken the picture of an asymmetric cellulose acetate membrane. The membrane was cast from a casting solution whose composition was 25 wt% polymer, 30 wt% formamide, and 25 wt% acetone. The asymmetric membrane prepared by the Manjikian method [57] was heat-treated before a carbon replica was made by shadowing the dense side of the membrane with Pt-C from a 9° angle. The electron micrographic observation of the replica also showed semiglobular unit cells 18 to 19 nm in diameter, compacted irregularly on the membrane surface.

On the basis of the above observation, Schultz and Asunmaa developed the following transport mechanism. They made an assumption that the low-density and the noncrystalline region of the polymer that fills the space between the circular cells is incorporated into the unit cell as its part. Those spaces (between the unit cells) were therefore assumed to be vacant. In reverse osmosis operation these vacant spaces are filled only with water, and this water is assumed to be more ordered than the ordinary water under strong influence from the polymeric material. This water flows by the viscous flow mechanism through channels that are formed by connecting the vacant spaces. Suppose r_p^* is the effective radius of this pore (m), n_p is the number of the pore in a unit area (1/m^2), p is the pressure drop across the membrane (Pa), η is the water viscosity (Pa s), L is the effective layer thickness (m), and τ is the tortuosity factor (–), the volumetric

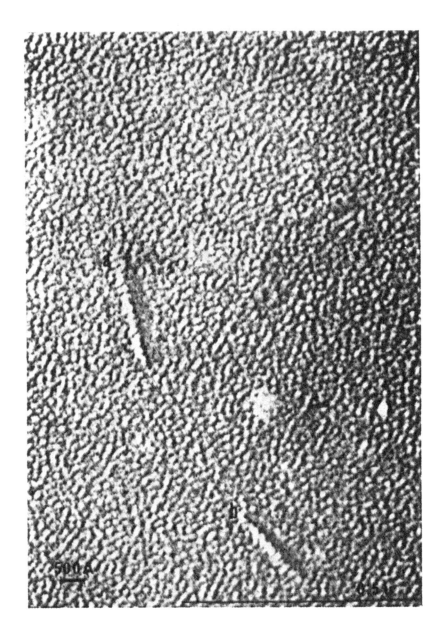

Figure 4.4. Electron micrographic picture of an ultrathin cellulose acetate membrane. (Reproduced from [56] with permission.)

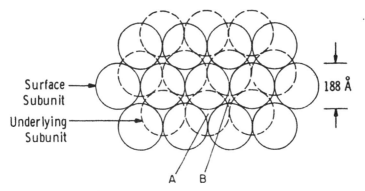

Surface —→ Subunit

Underlying —→ Subunit

188 Å

A B

Figure 4.5. Schematic representation of closely packed spherical polymer cells. (Reproduced from [56] with permission.)

flow rate of water can be written by Poiseuille's law as

$$J = n_p \pi p (r_p^*)^4 / 8\eta L\tau \qquad (4.1)$$

Corresponding to a channel bending and interconnecting the void spaces, $\tau = 2.5$ is appropriate due to Carman's analysis [58]. The closest packing of the spherical unit cells can be schematically illustrated in Figure 4.5. The area of the triangular void space can then be calculated to be $9.4^2(\sqrt{3} - \pi/2) = 14.30$ nm^2. This area corresponds to the effective radius of 2.13 nm. Assuming that the monolayer of water adsorbed to the channel wall is totally immobile, the thickness of the monolayer of water (0.28 nm) has to be subtracted from the effective radius. Then the r_p^* value to be used in Equation 4.1 becomes 1.85 nm. The numerical value for n_p can be calculated from the closest packing of the spherical unit cells, to be $6.5 \times 10^{15}/m^2$. Inserting the r_p^* and n_p values obtained above into Equation 4.1, η is calculated to be 0.035 Pa s. This is orders of magnitude higher than that of the ordinary water, indicating that the water that fills the membrane pore is far more viscous than ordinary water. The structure of water is usually considered to consist of two parts. One is clusters that are formed by strong hydrogen bonding working between water molecules. Water molecules are strongly ordered in a cluster. The other is a part where disorder prevails among water molecules. A water molecule fluctuates between these two parts of the water structure, and a thermodynamic equilibrium is maintained. The cluster size increases with a decrease in temperature, as shown in Figure 4.6 where the size of a cluster is given as n_{cl} [59], [60]. As mentioned earlier, pores with an effective radius of 2.13 nm are filled with special water of high viscosity. Therefore, assuming the cross-section of the pore is occupied by two water clusters of spherical shape, the diameter of such a cluster is 2.13

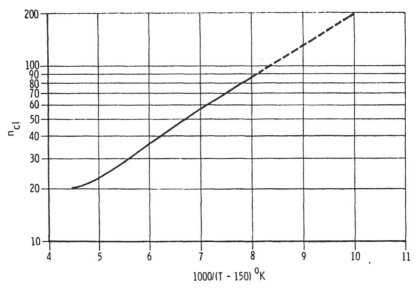

Figure 4.6. Number of water molecules in a cluster as a function of temperature. (Reproduced from [56] with permission.)

nm. Since the volume of one water molecule is 0.0299 nm^3, one cluster is occupied by 162 water molecules, corresponding to $T' = 255$ K in Figure 4.6. The structure of the water in the membrane pore is that of water (not ice) that was supercooled to 255 K. For the viscous flow of supercooled water, Miller proposed the following equation to calculate the activation energy $\Delta H^{\ddagger}$ (J/mol) [61]:

$$\Delta H^{\ddagger} = 4.27 \times 10^3 [T'/(T' - 150)]^2 \qquad (4.2)$$

Equation 4.2 yields 25.18×10^3 J/mol at 255 K. The latter value agrees very well with 24.91×10^3 J/mol, which was obtained experimentally by Keilin for a cellulose acetate membrane with 53.5% water content exhibiting 99.5% sodium chloride rejection.

Schultz and Asunmaa proceeded further to the theoretical calculation of the viscosity of supercooled water. According to Glasstone-Laidler-Eyring [62], the liquid viscosity can be given as

$$\eta = \frac{hN}{V} \exp \frac{\Delta F^{\ddagger}}{RT}$$

$$= \left[\frac{hN}{V} \exp \frac{-\Delta S^{\ddagger}}{R} \right] \left[\exp \frac{\Delta H^{\ddagger}}{RT} \right] \qquad (4.3)$$

where η is viscosity (Pa s), h is Planck's constant (6.626×10^{-34} J s), N is Avogadro's number (6.022×10^{23} mol^{-1}), V is molar volume of liquid (m^3/mol), $\Delta F^{\ddagger}$ is activation energy of viscous flow (J/mol), $\Delta S^{\ddagger}$ and $\Delta H^{\ddagger}$ are entropy and enthalpy of viscous flow, $\mathbf{R}$ is the gas constant (8.314 J/K mol), and T is absolute temperature (K). $\Delta F^{\ddagger}$ has been given by

$$E_{vap}/\Delta F^{\ddagger} = 2.45 \qquad (4.4)$$

for approximately a hundred nonmetal materials, including water and other liquids that can form molecular associations. Therefore, Equation 4.4 should also be applicable for water in the membrane pore. Furthermore, heat of vaporization, E_{vap} (J/mol), can be given by

$$E_{vap} \simeq E_w + C_w(T - T') \qquad (4.5)$$

where E_w is heat of vaporization (J/mol) at room temperature T (K), C_w is molar heat capacity of liquid (J/mol K), and T' corresponds to the temperature to which water in the pore is supercooled. According to Equation 4.5, E_{vap} is calculated to be 44.55×10^3 J/mol, and therefore $\Delta F^{\ddagger}$ is 18.18×10^3 J/mol, from Equation 4.4. Since $\Delta H^{\ddagger}$ is 25.18×10^3 J/mol, $\Delta S^{\ddagger}$ is calculated to be 23.57 J/mol K. Inserting all available numerical values into Equation 4.3, the viscosity of water in the pore is calculated to be 0.035 Pa s at 296 K. This value is exactly the same as the one obtained earlier from the Poiseuille equation and 37 times higher than 0.000936 Pa s, which is the viscosity of ordinary water at 296 K. Thus, on the basis of the microstructure of the dense surface layer of an asymmetric membrane, Schultz and Asunmaa concluded that water has a structure different from ordinary water, when it is present in the membrane. Similar structures were observed at the dense surface layer of aromatic polyamide membranes and of composite reverse osmosis membranes. The presence of water, with a different structure than ordinary water, in the membrane polymer was also confirmed by the DSC technique and by NMR.

A more detailed analysis of the membrane surface was made by Panar et al. They prepared a sample for the electron micrographic observation, by the freeze-cleave technique. A wet membrane of $1/8 \times 1/2$ in.^{2} was held in a small vise and subjected to a vacuum of 10^{-7} mmHg while the sample was being cooled in liquid nitrogen. The freeze-dried sample was then sliced with a knife cooled with liquid nitrogen, and the surface was shadowed with Pt/C by vacuum deposition from an angle of 40°. The replica so prepared was investigated under an electron microscope Zeiss EM 9S [63]. They also investigated the structure of the casting solution. A drop of the casting solution was frozen before it was sliced to observe the cross-section by electron micrograph. Figure 4.7 shows an electron microscopic picture of the cross-section of an aromatic polyamidehydrazide membrane. The surface of the membrane is covered by a closest monolayer packing of micelles with diameters from 40 to 80 nm. When X-ray diffraction was applied to this layer, no crystalline structure was detected. The above monolayer covers the surface of a support layer where the spherical micelles are irregularly packed with void spaces of 7.5 to 10.0 nm in size. In the surface layer the micelles are contacting each other, compressed, and deformed, and few void spaces exist between micelles. While the structural unit of both surface and support layers is macromolecular micelles, they are more densely packed at the surface layer. The surface layer seems to exhibit a mechanical property different from that of the support layer. For example, the surface layer can be removed from the support layer. In some extreme cases the surface layer can be totally isolated from the support layer while adhering to water. Figure 4.8 shows a section of the surface layer free of support layers. The grainy surface in this micrograph is the fracture surface of water surrounding the gel membrane. The reason for the formation of the monolayer on the surface is probably due to the surface tension that tends to decrease the area of the surface, resulting in the close packing of the macromolecular micelles at the surface of the casting solution. The structure observed for aromatic polyamidehydrazide membranes seems to be common for all membranes that have asymmetric structures. As mentioned earlier, the microstructure of the dense surface layer does not change during the freeze-drying of the membrane. Therefore, the pictures shown in Figures 4.7 and 4.8 can be considered as the microstructure of membranes in a wet state. It is also speculated that the macromolecular micelles found in the polymeric membrane have already been formed in the casting solution. Figure 4.9 shows an electron micrographic picture of aromatic polyamide-hydrazyde casting solution freeze-dried, and sliced with a microtome. The picture was taken from a 45° angle to the surface. The micellar structure is clearly observed in the picture. The surface monolayer, distinct from the micellar structure of the support layer underneath, is also clearly shown in the picture. Although only 60 s elapsed between the film casting and the freeze-drying, this period seems to be enough for an interfacial structure to be formed. Thus, Panar et al. showed that the microscopic structure of the membrane is formed instantly when the film is cast from a polymeric solution, and this becomes the precursor of the membrane structure. Of course, the structure is not the same as that of the membrane, since two steps involving solvent evaporation and polymer gelation have to be passed before the membrane is ultimately formed. Their observation, however, reveals that the structure of the membrane is affected significantly by that of the casting solution. Contrary to Schultz and Asunmaa, Paner et al. stated that there is no void space between the spherical micelles. The mass transport through the membrane, therefore, has to take place in spaces that are present in the polymer

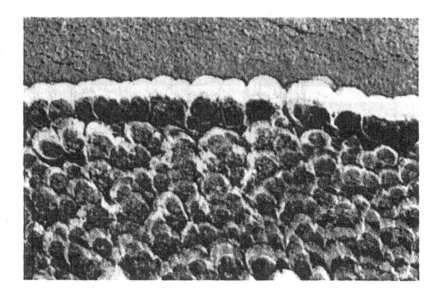

Figure 4.7. Electron micrographic picture of the cross-section of an asymmetric aromatic polyamide-hydrazide membrane. (Reproduced from [63] with permission.)

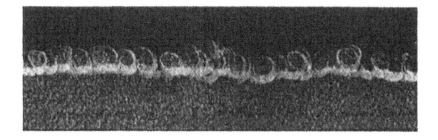

Figure 4.8. Electron micrographic picture of the surface skin of aromatic polyamide-hydrazide membrane. (Reproduced from [63] with permission.)

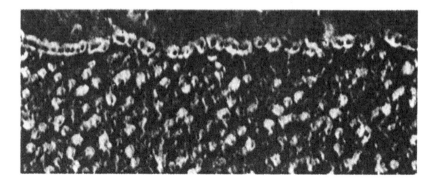

Figure 4.9.   Electron micrographic picture of aromatic polyamide-hydrazide casting solution freeze-dried, and sliced with a microtome. (Reproduced from [63] with permission.)

network of the micelle. A similar micellar structure of the macromolecules was also found in the surface layer of the composite FT 30 membrane (Figure 4.10) [64]. Investigation of the pore size on the selective skin layer was also conducted by Chan et al., with respect to aromatic polyamide membranes [65].

There are excellent summaries on the characterization of polymeric membranes, using methods other than the electron micrograph [66], [67]. Among others, the adsorption method was used by Ohya et al. [68], [69], Broens et al. [70], and Han et al [71]. Thermoporometry was used by Brun et al. [72], Zeman and Tkacik [73], and Cuperus et al [74]. The molecular probe method was used by Michaels [75] and Zeman and Wales [76]. Recently, Dietz et al. applied the impedance method to determine membrane porosity [77].

4.2   Structure of the Casting Solution

It was shown by an electron micrograph that a membrane consists of spherical micelles packed densely at the membrane surface and loosely in the interior of the membrane. It was also shown that the structure of the polymer in a film casting solution is essentially the same as that of the membrane. In other words, the microscopic structure of a membrane is intrinsically governed by that of the casting solution structure. It is therefore interesting to know about the structure of the polymer in a film casting solution, and particularly about the size of individual polymeric molecules and their packing mode.

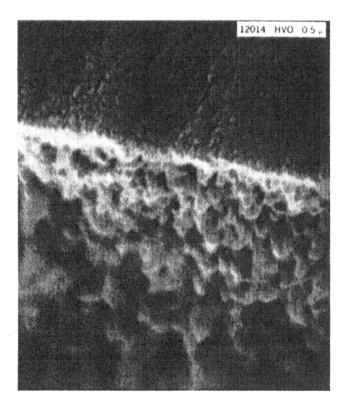

Figure 4.10. Electron micrographic picture of a composite FT-30 membrane. (Reproduced from [64] with permission.)

The intrinsic viscosity, $[\eta]$ (volume of solvent/weight of polymer), is a measure of the extension of a polymeric molecule in a solvent at infinite dilution. It is defined as

$$[\eta] = \lim_{g \to 0} \frac{1}{g} \left(\frac{\eta}{\eta_0} - 1 \right) \tag{4.6}$$

where η and η_0 are the viscosity (Pa s) of a polymer solution and that of the solvent, respectively, and g is the concentration given as weight of the polymer/volume of the solvent. The intrinsic viscosity $[\eta]$ is further related to the viscosity average molecular weight of the polymer, $\overline{M}_v$ by

$$[\eta] = K\overline{M}_v^\alpha \tag{4.7}$$

where K and α are called Mark-Houwink constants, which depend on the polymer, solvent, and temperature, but are independent of the mean molecular weight or molecular weight distribution over a reasonably broad range of these parameters. The value of α depends on the strength of the polymer-solvent interactions, ranging from 0.5 in poor (theta) solvents, through 0.65 in fair solvents, to 0.8 in good solvents. Values of $\overline{M}_v$ lie close to the weight-average molecular weight, $\overline{M}_w$, because of the greater contribution of larger molecules to viscosity. The macroscopic viscosity of a suspension of spherical particles in a medium with viscosity η_0 is given by the Einstein equation,

$$\eta = \eta_0(1 + 2.5\phi) \tag{4.8}$$

where ϕ is equal to the volume fraction of the spherical particles in the solution. In the derivation of the above equation, it is assumed that there is no interaction between the suspended particles. Therefore, the Einstein equation is applicable for the infinite dilution of polymer in the solvent, assuming polymers are spherical particles. The volume fraction ϕ can be given by

$$\phi = \frac{g}{M}N_0 v\varepsilon \tag{4.9}$$

where M is the molecular weight (kg/mol), v is the volume of an unswollen spherical polymer molecule (m³), N_0 is Avogadro's number (1/mol), and ε is an effective volume factor (–) to allow for swelling of the polymer by solvent. If the density of amorphous polymer at the solution temperature is ρ (kg/m³), v can be written as

$$v = M/\rho N_0 \tag{4.10}$$

Combining Equations 4.8 and 4.9 yields

$$\frac{\eta - \eta_0}{g\eta_0} = \frac{2.5N_0 v\varepsilon}{M} \tag{4.11}$$

If ε becomes ε_0 in the limit of zero concentration,

$$[\eta] = \lim_{g \to 0} \left(\frac{\eta - \eta_0}{g\eta_0} \right) \tag{4.12}$$

$$= \frac{2.5 N_0 v \varepsilon_0}{M} \tag{4.13}$$

With Equation 4.7

$$\varepsilon_0 v = \frac{KM^{\alpha+1}}{2.5 N_0} \tag{4.14}$$

or with Equation 4.10,

$$\varepsilon_0 = \frac{KM^{\alpha}\rho}{2.5} \tag{4.15}$$

Assuming that the shape of the polymer in the dilute solution is spherical and denoting the radius of the polymer as r_0,

$$\frac{4}{3}\pi r_0^3 = \varepsilon_0 v \tag{4.16}$$

Combining Equations 4.15 and 4.16,

$$r_0 = \left(\frac{3KM^{\alpha+1}}{10\pi N_0} \right)^{1/3} \tag{4.17}$$

or

$$r_0 = \left(\frac{3[\eta]M}{10\pi N_0} \right)^{1/3} \tag{4.18}$$

The above equation allows the calculation of the polymer radius at infinite dilution by knowing the intrinsic viscosity of the polymer solution [78].

When the polymer concentration is increased, the polymer molecules are desolvated and their size decreases. At the same time the number of polymer molecules in a unit volume of the polymer solution increases. As a result the distance between two polymer molecules decreases, the polymer molecules becoming more and more closely packed. When the polymer packing becomes a simple cubic packing, all polymers touch the neighboring six polymers. The volume fraction of the solvated polymer should be 0.524 when this critical point is reached, regardless of the size of the solvated polymer. Many properties of the polymer solution start to change when this critical point is reached. The most notable change is in the slope of two straight lines arising in the log-log plot of zero-shear solution viscosity against weight percent of polymer in the solution. The slope is reported to be much greater above a critical concentration that can be obtained from the intersect of two straight lines. According to Rudin and Johnston, the rapid increase in the Newtonian viscosity occurs when the volume fraction of the solvated polymer reaches 0.524 [79]. They also found that the viscosity data reported in the literature can be better explained by using a volume fraction of 0.507 rather than 0.524. The former value is used hereafter in this chapter, but it should be noted that this value is purely empirical. Then using Equation 4.9 and setting $\phi = 0.507$,

$$\phi_x = 0.507 \tag{4.19}$$

$$= \frac{g_x N_0 v \varepsilon_x}{M} \tag{4.20}$$

where subscript x denotes the critical conditions that will be called the limiting conditions hereafter. When Equation 4.20 is rearranged,

$$g_x v \varepsilon_x = \frac{0.507M}{N_0} \tag{4.21}$$

Combination of Equations 4.10 and 4.21 yields

$$g_x \varepsilon_x = 0.507 \rho \tag{4.22}$$

In order to calculate the value of g_x, we must first be able to evaluate ε_x. From the studies of Maron and Chiu on polymethylmethacrylate in various solvents, the following equation was obtained [80]:

$$\varepsilon_x = 2.60 + \left(0.34 \times 10^{-5}\right) M \tag{4.23}$$

Since the molecular weight of the repeat unit of polymethylmethacrylate is 100, and the repeat unit contains two atoms (carbon) that constitutes the main chain of the polymer, Rudin and Johnston generalized the above equation to

$$\varepsilon_x = 2.60 + 100 \left(\frac{0.34}{2}\right) \times 10^{-5} Z \tag{4.24}$$

where Z is the number of main chain atoms in the polymer molecule. Combining Equations 4.22 and 4.24, we obtain

$$g_x = 0.507\rho / \left(2.60 + 17 \times 10^{-5}Z\right) \tag{4.25}$$

The quantity for g_x can be calculated from the above equation. The volume of the solvated molecule at the critical point is $v\varepsilon_x$. Then the radius of the solvated polymer at the critical condition, r_x, is given as

$$\frac{4}{3}\pi r_x^3 = v\left(2.60 + 17 \times 10^{-5}Z\right) \tag{4.26}$$

$$r_x = \left[\frac{3v\left(2.6 + 17 \times 10^{-5}\right)Z}{4\pi}\right]^{1/3} \tag{4.27}$$

The size of the polymer at the critical point is thus known. When the polymer concentration is between infinite dilution and the critical concentration, the effective volume factor, ε, is given as a function of the polymer concentration, g:

$$\frac{1}{\varepsilon} = \frac{1}{\varepsilon_0} + \frac{g}{g_x}\left(\frac{\varepsilon_0 - \varepsilon_x}{\varepsilon_0\varepsilon_x}\right) \tag{4.28}$$

The numerical values for ε_0, ε_x, and g_x are given by Equations 4.15, 4.24 and 4.25. The radius of the polymer molecule can be calculated by [79]

$$r = \left(\frac{3v\varepsilon}{4\pi}\right)^{1/3} \tag{4.29}$$

Table 1 shows the data for the macromolecular radius, r, at infinite dilution, of aromatic polyamide polymers with different molecular weights [81]. The table also shows similar data for the case when $CaCl_2$ was added to the polymer solution as an electrolyte additive, calculated on the basis of the intrinsic viscosity data determined in the presence of the electrolyte additive. It is clear that the radius of the macromolecule increases as the polymer molecular weight increases. In the presence of the electrolyte additive, the radius is about twofold as compared to the radius in the absence of the electrolyte additive. The latter results can be interpreted either by inclusion of several macromolecules into a single macromolecular sphere due to complex formation, in which an electrolyte is shared by amide groups of several aromatic polyamide molecules, or by the solvation of electrolyte resulting in higher swelling of the spherical macromolecule.

Table 1. The Radius of the Aggregate of Polyamide Polymers
in Dilute Solution at 20°C [81]

Polymer	$\overline{M}_n$		Radius of the polymer molecule $\times 10^{10}$, m	
	without $CaCl_2$	with $CaCl_2^a$	without $CaCl_2$	with $CaCl_2^a$
PA-1	10,500	22,500	46.9	71.7
PA-2	18,400	47,850	63.0	112.7
PA-3	20,700	53,030	68.3	120.0
PA-4	29,500	90,330	86.7	164.7
PA-5	31,300	129,090	90.1	205.0

^{a} $CaCl_2$ to polymer weight ratio = 1:3.3.

Table 2. Polymer Size in Casting Solutions of
Different Polymer Concentrations at 95°C

Casting solution	Polymer size $\times 10^{10}$, m
PA- 7.0-0.3	72.3
PA- 9.0-0.3	66.1
PA-10.5-0.3	62.5
PA-12.5-0.3	58.6
PA-14.0-0.3	56.1

^{a} PA-7.0-0.3 indicates polymer concentration of
7 wt% and (nonsolvent $CaCl_2$/polymer) weight ratio
of 0.3. Molecular weight of polymer is 31,300.

Usually, the casting solution for aromatic polyamide reverse osmosis membranes includes the electrolyte additive, and the composition is near the critical concentration. Then the radius of macromolecular spheres should be nearly equal to 52 Å, as indicated by Table 2, and their packing fashion is close to the cubic packing. Approximating the square-shaped interstitial void area generated between four neighboring spheres (see Figure 4.11) by a circle of equal area, the effective radius of such interstitial void spaces is 27 Å.

This is considered too large for a reverse osmosis membrane pore. Presumably, a significant reduction in the size of the spherical macromolecule takes place during the evaporation and gelation steps. The macromolecular radius calculated above is, however, far smaller than the radius of the macromolecular nodule observed by Panar et al., under the electron microscope [63]. Kesting

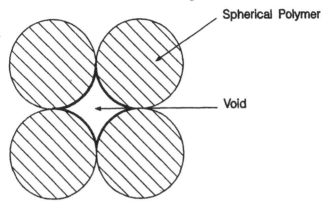

Figure 4.11. Schematic representation of a void space surrounded by four neighboring spheres.

suggested, therefore, that macromolecular nodules observed by Paner et al. were comprised of several macromolecules [82]. In other words, the nodule is an aggregate of the macromolecular spheres that are too small to be observed by the electron micrographic technique. While the interstitial void space created between individual macromolecular spheres corresponds to the pore of reverse osmosis membranes, the void spaces generated between aggregates of the macromolecular spheres should correspond to the pores of ultrafiltration membranes and those of microfiltration membranes. Therefore, it is expected that the pore sizes of the ultrafiltration and microfiltration membranes are far greater than the size of the interstitial void space calculated on the basis of the individual macromolecular sphere [82], [83]. The mode of the aggregate formation depends on the gelation path on the triangular nonsolvent-solvent-polymer diagram illustrated in Figure 3.24 and the speed of movement on the path. Thorough discussions were made, and the results were experimentally confirmed by the work of Kamide et al [84]. It is also interesting to note that the process of aggregate growth can be discussed as a fractal process; i.e., the individual macromolecular spheres agglomerate to form an aggregate (primary particle by Kamide); then the aggregates agglomerate into an aggregate of aggregates (secondary particle by Kamide); the aggregates of the aggregates further agglomerate to form an aggregate of aggregates of aggregates, and the size of the structural units increases progressively [85]. Accordingly, the possibility of a binodal, or even multinodal, pore size distribution arises.

The polymer nodules were observed not only on the surface of reverse osmosis membranes, but also on the surface of asymmetric membranes prepared for the separation of gas mixtures by Kazama et al [86]. While preparing hollow fiber membranes from their Cardo-type polyamide polymer, they investigated the microstructure of the membrane by an electron microscope. Skin layers

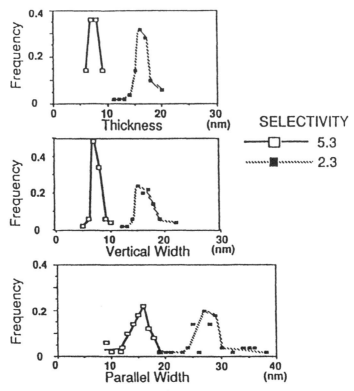

Figure 4.12. Selectivity vs. polymeric nodule size of the Cardo-Type polyamide membranes. (Reproduced from [86] with permission.)

were observed on both bore and shell sides of the hollow fiber. In particular, the topmost part of the inner skin layer consisted of two layers of closely packed polymer nodules, the thickness of each layer being 8 nm. Each polymer nodule was football shaped. The nodule width of 8 nm was observed when the membrane was sectioned vertically to the hollow fiber spinning direction, whereas the width was about 15 nm when the membrane was sectioned parallel to the spinning direction. The nodule size could be altered by changing the membrane casting conditions. The membranes were used for O_2/N_2 separation after drying. Figure 4.12 shows that a membrane of higher selectivity could be produced by reducing the size of the nodule. Kazama et al. thought that the space between the nodules could be closed more easily when the nodule was smaller, thus preventing the leak of the gas mixture through such void spaces.

The function of the additives in the casting solution is not very clear. In the above example the $CaCl_2$ additive increased the size of the polymer aggregate either by binding several molecules of aromatic polyamide polymer together or by intensifying the degree of swelling of the polymer molecule in the casting solution. It is postulated, on the other hand, that polyvinylpyrrolidone polymer, when added to the casting solution of polyethersulfone (Victrex), acts as a polymer surfactant. When no polyvinylpyrrolidone is added, water in the membrane has to be dispersed in a very hydrophobic environment of polyethersulfone polymer matrix, resulting in the formation of large water droplets during the gelation step in a continuous polymer phase. When polyvinylpyrrolidone is added, on the other hand, hydrophilic polyvinylpyrrolidone stands between the water and polyethersulfone, lowering the interfacial tension at the polymer/water interface. Smaller water droplets can be dispersed during the gelation step, resulting in smaller pore sizes in the dense skin layer [87]. It has also been found that polyvinylpyrrolidone polymer, although miscible in water, is not completely released into the gelation media (water) during the gelation step, since polyethersulfone and polyvinylpyrrolidone polymers are intermingled with each other. Miyano et al. measured the concentration of polyvinylpyrrolidone in the top surface layer of asymmetric polyethersulfone membranes, by the FTIR-IRS method, immediately after the gelation process and also after many hours of pure-water permeation experiments [88]. The infrared beam is supposed to penetrate to a depth of 0.5 μm. Table 3 clearly indicates that a large amount of polyvinylpyrrolidone polymer remains in the membrane surface even after a lengthy water permeation experiment. When the polyvinylpyrrolidone content in the casting solution is low (about 11%), the polyvinylpyrrolidone content in the polymer surface is almost the same as polyvinylpyrrolidone content in the casting solution. When the polyvinylpyrrolidone content in the casting solution is high (22%), the polyvinylpyrrolidone content in the polymer surface depends on the molecular weight of polyvinylpyrrolidone. With a molecular weight of 360,000, 17.6% remains in the membrane surface even after 73 h of the pure-water permeation experiment, whereas with a molecular weight of 10,000, only 12.4% remains. These results show clearly the effect of the polymer entanglement when the polyvinylpyrrolidone molecular weight is high.

The intrinsic viscosity data of the polymer solutions, and the size of the polymer and the polymer aggregate in the casting solutions, were correlated to the pore size and the pore size distribution on the membrane surface prepared from aromatic polyamide, polysulfone, and polyethersulfone polymers [89]–[93].

Example 1

Viscosities for polyamide solutions in dimethyl acetamide (DMA) solvent were measured with an Ubbelohde viscometer. The time required for the air-solution boundary to pass between two marks on the capillary was

Table 3. Change in PVP Content
During Pure-Water Permeation Experiment [88]

Pure-water permeation time (h)	Surface PVP concentration, C_m, (wt%)[a]
Film No. 13, $M_w = 10,000$, $C_s = 11.5$ wt%	
0	11.3
2	10.8
4.5	10.2
25	10.1
Film No. 15, $M_w = 10,000$, $C_s = 22.0$ wt%	
0	16.5
2	12.8
4.5	13.6
7.5	12.4
20	11.0
73	12.4
Film No. 27, $M_w = 360,000$, $C_s = 11.0$ wt%	
0	10.1
2	10.6
4.5	10.1
20	10.8
73	10.5
Film No. 28, $M_w = 360,000$, $C_s = 22.0$ wt%	
0	17.1
2	17.6
4.5	17.0
25	17.1
73	17.6

[a] The error limit is $\pm$ 0.5%.

Polymer concentration, g (g/dl of solvent)	Time, t (s)
0	77
0.2	89.8
0.25	93.2
0.4	104
0.5	111.6

Calculate the intrinsic viscosity using the equation

$$\eta_{sp} = \frac{\eta - \eta_0}{\eta_0} \simeq \frac{t - t_0}{t_0} \tag{4.30}$$

where η, η_0, t, and t_0 are solution viscosity, solvent viscosity, and time required for polymer solution and solvent to pass between two marks on the capillary tube, respectively; and the equation

$$[\eta] = \lim_{c \to 0} \eta_{sp}/g \tag{4.31}$$

where $[\eta]$ is intrinsic viscosity.

Since η_{sp}/g and g are linearly related, the following equation can be established:

$$\eta_{sp}/g = Ag + B \tag{4.32}$$

Then B is the intercept of η_{sp}/g at $g = 0$ and is equal to $[\eta]$. Solving Equation 4.32 for A and B by linear regression analysis,

$$A = 0.2262 \tag{4.33}$$
$$B = 0.7859 \tag{4.34}$$

therefore, $[\eta] = 0.7859$ (dl/g)

Example 2

With respect to an aromatic polyamide polymer with a molecular weight of 31,300, the apparent molecular weight of the polymer in the presence of $CaCl_2$ at different $CaCl_2$ ratios is given in Table 4. Calculate the limiting polymer concentration g_x. Calculate also the limiting values of the volume fraction of the polymer, nonsolvent ($CaCl_2$), and solvent.

Use the following structure of the polymer repeat unit for the calculation, and $\rho = 1.30 \times 10^3$ kg/m^3, and the density of $CaCl_2 = 2.51 \times 10^3$ kg/m^3.

Since the number of main-chain atoms and the molecular weight are the same for both isophthalic and terephthalic structures, only the isophthalic struc-

Table 4. Quantities Necessary for Calculation of g_x

CaCl$_2$/polymer	$(\overline{M}_n)_{CaCl_2}$	Z	ε_x	$g_x \times 10^{-3}$, kg/m^3 solvent
0	31,300	2104	2.958	0.1717
0.1	91,900	6178	3.650	0.1392
0.2	117,800	7919	3.946	0.1287
0.3	127,800	8592	4.061	0.1251
0.4	132,100	8881	4.110	0.1236
0.5	132,100	8881	4.110	0.1236
0.6	132,100	8881	4.110	0.1236

ture is considered. The number of main-chain atoms in the m-phenyleneiso-phthalamide structure $= 8(C) + 4(H) + 2(N) + 2(O) = 16$. The molecular weight of m-phenylene-isophthalamide structure $= 14 \times 12 + 10 \times 1 + 2 \times 14 + 2 \times 16 = 238$. The number of the main-chain atom in a polymer is

$$Z = \frac{16}{238} \times \overline{M}_n \tag{4.35}$$

ε_x and g_x can be calculated by Equations 4.24 and 4.25. The results are listed in Table 4 together with the data for the apparent molecular weight of the polymer corresponding to different amounts of CaCl$_2$ added to the polymer solution. Since the weight ratio CaCl$_2$/polymer is known, the limiting CaCl$_2$ concentration, designated as $(g_x)_{CaCl_2}$, is also known. Note that both g_x and $(g_x)_{CaCl_2}$ are based on kilograms of either polymer or CaCl$_2$ in 1 m^3 of solvent. Using the density data, we can calculate the limiting volumes of polymer, v_x, and CaCl$_2$, $(v_x)_{CaCl_2}$, in 1 m^3 of solvent. The volume fractions, ϕ, of polymer, CaCl$_2$, and solvent can thus be calculated at the limiting condition. All numerical values obtained are summarized in Table 5.

Example 3

Calculate the radius of the polymer aggregate r for an aromatic polyamide polymer of molecular weight 31,300 in DMA solvent when the weight ratio (CaCl$_2$/polymer) = 0.3. Use $[\eta] = 3.05 \times 10^{-1}$ m^3/kg.

Using Equation 4.15, the quantity ε_0 is calculated as

$$\varepsilon_0 = \frac{\left(3.05 \times 10^{-1}\right)\left(1.3 \times 10^3\right)}{(2.5)} = 1.586 \times 10^2$$

Table 5. Quantities Necessary for Calculation of Volume Fractions

CaCl$_2$/ polymer	$(g_x)_{CaCl_2}$ $\times 10^{-3}$, kg/m^3	v_x, m^3/m^3 solvent	$(v_x)_{CaCl_2}$, m^3/m^3 solvent	$\phi_{polymer}$	ϕ_{CaCl_2}	$\phi_{solvent}$
0	0	0.1321	0	0.1167	0	0.883
0.1	0.01392	0.1071	0.00555	0.0963	0.00499	0.899
0.2	0.02574	0.0990	0.01025	0.0892	0.00924	0.902
0.3	0.0375	0.0962	0.01494	0.0866	0.0135	0.900
0.4	0.04944	0.0951	0.01970	0.0853	0.0177	0.897
0.5	0.0618	0.0951	0.02462	0.0849	0.0220	0.893
0.6	0.07416	0.0951	0.03084	0.0845	0.0274	0.888

Table 6. Quantities Necessary for Calculation of r at 20°C

$g \times 10^{-3}$ kg/m^3 solvent	g/g_x	$\frac{g}{g_x} \cdot \frac{\varepsilon_0 - \varepsilon_x}{\varepsilon_0 \varepsilon_x}$	$\frac{1}{\varepsilon_0} + \frac{g}{g_x} \frac{\varepsilon_0 - \varepsilon_x}{\varepsilon_0 \varepsilon_x}$	ε	$r \times 10^{10}$ m
0.07	0.5596	0.1342	0.1405	7.117	65.17
0.09	0.7194	0.1726	0.1789	5.590	60.15
0.105	0.8393	0.2013	0.2076	4.831	57.27
0.125	0.9992	0.2397	0.2460	4.065	54.09
0.140	1.1192	0.2685	0.2748	3.639	52.12

From Table 4, $\varepsilon_x = 4.061$ and $g_x = 0.1251 \times 10^3$ kg/m^3 solvent. Then ε can be obtained for different values of g by Equation 4.28. The results are summarized in Table 6.

From Equation 4.10,

$$v = \frac{(127, 800)}{(1.3 \times 10^3)(6.023 \times 10^{26})}$$
$$= 163,220 \times 10^{-30} m^3$$

Since ε and v are known, r can be obtained by using Equation 4.29. The results of the calculation are summarized in Table 6.

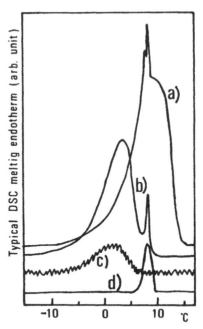

Figure 4.13. Typical thermograms by differential scanning calorimetry. [(a) 57.6%, (b) 32.6%, (c) 14.5%, (d) pure water.] (Reproduced from [95] with permission.)

4.3   Structure of the Permeant in the Membrane

The size of the pore at the surface of the reverse osmosis membrane is so small that the permeant molecule in the pore is under a strong interaction force exerted from the polymeric material that constitutes the pore wall. Hence, it is expected that the structure of the permeant is different from that of the permeant in the bulk phase.

The structure of water in cellulose acetate membranes was investigated using Differential Scanning Calorimetry (DSC) [94], [95]. In particular, Taniguchi and Horigome have prepared asymmetric membranes and dense symmetric membranes with different water contents and made DSC observation. Some of their results are shown in Figure 4.13. The figure shows that a very sharp peak corresponding to the endotherm at the melting point of ordinary water (d) decreases in area with a decrease in membrane water content from (a) to (c) and disappears completely at (c). A broader peak appears in the membrane, the area of which decreases also from (a) to (c). However, it does not completely disappear even for (c). As water content in the membrane decreases, the peak moves towards a lower temperature. Taniguchi and Horigome thought that the sharp peak of the endotherm originated from perfectly free water, whereas the

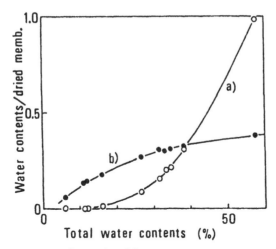

Figure 4.14. Amount of bound and free water vs. water content in the membrane. [(a) Free water, (b) bound water.] (Reproduced from [95] with permission.)

broader peak was due to the formation of water bound weakly to the macromolecule. The reason for the shift of the broader peak from the sharp endotherm of free water is that the ordered structure in ordinary water is broken by interaction with the macromolecule; i.e., a greater entropy change is expected when a highly ordered ice structure is formed from less-ordered water that is under the influence of the interaction force. An increase in entropy changes results by lowering the freezing point. As mentioned earlier, the sharp peak that appeared in Figure 4.13 is that of free water, whereas the broader peak is considered to be that of water weakly interacting with the polymer. The sum of the areas of both peaks corresponds to the total amount of free water in the membrane. Subtracting the latter quantity from the total amount of water in the membrane, the amount of bound water can be calculated. The amount of free water and bound water so obtained are plotted vs. water content in the membrane, in Figure 4.14. Both types of water decrease with a decrease in the water content. In particular, the amount of free water becomes zero when the water content becomes 12 to 13%. The membrane has the sorption capacity for sodium chloride even when the water content of the membrane is as low as 12 to 13%; i.e., the distribution coefficient of sodium chloride between aqueous solution and the membrane of the above water content is 22%. This suggests that even bound water can dissolve sodium chloride solute. Taniguchi and Horigome further concluded from extrapolation of the correlation between the distribution coefficient and the water content that the bound water has no dissolving power on sodium chloride solute when the membrane water content becomes as low as 7.8%. The above experimental results suggest that there are four types of water in cellulose ac-

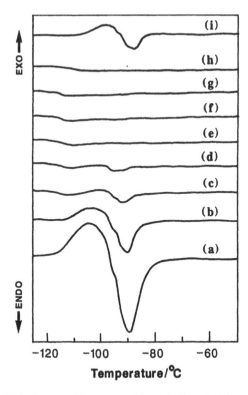

Figure 4.15. Endotherms of 2-propanol in polydimethylsilsiloxane membrane. (Reproduced from [96] with permission.)

etate membranes. They are (1) perfect free water, (2) free water that is weakly interacting with polymeric material, (3) bound water that has dissolving power on sodium chloride solute, and (4) bound water that has no dissolving power on sodium chloride solute. The amount of each water depends on the total amount of water in the cellulose acetate membrane. The interfacial pure-water layer proposed by Sourirajan is considered to be the same as the fourth type of water proposed by Taniguchi and Horigome. The interfacial water proposed by Sourirajan is driven from the inside of the membrane, under a high pressure drop across the membrane in the reverse osmosis process.

Water and also organic liquids can exist in the membrane in bound states [96]. Figure 4.15 shows melting endotherms of 2-propanol in polydimethyl-siloxane membrane, of dry polydimethylsiloxane membrane, and of bulk 2-propanol. The areas of the endothermic curve decrease with decreasing 2-propanol content in the membrane (Figure 4.15(a)–(d)). In Figure 4.15(e)–(g), endotherms of 2-propanol were scarcely observed around its melting point of −88.0°C, even though 2-propanol was present in the polydimethylsiloxane

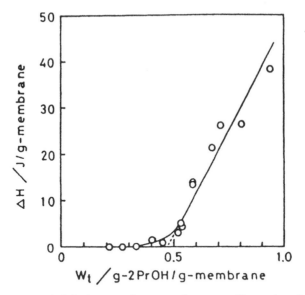

Figure 4.16. Enthalpic heat vs. 2-propanol content. (Reproduced from [96] with permission.)

membrane. No other new endotherm was observed in the temperature range $-160°C$ to $20°C$. The membrane in Figure 4.15(g) contained 3.7×10^{-3} g of 2-propanol. The endotherm of 4.4×10^{-3} g of bulk 2-propanol is shown in Figure 4.15(i). If most of the 2-propanol in the membrane could be assigned as bulk 2-propanol, we should be able to observe an endotherm in Figure 4.15(e)–(g). Figure 4.16 shows the relationship between enthalpic heat and 2-propanol content. The slope of the straight line should correspond to the heat of melting of bulk 2-propanol ($\Delta H_M = 89.9$ J /g). Extrapolation to $\Delta H_M = 0$ should intercept the 2-propanol content axis at a point that is equal to the total bound 2-propanol. From Figure 4.16 it was concluded that the total bound 2-propanol content in the polydimethylsiloxane membrane was 0.47g 2-propanol/g membrane. The presence of sorbed molecules in the bound state in the membrane should exert significant influence on the membrane transport of such molecules in the pervaporation and vapor permeation processes. Investigations were made for the state of permeant in polydimethylsiloxane membranes, with respect to various organic solvents [97]–[99].

Nomenclature for Chapter 4

4.1 Microscopic Structure of the Membrane

$$C_w$$ = molar heat capacity of liquid, J/mol K

$$E_{vap}$$ = heat of vaporization, J/mol

$$E_w$$ = heat of vaporization at temperature T, K

$$\Delta F^{\ddagger}$$ = activation energy of viscous flow, J/mol

$$\Delta H^{\ddagger}$$ = enthalpy of viscous flow, J/mol

$$h$$ = Planck's constant, 6.626×10^{-34} J s

$$J$$ = volumetric flow rate of water, m^3/s

$$L$$ = effective film thickness, m

$$N$$ = Avogadro's number, 6.022×10^{23} mol^{-1}

$$n_p$$ = number of pores in a unit area, $1/m^2$

$$p$$ = pressure drop across the membrane, Pa

$$R$$ = gas constant, 8.314 J/mol K

$$r_p^*$$ = effective radius of the pore, m

$$\Delta S^{\ddagger}$$ = entropy of viscous flow, J/mol K

$$T$$ = absolute temperature, K

$$T'$$ = cluster temperature, K

$$V$$ = molar volume of liquid, m^3/mol

$$\eta$$ = viscosity of water, Pa s

$$\tau$$ = tortuosity factor, —

4.2 Structure of the Casting Solution

$$g$$ = polymer concentration, (weight of polymer/volume of solvent)

$$K$$ = Mark-Houwink constant defined by Equation 4.7

$$M$$ = molecular weight of polymer

$$M_v$$ = average molecular weight of polymer

$$N_0$$ = Avogadro's number, 1/mol

$$r$$ = radius of spherical polymer, m

$$r_0$$ = r when concentration is zero, m

$$v$$ = volume of an unswollen spherical molecule, m^3

$$Z$$ = number of main-chain atoms in a polymer molecule

$$\alpha$$ = Mark-Houwink constant defined by Equation 4.7

$$\varepsilon$$ = effective volume factor, —

$$\varepsilon_0$$ = ε when concentration is zero, —

$$\eta$$ = viscosity of solution, Pa s

$$\eta_0$$ = viscosity of solvent, Pa s

$[\eta]$ = intrinsic viscosity, (volume/weight)

ρ = density of amorphous polymer, kg/m^3

ϕ = volume fraction of polymer in the solution, —

Subscript

x = quantities corresponding to the limiting condition

5

Membrane Transport/Solution-Diffusion Model

5.1 Reverse Osmosis

5.1.1 Reverse Osmosis Transport

The solution-diffusion model is currently being used by the majority of the membrane community. An excellent summary of the applicability of this model for various membrane transport phenomena is given in the classic book by Crank and Park [100]. Lonsdale applied this model for reverse osmosis transport [101] when the latter process emerged as an important desalination process.

The most general description of the mass transport across a membrane is based on irreversible thermodynamics.

$$J_i = L_{ii}X_i + \sum L_{ij}X_j \tag{5.1}$$

where J_i is the mass flux of the component i, L_{ii} and L_{ij} are phenomenological coefficients, and X_i and X_j are the forces under which the mass transfer takes place. The first term of Equation 5.1 indicates that the flux of the ith component is caused by a force, X_i, that is acting on the ith component. The second term expresses the relation between the flux of the ith component and the forces acting on components other than ith component. L_{ii} is always positive, and according to Onsager's reciprocal relationship,

$$L_{ij} = L_{ji} \tag{5.2}$$

131

Furthermore, thermodynamic conditions require that

$$L_{ii}L_{jj} - L_{ij}^2 \geq 0 \tag{5.3}$$

The most appropriate choice of the force X for the molecular diffusion through the membrane under isothermal conditions without external forces being applied to the mass transfer of the ith component is the chemical potential gradient and

$$X_i = -\nabla \mu_i \tag{5.4}$$

Then the flux is

$$J_i = -L_{ii} \nabla \mu_i \tag{5.5}$$

In the above equation the second term of Equation 5.1 is eliminated, which means the flow of the ith component is totally decoupled from the flow of other components.

In reverse osmosis transport both concentration and pressure gradients contribute to the chemical potential gradient by

$$\nabla \mu_i = \mathbf{R}T \nabla \ln a_{im} + v_i \nabla p \tag{5.6}$$

where a_{im} and v_i are the activity in the membrane (mol/m^3) and the partial molar volume (m^3/mol) of the ith component, respectively. The reverse osmosis process can be considered as a binary system where the transport of solvent, usually water, and that of solute are involved. Designating solute and solvent by subscripts A and B, respectively, Equation 5.6 can be rewritten for solvent as

$$\nabla \mu_B = \mathbf{R}T \nabla \ln a_{Bm} + v_B \nabla p \tag{5.7}$$

Integrating Equation 5.7 with an assumption, $v_B = \text{const}$, from the high-pressure to the low-pressure side of the membrane,

$$\Delta \mu_B = v_B \left(\frac{\mathbf{R}T}{v_B} \Delta \ln a_{Bm} + \Delta p \right) \tag{5.8}$$

It should be noted that the integration was carried out inside the membrane from the high-pressure side to the low-pressure side of the membrane. Therefore, Δ is defined as the quantity at the low-pressure side of the membrane minus the quantity at the high-pressure side of the membrane.

Assuming that thermodynamic equilibrium is established at both sides of the membrane, $a_{Bm} = a_B$ at the membrane-solution boundaries on both sides of the membrane, where a_B is the activity of solvent (mol/m^3) outside of the membrane.

Since osmotic pressure, π (Pa), is defined as

$$\pi = -\frac{RT}{v_B} \ln a_B \tag{5.9}$$

Equation 5.8 can be written as

$$\Delta \mu_B = v_B(\Delta p - \Delta \pi) \tag{5.10}$$

It should be noted that the osmotic pressure in the above equation is based on the solution outside the membrane. Therefore, $\Delta \pi$ is the difference in the osmotic pressures of solutions that are outside the membrane and in contact with the low-pressure side and the high-pressure side of the membrane.

As for the solute with subscript A, an assumption is made that the solution behaves as a dilute solution, and the activity coefficient remains constant. Then Equation 5.6 becomes

$$\nabla \mu_A = RT \nabla \ln a_{Am} + v_A \nabla p \tag{5.11}$$
$$= RT \nabla \ln c_{Am} + v_A \nabla p \tag{5.12}$$

where a_{Am}, c_{Am}, and v_A designate the activity of solute in the membrane (mol/m^3), the concentration of solute in the membrane (mol/m^3), and the molar volume of solute (m^3/mol), respectively. Again, integrating from the high-pressure to the low-pressure side of the membrane, Equation 5.12 becomes

$$\Delta \mu_A = RT\Delta \ln c_{Am} + v_A \Delta p \tag{5.13}$$

Since $\Delta \ln c_{Am} = \ln(c_{Am}$ at the low-pressure side of the membrane/c_{Am} at the high-pressure side of the membrane), it is easy to compare the contribution of the first term and that of the second term of the right side of Equation 5.13 to $\Delta \mu_A$. The second term can easily be ignored when the solute concentration at the low-pressure side of the membrane is less than 90% of that at the high-pressure side of the membrane, which is usually the case in reverse osmosis. Then,

$$\Delta \mu_A = RT\Delta \ln c_{Am} \tag{5.14}$$

The following physical meaning is usually given to the phenomenological coefficient L_{11}:

$$L_{BB} = \frac{c_{Bm}}{f_{Bm}} \qquad (5.15)$$

where c_{Bm} and f_{Bm} are the concentration of the solvent (mol/m^3) in the membrane and the friction $(\text{J s/m}^2 \text{ mol})$ working between a unit mole of the solute and the membrane material. Assuming that the flow of solvent is totally decoupled from that of the solute and the chemical potential gradient is constant across the membrane, we can combine Equations 5.5 and 5.15,

$$J_B = -\frac{c_{Bm}}{f_{Bm}} \frac{\Delta \mu_B}{\delta} \qquad (5.16)$$

where δ is the membrane thickness. Combining Equations 5.10 and 5.16,

$$J_B = -\frac{c_{Bm}}{f_{Bm}} v_B \frac{\Delta p - \Delta \pi}{\delta} \qquad (5.17)$$

Since the friction f_{Bm} can be given as

$$f_{Bm} = \frac{RT}{D_{Bm}} \qquad (5.18)$$

where D_{Bm} is diffusion coefficient (m^2/s) of solvent in the membrane, Equation 5.17 becomes

$$J_B = -\frac{c_{Bm} D_{Bm} v_B}{RT\delta} (\Delta p - \Delta \pi) \qquad (5.19)$$

As for the solute, the phenomenological coefficient L_{AA} can be given as

$$L_{AA} = \frac{c_{Am}}{f_{Am}} \qquad (5.20)$$

Then the solute flux can be written as

$$J_A = -\frac{c_{Am}}{f_{Am}} \frac{\Delta \mu_A}{\delta} \qquad (5.21)$$

Combining Equations 5.14 and 5.21,

$$J_A = -\frac{c_{Am}}{f_{Am}} \frac{RT\Delta \ln c_{Am}}{\delta} \qquad (5.22)$$

Approximating $\ln c_{Am}$ by $(\Delta c_{Am})/c_{Am}$ and using the relation $f_{Am} = RT/D_{Am}$,

$$J_A = -D_{Am}\frac{\Delta c_{Am}}{\delta} \tag{5.23}$$

Assuming thermodynamic equilibrium is established at both sides of the membrane, the relation between the concentration inside the membrane, c_{Am}, and outside the membrane, c_A, can be written as

$$c_{Am} = K_A c_A \tag{5.24}$$

where K_A is the distribution constant. Equation 5.23 can then be written as

$$J_A = -D_{Am}K_A\frac{\Delta c_A}{\delta} \tag{5.25}$$

Equations 5.19 and 5.25 describe the fluxes of the solvent and solute, according to the solution-diffusion model.

The reverse osmosis performance data are often compared using the solvent flux, J_B, and a quantity called solute separation, f', defined as

$$f' = 1 - \frac{c_{A3}}{c_{A2}} \tag{5.26}$$

where subscripts 2 and 3 indicate the solution (outside the membrane) on the high-pressure and low-pressure sides of the membrane. Since

$$\frac{J_A}{J_B} = \frac{c_{A3}}{c_{B3}} \tag{5.27}$$

$$f' = 1 - \frac{J_A c_{B3}}{J_B c_{A2}} \tag{5.28}$$

$$= 1 - \frac{D_{Am}K_A RT(c_{A2} - c_{A3})c_{B3}}{D_{Bm}c_{Bm}V_B(p_2 - p_3 - \pi_2 + \pi_3)c_{A2}} \tag{5.29}$$

Rearranging,

$$f' = \frac{1}{1 + \dfrac{D_{Am}K_A RT c_{B3}}{D_{Bm}c_{Bm}V_B(p_2 - p_3 - \pi_2 + \pi_3)}} \tag{5.30}$$

Equations 5.19 and 5.30 describe the performance of a reverse osmosis membrane.

Example 1

The following numerical values were given by Lonsdale for parameters involved in his transport equations (Equations 5.19 and 5.30) [101]: $D_{Bm}C_{Bm} = 2.7 \times 10^{-8}$ kg/m s, and $D_{Am}K_A = 4.2 \times 10^{-14}$ m²/s.

Calculate the solute separation and the solvent flux, when the feed sodium chloride molality is 0.1 and the operating pressure is 4.134×10^6 Pa (gauge). The thickness of the selective layer is assumed to be 10^{-7} m.

The following numerical values can be used: $RT = 2.479 \times 10^3$ J/mol at 25°C, $c_{B3} = 10^3$ kg/m³, and $v_B = 18.02 \times 10^{-6}$ m³/mol. The coefficient for osmotic pressure $= 2.5645 \times 10^8$ Pa per mole fraction.

Since the feed mole fraction is

$$\frac{0.1}{0.1 + (1000/18.02)} = 1.799 \times 10^{-3} \qquad (5.31)$$

the feed osmotic pressure (Pa) is

$$(2.5645 \times 10^8) \times (1.799 \times 10^{-3}) = 0.461 \times 10^6 \qquad (5.32)$$

Approximating $\pi_2 - \pi_3 = 0.461 \times 10^6$ Pa, Equation 5.30 becomes

$$f' = \left[1 + \frac{(4.2 \times 10^{-14})(2.479 \times 10^3)(10^3)}{(2.7 \times 10^{-8})(18.02 \times 10^{-6})(4.134 \times 10^6 - 0.461 \times 10^6)} \right]^{-1}$$
$$= 0.945$$

Then the sodium chloride molality of the permeate becomes

$$0.1 \times (1 - 0.945) = 0.0055 \qquad (5.33)$$

The permeate sodium chloride mole fraction is

$$\frac{0.0055}{0.0055 + \frac{1000}{18.02}} = 9.910 \times 10^{-5} \qquad (5.34)$$

The osmotic pressure (Pa) of the permeate is

$$(2.5645 \times 10^8) \times (9.910 \times 10^{-5}) = 0.0254 \times 10^6 \qquad (5.35)$$

and $\pi_2 - \pi_3 = (0.461 - 0.0254) \times 10^6 = 0.4356 \times 10^6$ Pa. The solute separation is recalculated by using the above value for $\pi_2 - \pi_3$, as

$$f' = \left[1 + \frac{(4.2 \times 10^{-14})(2.479 \times 10^3)(10^3)}{(2.7 \times 10^{-8})(18.02 \times 10^{-6})(4.134 \times 10^6 - 0.4356 \times 10^6)}\right]^{-1}$$
$$= 0.945$$

$f' = 0.945$ is therefore a sufficiently accurate answer when three digits after the decimal point are required. From Equation 5.19, the flux (kg/m^2 s) is

$$J_B = \frac{(2.7 \times 10^{-8})(18.02 \times 10^{-6})(4.314 \times 10^6 - 0.4356 \times 10^6)}{(2.479 \times 10^3)(10^{-7})}$$
$$= 76.11 \times 10^{-4}$$

When there is no solute, there is no osmotic pressure effect. Therefore,

$$J_B = \frac{(2.7 \times 10^{-8})(18.02 \times 10^{-6})(4.314 \times 10^6)}{(2.479 \times 10^3)(10^{-7})}$$
$$= 81.14 \times 10^{-4}$$

Equations 5.30 and 5.19 indicate that when $p_2 - p_3 = -\Delta p$ approaches infinity, f' and J_B approach 1.0 and infinity, respectively.

5.1.2 Concentration Polarization

When a complete separation of a solution takes place by a reverse osmosis membrane, only one component of the solution (usually the solvent) goes through the membrane and is enriched in the permeate. The other component (usually the solute), on the other hand, is left behind on the feed solution side of the membrane. Unless the solute diffuses back to the main body of the feed solution quickly, the solute concentration in the vicinity of the membrane increases until a steady state is reached where the diffusion rate, enhanced by the high concentration near the membrane surface, counterbalances the rate of the solute accumulation near the membrane surface. This phenomenon is called concentration polarization and may occur even when the separation of the solute from the solvent is not perfect. It is known that the concentration polarization exerts an unfavorable effect on the performance of reverse osmosis membranes. For example, the experimentally observable solute separation defined later by Equation 5.46 is lowered, since the concentration in the permeate, c_{A3}, is governed not by the concentration of the solute in the main body of the feed solution, but

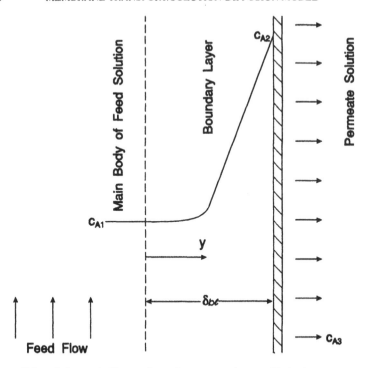

Figure 5.1. Schematic illustration of concentration profile in the concentrated boundary layer.

by the concentration near the membrane surface. As the latter concentration increases, so does the concentration in the permeate. The high concentration near the membrane surface also increases the osmotic pressure of the feed solution, which results in a decrease in the permeation rate. Many attempts were made to describe this phenomenon, using various transport equations. The simplest and most popular attempts are based on the boundary film theory in which the presence of a boundary layer of thickness δ_{bl} is assumed. The diffusion of the solute from the solution at the membrane surface to the main body of the feed solution occurs at this boundary layer. The concentration profile developed across this layer is schematically illustrated in Figure 5.1. On the basis of the above assumption, the net flow rate of the solute in the boundary layer is the sum of the diffusive flow and the convective flow and should be equal to the solute permeation rate through the membrane; otherwise, accumulation of the solute would occur. Therefore,

$$-D_{AB}\frac{dc_A}{dy} + vc_A = vc_{A3} \qquad (5.36)$$

where D_{AB} is the diffusion coefficient of solute A in solvent B at the boundary layer (m²/s), c_A is the solute concentration (mol/m³) as a function of distance (y) (see Figure 5.1), and v is the solution velocity (m/s). Rearrangement of Equation 5.36 yields

$$\frac{dc_A}{dy} = \frac{v}{D_{AB}}(c_A - c_{A3}) \tag{5.37}$$

and

$$\frac{dc_A}{c_A - c_{A3}} = \frac{v}{D_{AB}}dy \tag{5.38}$$

Integrating Equation 5.38,

$$\ln(c_A - c_{A3}) = \frac{v}{D_{AB}}y + C \tag{5.39}$$

where C is the integral constant. Since $c = c_{A1}$ when $y = 0$,

$$\ln(c_{A1} - c_{A3}) = C \tag{5.40}$$

Substituting C in Equation 5.40 for Equation 5.39,

$$\ln\frac{(c_A - c_{A3})}{(c_{A1} - c_{A3})} = \frac{v}{D_{AB}}y \tag{5.41}$$

Since $c_A = c_{A2}$ when $y = \delta_{bl}$, where δ_{bl} is the thickness of the boundary layer,

$$\ln\frac{(c_{A2} - c_{A3})}{(c_{A1} - c_{A3})} = \frac{v}{D_{AB}}\delta_{bl} \tag{5.42}$$

and

$$\frac{c_{A2} - c_{A3}}{c_{A1} - c_{A3}} = \exp\left(\frac{v}{D_{AB}}\delta_{bl}\right) \tag{5.43}$$

Defining the mass transfer coefficient, k, as

$$k = \frac{D_{AB}}{\delta_{bl}} \tag{5.44}$$

$$\frac{c_{A2} - c_{A3}}{c_{A1} - c_{A3}} = \exp\left(\frac{v}{k}\right) \tag{5.45}$$

The above equation is a standard equation to describe the concentration polarization to calculate the boundary concentration, c_{A2} (mol/m³), from the feed concentration, c_{A1} (mol/m^3), the permeate concentration, c_{A3} (mol/m^3), the permeation velocity of the solution, v (m/s), and the mass transfer coefficient, k (m/s). The last parameter depends on the membrane configuration and the hydrodynamic conditions of the feed solution, and many techniques were proposed to evaluate this parameter numerically [102].

It is impossible to measure the solute concentration, c_{A2}, in the vicinity of the membrane surface, by experiment. Therefore, the solute separation, f', cannot be experimentally determined, although, rigorously speaking, f' is the solute separation inherent in the membrane. The solute concentration in the bulk feed solution, c_{A1}, on the other hand, can be experimentally determined. Therefore, the solute separation defined on the basis of c_{A1} as

$$f = 1 - \frac{c_{A3}}{c_{A1}} \qquad (5.46)$$

is an experimentally observable solute separation. These two solute separations, one defined by Equation 5.26 and the other defined by Equation 5.46, are different. While the former value is inherent in the membrane, the latter value is affected by the concentration polarization.

Usually it is believed that the concentration polarization is insignificant for the membrane gas separation, because of the high diffusivity of molecules in the gaseous phase.

5.2   Prediction of Reverse Osmosis Membrane Performance

5.2.1   Prediction of Membrane Performance Involving Single Solute

It is possible to predict reverse osmosis performance data under different operating conditions. The method can be extended to the prediction for the separation of different solutes, either organic or inorganic, if the separation data of sodium chloride solute are known for a membrane. Furthermore, the separation of individual ions involved in the mixture of electrolyte solutes can also be predicted on the basis of the separation data of sodium chloride. The method was established by Sourirajan and co-workers for asymmetric cellulose acetate and aromatic polyamide membranes, but should be applicable for other types of membranes. The prediction is made on the basis of a set of equations derived by Kimura and Sourirajan [51],[103]–[109].

As will be shown later, the effects of the pressure and concentration on the solvent and solute flux are exactly the same in Sourirajan and co-workers' transport equations as in Lonsdale's transport equations. Therefore, Sourirajan and co-workers' works are included in this chapter. However, there is an important difference between these two approaches. In Chapter 4 we learned that different reverse osmosis membranes can be produced from a given polymeric material, depending on the size of the "pores" generated on the surface of the membrane. While Sourirajan and co-workers considered that the transport equations were applicable for membranes of different pore sizes, prepared from a given polymeric material, Lonsdale considered that they were applicable only for a perfectly dense membrane without any pores. As a result, many sets of numerical values are allowed for transport parameters, depending on the pore size, according to Sourirajan and co-workers' approach, whereas there is only one single set of numerical parameters for a given polymeric material, according to Lonsdale's approach. The latter set of transport parameters is considered intrinsic to the polymeric material. Apparently, Sourirajan and co-workers assumed the presence of pores in any reverse osmosis membranes for which they applied their transport equations, whereas, for Lonsdale, pores were considered as defects that prevented membranes from achieving the highest solute rejection that is the intrinsic property of the polymeric material.

There seems to be a historical reason for the difference in these two approaches. The solution-diffusion approach was established for the permeation of liquid and gas through the membrane before reverse osmosis membranes of practical usefulness were developed by the phase-inversion technique. Dense membranes without asymmetricity were prepared from polymeric materials, and their transport properties were measured, assuming that the membranes were defect-free and the transport parameters so produced were the values intrinsic to the material. The membrane with the highest separation capacity for a given polymer was believed to be that which could exhibit the transport properties intrinsic to the polymer. The goal of membrane production engineering was to ensure the membrane intrinsic transport properties of the polymeric material. This approach is still popular in the membrane manufacturing industry.

The membranes of practical usefulness are, on the other hand, prepared by the phase-inversion technique. We have learned that the top selective layer of the asymmetric membrane prepared by the phase-inversion technique has a microscopic structure that depends on the conditions under which the membranes are prepared. Obviously, the numerical parameters associated with the membrane transport equations should depend on the microscopic structure of the membrane and also on the conditions under which membranes are prepared.

According to Kimura and Sourirajan's approach, the solvent flux is written as

$$J_B = A \times (p_2 - p_3 - \pi_2 + \pi_3) \tag{5.47}$$

where A is a proportionality constant and is called the pure-water permeability constant. The dimension is mol/m² s Pa. Equation 5.47 is the same mathematical formula as Equation 5.19 if

$$A = \frac{c_{Bm}D_{Bm}v_B}{RT\delta} \tag{5.48}$$

When the feed liquid is pure water, both π_2 and π_3 are zero, and therefore,

$$J_B = A \times (p_2 - p_3) \tag{5.49}$$

Solute permeation flux, J_A, was given as

$$J_A = \left(\frac{D_{Am}K_A}{\delta}\right)(c_{A2} - c_{A3}) \tag{5.50}$$

This equation is also the same mathematical formula as Equation 5.25. The quantity $D_{Am}K_A/\delta$ is called solute transport parameter and its dimension is meters per second (m/s). Furthermore,

$$c_{A1} = c_1 X_{A1} \tag{5.51}$$
$$c_{A2} = c_2 X_{A2} \tag{5.52}$$
$$c_{A3} = c_3 X_{A3} \tag{5.53}$$

where c is the total molar concentration (mol/m³) including solute and solvent, and X_A is the mole fraction of the solute. The meanings of subscripts 1, 2, and 3 are given in the nomenclature.

Substituting Equations 5.52 and 5.53 for c_{A2} and c_{A3} in Equation 5.50,

$$J_A = \left(\frac{D_{Am}K_A}{\delta}\right)(c_2 X_{A2} - c_3 X_{A3}) \tag{5.54}$$

Also, using the relation

$$\frac{J_A}{J_A + J_B} = X_{A3} \tag{5.55}$$

Equation 5.54 can be written as

$$J_B = \frac{D_{Am}K_A}{\delta}\left(\frac{1 - X_{A3}}{X_{A3}}\right)(c_2 X_{A2} - c_3 X_{A3}) \tag{5.56}$$

The equation for the concentration polarization was given by Equation 5.45 as

$$\frac{c_{A2} - c_{A3}}{c_{A1} - c_{A3}} = \exp\left(\frac{v}{k}\right) \tag{5.57}$$

Using the relations given in Equations 5.51, 5.52, and 5.53, with an assumption of a constant total molar concentration,

$$c_1 = c_2 = c_3 = c \tag{5.58}$$

and also the relation

$$v = \frac{J_A + J_B}{c_1} \tag{5.59}$$

Equation 5.57 can be rearranged to

$$J_B = c_1 k (1 - X_{A3}) \ln \frac{X_{A2} - X_{A3}}{X_{A1} - X_{A3}} \tag{5.60}$$

Equations 5.47, 5.49, 5.56, and 5.60 are basic transport equations proposed by Kimura and Sourirajan and enable the calculation of the flux of water, the flow of solute, and the degree of concentration polarization.

With respect to cellulose acetate (acetyl content 39.8%) membrane, the following correlations were established experimentally [51]:

Correlation of A:

· At a given temperature, $A = A_0 \exp(\alpha p)$ where $\underline{\alpha}$ is a constant.

· At a given pressure, $A\mu_w$ = constant where μ_w is the viscosity of water (Pa s).

Correlation of $(D_{Am} K_A / \delta)$:

· At a given temperature, $(D_{Am}K_A/\delta)_{NaCl} \propto (p_2 - p_3)^{\underline{\beta}}$ where $\underline{\beta}$ is a constant.

· At a given pressure, $(D_{Am}K_A/\delta)_{NaCl} \propto \exp(0.005T)$.

· At a given temperature and pressure, $D_{Am}K_A/\delta$ is independent of X_{A2} for completely ionized inorganic solutes and also for completely nonionized polar organic solutes.

Correlation of k

· At a given temperature and pressure, $k_{NaCl} \propto Q^n$ where Q is the flow rate of feed solution (m^3/s).

· At a given pressure, $k_{NaCl} \propto \exp(0.005T)$.

· At any given temperature, pressure, and feed flow rate,

$$k_{solute} = k_{NaCl}a[(D_{AB})_{solute}/(D_{AB})_{NaCl}]^{2/3} \qquad (5.61)$$

where $(D_{AB})_{solute}$ and $(D_{AB})_{NaCl}$ are diffusion coefficients (m²/s) of a solute and sodium chloride, respectively. Equation 5.61 is valid, provided that the viscosities involved are not too different from the viscosity of pure water, and enables the calculation of the mass transfer coefficient (m/s) for many solutes, using the mass transfer coefficient of sodium chloride.

Example 2

Under the following operating conditions for reverse osmosis,

· Feed solution: NaCl-H₂O

· Feed molality: 0.6

· Operating pressure: 10,335 kPa (gauge)

· Effective membrane area: 13.2 cm²

the following experimental data were obtained: Pure water permeation rate $= 159.8 \times 10^{-3}$ kg/h; permeation rate in the presence of NaCl solute in the feed $= 122.9 \times 10^{-3}$ kg/h; and solute separation on the basis of molality $= 81.2\%$. Calculate the parameters A, $D_{Am}K_A/\delta$, and k. The following numerical values can be used:

· $c_1 = c_2 = c_3 = c = 55.3$ kg-mol/m³

· Molecular weight of NaCl $= 58.45$

The calculation of the pure water flux from the data on the pure-water permeation rate is

$$J_B = \frac{(159.8 \times 10^{-3})}{(18.02)(13.2 \times 10^{-4})(3600)}$$
$$= 1.867 \times 10^{-3} \text{ kg-mol/m}^2 \text{ s}$$

From Equation 5.49,
$$A = 1.867 \times 10^{-3}/10,335 = 1.806 \times 10^{-7} \text{kg} - \text{mol/m}^2 \text{ s kPa}$$

The calculation of the permeation flux in the presence of NaCl solute in the feed is

$$J_B = \frac{(122.9 \times 10^{-3})}{(18.02)(13.2 \times 10^{-4})(3600)\left[1 + \frac{(0.6)(1-0.812)(58.45)}{1000}\right]}$$
$$= 1.426 \times 10^{-3} \text{ kg-mol/m}^2 \text{ s}$$

Since the separation is 0.812, the molality of the permeate solution is

$$(0.6)(1 - 0.812) = 0.1128 \text{ molal}$$

Table 1. Osmotic Pressure Data
Pertinent to Different Electrolyte Solutes
at 25°C (kPa)

Molality	NaCl	LiCl	KNO$_3$	MgCl$_2$	CuSO$_4$
0	0	0	0	0	0
0.1	462	462	448	641	276
0.2	917	931	862	1303	510
0.3	1372	1407	1262	1999	731
0.4	1820	1889	1648	2737	945
0.5	2282	2386	2020	3523	1165
0.6	2744	2889	2379	4357	1379
0.7	3213	3413	2737	5233	1593
0.8	3682	3944	3082	6178	1813
0.9	4158	4482	3427	7191	2055
1.0	4640	5040	3750	8266	2302
1.2	5612	6191	4385	10611	2834
1.4	6612	7398	4992	13231	3434
1.6	7646	8646	5557	16127	—

From Table 1 the osmotic pressure of the permeate solution is 520 kPa. From Equation 5.47,

$$\pi(X_{A2}) = p_2 - p_3 + \pi(X_{A3}) - \frac{J_B}{A} \tag{5.62}$$

Inserting numerical values,

$$\pi(X_{A2}) = 10{,}335 + 520 - \frac{1.426 \times 10^{-3}}{1.806 \times 10^{-7}} = 2957 \text{ kPa}$$

From Table 1 the molality of the concentrated boundary solution is 0.6459. Therefore,

$$X_{A1} = \frac{0.6}{0.6 + (1000/18.02)} = 0.01070$$

$$X_{A2} = \frac{0.6459}{0.6459 + (1000/18.02)} = 0.01150$$

$$X_{A3} = \frac{0.1128}{0.1128 + (1000/18.02)} = 0.002029 \tag{5.63}$$

Rearranging Equation 5.56 with an approximation $c_1 = c_2 = c_3 = c$,

$$\frac{D_{Am}K_A}{\delta} = \frac{J_B}{c\left[(1 - X_{A3})/(X_{A3})\right](X_{A2} - X_{A3})} \tag{5.64}$$

Inserting numerical values,

$$\frac{D_{Am}K_A}{\delta} = \frac{(1.426 \times 10^{-3})}{(55.3)\left(1 - 0.002029/0.002029\right)(0.01150 - 0.002029)}$$
$$= 5.536 \times 10^{-6} \text{ m/s}$$

Rearranging Equation 5.60,

$$k = \frac{J_B}{c(1 - X_{A3})\ln\left[(X_{A2} - X_{A3})/(X_{A1} - X_{A3})\right]} \tag{5.65}$$

Inserting numerical values,

$$k = \frac{(1.426 \times 10^{-3})}{(55.3)(1 - 0.002029)\ln(0.01150 - 0.002029)/(0.01070 - 0.002029)}$$
$$= 292.8 \times 10^{-6} \text{ m/s}$$

Example 3
For a given set of parameters,

$$A = 3.04 \times 10^{-7} \text{ kg-mol/m}^2\text{ s kPa}$$
$$D_{Am}K_A/\delta = 8.03 \times 10^{-7}\text{m/s}$$
$$k_{NaCl} = 22 \times 10^{-6}\text{m/s}$$

calculate the solute separation, f, the pure-water flux, and the permeate flux (in the presence of solute in the feed solution), when the feed NaCl concentration is 0.6 molal and the operating pressure is 6895 kPa (gauge). Assume that the osmotic pressure is proportional to the sodium chloride mole fraction.

Using the assumption given above and combining Equations 5.47 and 5.56,

$$A(p_2 - p_3) - AB(X_{A2} - X_{A3}) = (fracD_{Am}K_A\delta)\,c\frac{1 - X_{A3}}{X_{A3}}(X_{A2} - X_{A3}) \tag{5.66}$$

Rearranging,

$$X_{A2} - X_{A3} = \frac{A(p_2 - p_3)}{AB + \left[(D_{Am}K_A)/\delta\right]c\left[(1 - X_{A3})/X_{A3}\right]} \tag{5.67}$$

Combining Equations 5.47 and 5.60,

$$A(p_2 - p_3) - AB(X_{A2} - X_{A3}) = kc(1 - X_{A3}) \ln \frac{X_{A2} - X_{A3}}{X_{A1} - X_{A3}} \qquad (5.68)$$

Inserting numerical values,

$$A(p_2 - p_3) = (3.04 \times 10^{-7})(6895) = 20{,}961 \times 10^{-7} \text{kg} - \text{mol/m}^2\text{s}$$

Since 0.6 molal NaCl solution ($X_A = 0.0107$) has an osmotic pressure of 2744 kPa (see Table 1),

$$B = (2744)/(0.0107) = 256{,}449 \text{ kPa}$$

Therefore,

$$AB = (3.04 \times 10^{-7})(256{,}449) = 779{,}600 \times 10^{-7} \text{kg} - \text{mol/m}^2\text{s}$$

Furthermore,

$$\left(\frac{D_{Am}K_A}{\delta}\right)c = (8.03 \times 10^{-7})(55.3) = 444.06 \times 10^{-7} \text{kg} - \text{mol/m}^2\text{s}$$

$$kc = (22 \times 10^{-6})(55.3) = 12{,}166 \times 10^{-7} \text{kg} - \text{mol/m}^2\text{s}$$

Inserting the above numerical values into Equation 5.67,

$$X_{A2} - X_{A3} = \frac{(20{,}961 \times 10^{-7})}{(779{,}600 \times 10^{-7}) + (444.06 \times 10^{-7})\left[(1 - X_{A3})/X_{A3}\right]} \qquad (5.69)$$

Defining the quantities α and β as

$$\alpha = A(p_2 - p_3) - AB(X_{A2} - X_{A3}) \qquad (5.70)$$

$$\beta = kc(1 - X_{A3}) \ln\left(\frac{X_{A2} - X_{A3}}{X_{A1} - X_{A3}}\right) \qquad (5.71)$$

and inserting numerical values,

$$\alpha = 20{,}961 \times 10^{-7} - 779{,}600 \times 10^{-7}(X_{A2} - X_{A3}) \qquad (5.72)$$

$$\beta = 12{,}166 \times 10^{-7}(1 - X_{A3}) \ln\left(\frac{X_{A2} - X_{A3}}{X_{A1} - X_{A3}}\right) \qquad (5.73)$$

From Equation 5.68,

$$\alpha = \beta \tag{5.74}$$

The numerical values for X_{A3} and $X_{A2} - X_{A3}$ can be solved from Equation 5.69 and Equations 5.72 to 5.74 following steps 1 to 5.

· Step 1. Assume X_{A3}.

· Step 2. Solve Equation 5.69 for $X_{A2} - X_{A3}$, using the numerical value for X_{A3} assumed in step 1.

· Step 3. Calculate α using the numerical value for $X_{A2} - X_{A3}$ obtained in step 2.

· Step 4. Calculate β.

· Step 5. If $\alpha = \beta$, the numerical value assumed for X_{A3} in step 1 and the numerical value obtained in step 2 for $X_{A2} - X_{A3}$ are correct; otherwise go back to step 1.

Table 2 summarizes the values for $X_{A2} - X_{A3}$, α, and β, obtained in the intermediate calculation steps. From Table 2, $X_{A3} = 0.00107$ seems to be the best solution. Correspondingly, $X_{A2} - X_{A3} = 0.01755$.

Table 2. X_{A2} - X_{A3}, α and β
Values Corresponding to Different X_{A3}

X_{A3}	$X_{A2} - X_{A3}$	$\alpha \times 10^7$	$\beta \times 10^7$
0.001	0.01714	7599	6919
0.002	0.02094	4637	10664
0.003	0.02261	3343	13065
0.0015	0.01951	5759	9126
0.0011	0.01772	7147	7449
0.00105	0.01744	7365	7192
0.00107	0.01755	7280	7294
0.00106	0.01750	7319	7249

Then

$$f = \frac{0.0107 - 0.00107}{0.0107} = 0.90 \tag{5.75}$$

Since

$$J_B = A(p_2 - p_3) - AB(X_{A2} - X_{A3})$$

$$= \alpha$$

$$= 7280 \times 10^{-7} \text{ kg-mol/m}^2 \text{ s}$$

A method to predict reverse osmosis performance data for different solutes, when the performance data for a reference NaCl solute is known, has been developed [110]–[127]. The following equation can easily be derived from Equations 5.49, 5.56, and 5.60 when $J_A \ll J_B$ (or $X_{A3} \ll 1$).

$$f = \frac{1}{1 + [(D_{Am}K_A/\delta)/v] \exp(v/k)} \tag{5.76}$$

From the above equation it is obvious that the solute separation, f, is governed by the quantity $D_{Am}K_A/\delta$ when v and k are nearly equal for different solutes. Therefore, if we are able to know the value of $D_{Am}K_A/\delta$ for a solute from that of sodium chloride, we can predict the separation of the given solute on the basis of the separation data for sodium chloride.

Relationships between $D_{Am}K_A/\delta$ for sodium chloride and $D_{Am}K_A/\delta$ for other solutes were therefore established. For completely ionized electrolyte solutes, the following correlation was found [113]:

$$(D_{Am}K_A/\delta)_{\text{solute}} \propto \exp\left[n_c \left(-\frac{\Delta\Delta G}{RT} \right)_{\text{cation}} + n_a \left(-\frac{\Delta\Delta G}{RT} \right)_{\text{anion}} \right] \tag{5.77}$$

where n_c and n_a represent the number of moles of cation and anion, respectively, in 1 mol of the ionized solute, and $(\Delta\Delta G/RT)_{\text{ion}}$ is interpreted as the free energy required to bring an ion from the bulk water to the interfacial water (J/mol). Applying Equation 5.77 to $(D_{Am}K_A/\delta)_{\text{NaCl}}$,

$$\ln(D_{Am}K_A/\delta)_{\text{NaCl}} = \ln C^*_{\text{NaCl}} + \left[\left(-\frac{\Delta\Delta G}{RT} \right)_{\text{Na}^+} + \left(-\frac{\Delta\Delta G}{RT} \right)_{\text{Cl}^-} \right] \tag{5.78}$$

where $\ln C^*_{\text{NaCl}}$ is a constant representing the porous structure of the membrane surface, in terms of $D_{Am}K_A/\delta$. Whereas $(-\Delta\Delta G/RT)_{\text{ion}}$ depends on the ionic species and the polymeric material from which membranes are produced, $\ln C^*_{\text{NaCl}}$ depends on the porous structure of the membrane.

Using the data on $-\Delta\Delta G/RT$ for Na^+ and Cl^- ions for the membrane material-solution involved, the value of $\ln C^*_{\text{NaCl}}$ for a particular membrane can be calculated from the value of $(D_{Am}K_A/\delta)_{\text{NaCl}}$ specified for the membrane. Also, using the value of $\ln C^*_{\text{NaCl}}$ so obtained, the corresponding value of $D_{Am}K_A/\delta$ for any completely ionized solutes can be obtained from the equation

$$\ln(D_{Am}K_A/\delta)_{\text{solute}} =$$
$$\ln C_{\text{NaCl}}^* + \left\{ n_c \left[-(\Delta\Delta G/RT) \right]_{\text{cation}} + n_a \left[-(\Delta\Delta G/RT \right]_{\text{anion}} \right\} \quad (5.79)$$

Thus, for any specified values of $(D_{Am}K_A/\delta)_{\text{NaCl}}$, the corresponding values of $D_{Am}K_A/\delta$ for a large number of completely ionized solutes can be obtained from Equation 5.79, using data on $-\Delta\Delta G/RT$ for the ions involved. Available data on $-\Delta\Delta G/RT$ for different ions applicable for cellulose acetate (acetyl content, 39.8%) membrane-aqueous solution system are listed in Tables 3 and 4.

With respect to electrolyte inorganic solutes, a few special cases arise. For a solution system involving ions and ion pairs, Equation 5.79 can be written as

$$\ln(D_{Am}K_A/\delta)_{\text{solute}}$$
$$= \ln C_{\text{NaCl}}^* + \alpha_D \left[n_c \left(\frac{-\Delta\Delta G}{RT} \right)_{\text{cation}} + n_a \left(\frac{-\Delta\Delta G}{RT} \right)_{\text{anion}} \right] +$$
$$(1 - \alpha_D) \left(\frac{-\Delta\Delta G}{RT} \right)_{ip} \quad (5.80)$$

where α_D represents the degree of dissociation, and the subscript ip refers to the ion pair formed; for the particular case where the ion pair itself is an ion, Equation 5.79 assumes the more general form,

$$\ln(D_{Am}K_A/\delta)_{\text{solute}} = \ln C_{\text{NaCl}}^* +$$
$$\alpha_D \left[n_c \left(\frac{-\Delta\Delta G}{RT} \right)_{\text{cation}} + n_a \left(\frac{-\Delta\Delta G}{RT} \right)_{\text{anion}} \right] +$$
$$(1 - \alpha_D) \left(\frac{-\Delta\Delta G}{RT} \right)_{ip} +$$
$$(1 - \alpha_D)(n_c - n_{ipc}) \left(\frac{-\Delta\Delta G}{RT} \right)_{\text{cation}} +$$
$$(1 - \alpha_D)(n_a - n_{ipa}) \left(\frac{-\Delta\Delta G}{RT} \right)_{\text{anion}} \quad (5.81)$$

where n_{ipc} and n_{ipa} represent the number of moles of cation and anion, respectively, involved in 1 mol of the ion pair. Data on $-\Delta\Delta G/RT$ for several inorganic ion pairs are also included in Table 3. For the case of a feed solution

Table 3. Data on Free Energy Parameters for
Some Inorganic Ions and Ion Pairs[a] at 25°C[5]

Species	$(-\Delta\Delta G/RT)_i$	Species	$(-\Delta\Delta G/RT)_i$
Inorganic cations		Inorganic anions	
H^+	6.34	OH^-	−6.18
Li^+	5.77	F^-	−4.91
Na^+	5.79	Cl^-	−4.42
K^+	5.91	Br^-	−4.25
Rb^+	5.86	I^-	−3.98
Cs^+	5.72	IO_3^-	−5.69
NH_4^+	5.97	$H_2PO_4^-$	−6.16
Mg^{2+}	8.72	BrO_3^-	−4.89
Ca^{2+}	8.88	NO_2^-	−3.85
Sr^{2+}	8.76	NO_3^-	−3.66
Ba^{2+}	8.50	ClO_3^-	−4.10
Mn^{2+}	8.58	ClO_4^-	−3.60
Co^{2+}	8.76	HCO_3^-	−5.32
Ni^{2+}	8.47	HSO_4^-	−6.21
Cu^{2+}	8.41	SO_4^{2-}	−13.20
Zn^{2+}	8.76	$S_2O_3^{2-}$	−14.03
Cd^{2+}	8.71	SO_3^{2-}	−13.12
Pb^{2+}	8.40	CrO_4^{2-}	−13.69
Fe^{2+}	9.33	$Cr_2O_7^{2-}$	−11.16
Fe^{3+}	9.82	CO_3^{2-}	−13.22
Al^{3+}	10.41	$Fe(CN)_6^{3-}$	−20.87
Ce^{3+}	10.62	$Fe(CN)_6^{4-}$	−26.83
Cr^{3+}	11.28		
La^{3+}	12.89		
Th^{4+}	12.42		
Ion Pairs			
$MgSO_4$		3.45	
$CoSO_4$		3.41	
$ZnSO_4$		2.46	
$MnSO_4$		2.48	
$CuSO_4$		2.85	
$CdSO_4$		3.04	
$NiSO_4$		2.18	
$KFe(CN)_6^{2-}$		−2.53	
$KFe(CN)_6^{3-}$		−17.18	

[a]System: aqueous solution/cellulose acetate (CA-398).

that is subject to partial hydrolysis, Equation 5.79 becomes

$$\ln(D_{Am}K_A/\delta)_{solute} = \ln C^*_{NaCl} + (1 - \alpha_H) \left[n_c \left(\frac{-\Delta\Delta G}{RT} \right)_{cation} + \right.$$

$$n_a \left. \left(\frac{-\Delta\Delta G}{RT} \right)_{anion} \right] +$$

$$\alpha_H \left[n_{hy} \left(\frac{-\Delta\Delta G}{RT} \right)_{hy} + \right.$$

$$\left. \left(\frac{-\Delta\Delta G}{RT} \right)_{OH^- \text{ or } H^+} \right] \qquad (5.82)$$

where α_H represents the degree of hydrolysis, the subscript *hy* refers to the hydrolyzed species resulting from the hydrolysis reaction, and subscripts OH^- and H^+ represent the hydroxyl and hydrogen ions, respectively. In Equations 5.80 and 5.82, the applicable values of α_D and α_H are those corresponding to the boundary conditions X_{A2}.

Example 4

Using the same values for A, $D_{Am}K_A/\delta$, and k as in Example 3, calculate the solute separation, f, at the operating pressure of 1724 kPa (gauge) (250 psig) when the feed solution is dilute. Use the relation

$$v = Ap/c \qquad (5.83)$$

From Equation 5.83,

$$v = (3.04 \times 10^{-7})(1724)/(55.3)$$
$$= 9.447 \times 10^{-6} \text{ m/s}$$

Using Equation 5.76,

$$f = \cfrac{1}{1 + \cfrac{(8.03 \times 10^{-7}) \times \exp\left(\cfrac{9.477 \times 10^{-6}}{22 \times 10^{-6}} \right)}{9.477 \times 10^{-6}}}$$
$$= 0.855$$

Table 4. Data on Free Energy Parameters for Some Organic Ions[a] at 25°C [5]

Species	$(-\Delta\Delta G/RT)_i$	Species	$(-\Delta\Delta G/RT)_i$
$HCOO^-$	-4.78	$m\text{-}CH_3C_6H_4COO^-$	-5.67
H Phthalate$^-$	-4.63	$m\text{-}OHC_6H_4COO^-$	-5.64
$C_2O_4^{2-}$	-14.06	$p\text{-}ClC_6H_4COO^-$	-5.63
$t\text{-}C_4H_9COO^-$	-6.90	$m\text{-}NO_2C_6H_4COO^-$	-5.92
$i\text{-}C_3H_7COO^-$	-6.11	$p\text{-}NO_2C_6H_4COO^-$	-5.93
cyclo-$C_6H_{11}COO^-$	-6.24	$o\text{-}ClC_6H_4COO^-$	-6.41
$n\text{-}C_4H_9COO^-$	-6.11	$o\text{-}NO_2C_6H_4COO^-$	-6.61
$n\text{-}C_3H_7COO^-$	-6.06	$HOOCCOO^-$	-6.60
$C_2H_5COO^-$	-6.14	$HOOCCH_2COO^-$	-6.46
CH_3COO^-	-5.95	$HOOC(CH_2)_2COO^-$	-5.65
$C_6H_5(CH_2)_3COO^-$	-5.93	$CH_3CHOHCOO^-$	-6.30
$C_6H_5(CH_2)_2COO^-$	-5.86	$HOOCCH(OH)CH_2COO^-$	-5.97
$C_6H_5(CH_2)COO^-$	-5.69	$HOOCCH(OH)CH(OH)\text{-}$	
$C_6H_5COO^-$	-5.66	COO^-	-6.40
$p\text{-}CH_3OC_6H_4COO^-$	-5.74	$HOOCCH_2C(OH)\text{-}$	
		$(COOH)CH_2COO^-$	-6.24

[a]System: aqueous solution/cellulose acetate (CA-398).

Example 5

Using the same membrane as in Example 4, and under the same operating conditions, calculate the solute separation of KCl, Na_2SO_4, $Mg(NO_3)_2$, and $CaCl_2$. Use the following data for the diffusivity of the above solutes:

$$
\begin{aligned}
(D_{AB})_{NaCl} &= 1.61 \times 10^{-9} m^2/s \\
(D_{AB})_{KCl} &= 1.995 \times 10^{-9} m^2/s \\
(D_{AB})_{Na_2SO_4} &= 1.230 \times 10^{-9} m^2/s \\
(D_{AB})_{Mg(NO_3)_2} &= 1.216 \times 10^{-9} m^2/s \\
(D_{AB})_{CaCl_2} &= 1.336 \times 10^{-9} m^2/s
\end{aligned}
$$

Since $(D_{Am}K_A/\delta)_{NaCl} = 8.03 \times 10^{-7}$ m/s, $\ln(D_{Am}K_A/\delta)_{NaCl} = -14.04$.

Using Equation 5.78 and $-\Delta\Delta G/RT$ values listed in Table 3, $\ln C^*_{NaCl}$ can be calculated as

$$
\ln C^*_{NaCl} = -14.04 - [5.79 + (-4.42)] = -15.41 \qquad (5.84)
$$

$\ln(D_{Am}K_A/\delta)$ and $D_{Am}K_A/\delta$ values for electrolytes other than sodium chloride can be obtained by using Equation 5.79 and $-\Delta\Delta G/RT$ values for ions involved in the electrolyte. The mass transfer coefficient for each electrolyte can also be obtained on the basis of k_{NaCl}, using Equation 5.61 together with the diffusion coefficient for the respective electrolyte. All numerical values involved in the calculation are listed in Table 5. Solute separation for each electrolyte can be calculated using Equation 5.76 and $v = 9.477 \times 10^{-6}$ m/s. The results are as follows:

$$
f_{KCl} = \cfrac{1}{1 + (9.008 \times 10^{-7}) \times \cfrac{\exp\left(\frac{9.477 \times 10^{-6}}{25.38 \times 10^{-6}}\right)}{(9.477 \times 10^{-6})}}
$$
$$
= 0.879
$$

$$
f_{Na_2SO_4} = \cfrac{1}{1 + (0.402 \times 10^{-7}) \times \cfrac{\exp\left(\frac{9.477 \times 10^{-6}}{18.39 \times 10^{-6}}\right)}{(9.477 \times 10^{-6})}}
$$
$$
= 0.993
$$

$$
f_{Mg(NO_3)_2} = \cfrac{1}{1 + (8.233 \times 10^{-7}) \times \cfrac{\exp\left(\frac{9.477 \times 10^{-6}}{18.25 \times 10^{-6}}\right)}{(9.477 \times 10^{-6})}}
$$
$$
= 0.873
$$

Table 5. $\Sigma(-\Delta\Delta G/RT)_i$, $\ln(D_{Am}K_A/\delta)$, $(D_{Am}K_A/\delta)$, and k Values for Different Electrolytes

Electrolyte	$\Sigma(-\Delta\Delta G/RT)_i$	$\ln(D_{Am}K_A/\delta)$	$(D_{Am}K_A/\delta) \times 10^7$ (m/s)	$k \times 10^6$ (m/s)
KCl	1.49	−13.92	9.008	25.38
Na_2SO_4	−1.62	−17.03	0.402	18.39
$Mg(NO_3)_2$	1.40	−14.01	8.233	18.25
$CaCl_2$	0.04	−15.37	2.113	19.43

$$f_{CaCl_2} = \cfrac{1}{1 + (2.113 \times 10^{-7}) \times \cfrac{\exp\left(\frac{9.477 \times 10^{-6}}{19.43 \times 10^{-6}}\right)}{(9.477 \times 10^{-6})}}$$

$$= 0.965$$

Example 6

A linear free-energy relationship was found applicable to the solute transport parameter $D_{Am}K_A/\delta$ of organic solutes in a homologous group in the form [115]

$$\ln(D_{Am}K_A/\delta) = \rho^* \Sigma\sigma^* + \delta^* \Sigma E_s + \ln C^* \tag{5.85}$$

where $\Sigma\sigma^*$ and ΣE_s are Taft's polar and steric constants characteristic to the substituent group. ρ^* and δ^* are the constants associated with $\Sigma\sigma^*$ and ΣE_s, and $\ln C^*$ is a constant representing the porosity of the membrane (similar to $\ln C^*_{NaCl}$ defined by Equation 5.79 for electrolytes). With respect to the homologous series of ketone solutes, $D_{Am}K_A/\delta$ values obtained experimentally for a cellulose acetate membrane are listed in Table 6, together with $\Sigma\sigma^*$ and ΣE_s values applicable to substituent groups involved in the ketone solutes. Using the relationship given by Equation 5.85, calculate ρ^*, δ^*, and $\ln C^*$, applying linear regression analysis.

Setting

$$x_i = \Sigma\sigma^* \tag{5.86}$$

$$y_i = \Sigma E_s \tag{5.87}$$

$$z_i = \ln(D_{Am}K_A/\delta) \tag{5.88}$$

$$a = \rho^* \tag{5.89}$$

$$b = \delta^* \tag{5.90}$$

$$c = \ln C^* \tag{5.91}$$

Equation 5.85 can be written as

$$z_i = ax_i + by_i + c \tag{5.92}$$

Table 6. $D_{Am}K_A/\delta$, $\Sigma\sigma^*$, and ΣE_s Values for Different Ketones

Ketones	Substituents	$\ln(D_{Am}K_A/\delta)$	$\Sigma\sigma^*$	ΣE_s
Methyl ethyl ketone	CH_3, C_2H_5	−13.407	−0.100	−0.07
Cyclopentanone	cyclo-C_4H_8	−13.724	−0.250	−0.51
Diisopropyl ketone	i-C_3H_7, i-C_3H_7	−14.374	0.380	−1.40
Diisobutyl ketone	i-C_4H_9, i-C_4H_9	−15.925	0.400	−1.86
Cyclohexanone	cyclo-C_5H_{10}	−13.942	−0.180	−0.79
Benzyl methyl ketone	$C_6H_5CH_2$, CH_3	−13.057	+0.215	−0.38
Acetophenone	C_6H_5, CH_3	−12.852	+0.600	−0.06
Methyl isopropyl ketone	CH_3, i-C_3H_7	−14.294	−0.190	−0.70
Methyl isobutyl ketone	CH_3, i-C_4H_9	−14.325	−0.200	−0.93
Acetone	CH_3, CH_3	−14.094	0	0

The best values for a, b, and c can be obtained by simultaneously solving the equations

$$\partial \sum (ax_i + by_i + c - z_i)^2 / \partial a = 0 \qquad (5.93)$$

$$\partial \sum (ax_i + by_i + c - z_i)^2 / \partial b = 0 \qquad (5.94)$$

$$\partial \sum (ax_i + by_i + c - z_i)^2 / \partial c = 0 \qquad (5.95)$$

Therefore,

$$\left(\sum x_i^2\right) a + \left(\sum x_i y_i\right) b + \left(\sum x_i\right) c - \sum (x_i z_i) = 0 \qquad (5.96)$$

$$\left(\sum x_i y_i\right) a + \left(\sum y_i^2\right) b + \left(\sum y_i\right) c - \sum (y_i z_i) = 0 \quad (5.97)$$

$$\left(\sum x_i\right) a + \left(\sum y_i\right) b + \sum c - \left(\sum z_i\right) = 0 \quad (5.98)$$

Using the given numerical values,

$$0.8916a + 1.754b - 0.885c - 14.1758 = 0 \quad (5.99)$$

$$1.754a + 7.8119b - 6.70c - 97.7569 = 0 \quad (5.100)$$

$$-0.885a - 6.70b + 10.0c + 139.947 = 0 \quad (5.101)$$

Solving the above linear equations simultaneously,

$$\ln(D_{Am}K_A/\delta) = 0.971\Sigma\sigma^* + 0.862\Sigma E_s - 13.33 \quad (5.102)$$

5.2.2 Prediction of Membrane Performance Involving Two Mixed Uni-Univalent Inorganic Electrolyte Solutes with No Common Ions

The prediction of the membrane performance involving more than one solute has been extensively discussed in the literature [125]-[132]. In particular, the mixed solute feed solutions we are concerned with here are of the type NaBr-KCl-H$_2$O [127]. Our objective is to predict the membrane performance (i.e., solute separation with respect to each ion, and also the permeation rate) from experimental RO data for a reference NaCl-H$_2$O feed solution only, under a given set of operating conditions. Although the following example is limited to a solution that includes a common electrolyte, the method can be generalized for cases without any common ions. The approach can also be easily extended to the system with any number of electrolytes and any number of valences involved in the electrolytes, as long as there is no association between cations and anions.

Nomenclature. The following special nomenclature will be used for this analysis. Subscripts A,B, and M refer to solute, solvent water, and membrane, respectively. All quantities with an asterisk refer to ionic properties; with respect to such quantities, the first subscript of boldface letters **i** (**1,3**,...,**2,4**,...) refers to the indicated ion, and the second subscript, m, 1, 2, or 3, refers to the indicated phase (m = membrane phase, 1 = bulk solution phase on the high-pressure side of the membrane, 2 = concentrated boundary solution phase, and 3 = permeate solution phase). Ions **1** and **3** (odd numbers) are different cations, and **2** and **4** (even numbers) are different anions; all ions are univalent. Quantities that do not refer specifically to ions have the subscript m, 1, 2, or 3, which refer to the indicated phase. Numerical subscripts **12**, **14**, **32**, **34** refer to single salts, with ions indicated by each number. For example the quantities $X_i^*, X_{i_1}^*, X_1^*, X_{1_m}^*, X_{13}^*, X_{Am}, X_{Am3}$, and X_{A3} represent mole fractions of ion **i**, ion **i** in bulk solution phase, ion **1**, ion **1** in membrane phase, ion **1** in permeate solution phase, solute A in membrane phase, solute A in membrane phase in equilibrium with X_{A3}, and solute A in permeate solution phase, respectively. The quantities c_2 and c_{m2} represent molar densities of solution in phase 2 and in membrane phase in equilibrium with X_{A2}, respectively. The quantity $(D_{Am}K_A/\delta)_{12}$ represents the solute transport parameter for the single solute **12**.

Basic transport equations are applied to a mixed solute system involving several ions in aqueous solutions. Equations 5.47, 5.50, 5.24, 5.60, and 5.55, given earlier for single-solute systems, can be rewritten in analogous forms for the above mixed-solute systems, as follows:

$$J_B = A\left[(p_2 - p_3) - \pi\left(\sum X_{i2}^*\right)_{mixt} + \pi\left(\sum X_{i3}^*\right)_{mixt}\right] \quad (5.103)$$

$$J_i^* = \frac{D_{im}^* K_{i2}^*}{\delta}c_2 X_{i2}^* - \frac{D_{im}^* K_{i3}^*}{\delta}c_3 X_{i3}^* \quad (5.104)$$

$$K_i^* = \frac{c_m X_{im}^*}{c X_i^*} \quad (5.105)$$

$$X_{i2}^* = X_{i3}^* + (X_{i1}^* - X_{i3}^*)\exp\left(\frac{J_B + \sum J_i^*}{kc}\right) \quad (5.106)$$

and

$$X_{i3}^* = J_i^* \Big/ \left(J_B + \sum J_i^*\right) \quad (5.107)$$

In addition, the conditions for overall electroneutrality prevail in each phase, so that

$$\sum z_i^* X_{i1}^* = 0 \quad (5.108)$$

$$\sum z_i^* X_{i2}^* = 0 \quad (5.109)$$

$$\sum z_i^* X_{i3}^* = 0 \quad (5.110)$$

$$\sum z_i^* X_m^* = 0 \quad (5.111)$$

and

$$\sum z_i^* J_i^* = 0 \quad (5.112)$$

where z_i^* represents the valence of ion **i**. Referring to Equation 5.24, the quantity K is a true equilibrium constant independent of salt concentration in a single-solute system. For the mixed-solute system under analysis, the corresponding quantity is K_i^*, which is, unlike K, composition dependent, and this dependency arises because of the competing effects of different ions on the distribution between the aqueous and the membrane phases.

For simplicity of analysis, the following assumptions are made:

1. $c = c_1 = c_2 = c_3$

2. $J_B \gg \sum_i J_i^*$

3. $\pi \left(\sum_i X_i^* \right)_{mixt} = B_{av} \left(\sum_i X_i^* \right)_{mixt}$

4. $k = k_{av} =$ constant for a given operating condition

5. For any salt and ion, the ratio of diffusivity through the membrane to that in water is constant, i.e.,

$$\frac{D_{Am}}{D_{AB}} = \frac{D_{im}^*}{D_i^*} \qquad (5.113)$$

On the basis of assumptions (1) to (5) above, and defining

$$\alpha = \exp(J_B/k_{av}c) \qquad (5.114)$$

Equations 5.103, 5.104, 5.106, and 5.107 can be written as

$$J_B = A(p_2 - p_3) - AB_{av} \left[\left(\sum X_{i1}^* \right)_{mixt} - \left(\sum X_{i3}^* \right)_{mixt} \right] \alpha \qquad (5.115)$$

$$J_i^* = \frac{D_{im}^* K_{i2}^*}{\delta} c X_{i2}^* - \frac{D_{im}^* K_{i3}^*}{\delta} c X_{i3}^* \qquad (5.116)$$

$$X_{i2}^* = X_{i3}^* + (X_{i1}^* - X_{i3}^*)\alpha \qquad (5.117)$$

$$X_{13}^* = J_1^*/J_B \qquad (5.118)$$

$$X_{33}^* = J_3^*/J_B \qquad (5.119)$$

$$X_{23}^* = J_2^*/J_B \qquad (5.120)$$

Expression for Water Flux

Let B_{12}, B_{32}, B_{14}, and B_{34} represent the slope of the mole fraction (X_A) vs. osmotic pressure (kPa) plots for the single solutes **12**, **32**, **14**, and **34**, respectively. Then the proportionality constant B_{av} in Equation 5.115 can be expressed as

$$B_{av} = \frac{B_{12} + B_{32} + B_{14} + B_{34}}{8} \qquad (5.121)$$

since there are eight ions in the four salts involved. Equation 5.121 enables the use of osmotic pressure data available in the literature [51] for calculating B_{av} in Equation 5.115. Furthermore, from the electroneutrality condition in each phase,

$$\sum (X_{i1}^*)_{mixt} = X_{11}^* + X_{31}^* + X_{21}^* + X_{41}^* = 2\left(X_{11}^* + X_{31}^*\right) \tag{5.122}$$

and

$$\sum (X_{i3}^*)_{mixt} = X_{13}^* + X_{33}^* + X_{23}^* + X_{43}^* = 2\left(X_{13}^* + X_{33}^*\right) \tag{5.123}$$

Then, Equation 5.115 becomes

$$J_B = A(p_2 - p_3) - 2AB_{av}(X_{11}^* - X_{13}^* + X_{31}^* - X_{33}^*)\alpha \tag{5.124}$$

Expression for Equilibrium of Solute

For a given solute system, when the concentration of the solution is expressed in terms of salt concentration, the equilibrium constant K for the salt is given by Equation 5.24, so that

$$K = c_m(X_{salt})_m/cX_{salt} \tag{5.125}$$

When the concentration of the same solution is expressed in terms of ionic concentrations, the Donnan equilibrium for a neutral membrane can be written as

$$K^{\pm} = \frac{c_m(X_{cation}^*)_m c_m(X_{anion}^*)_m}{c(X_{cation}^*)c(X_{anion}^*)} \tag{5.126}$$

Since

$$cX_{cation}^* = cX_{anion}^* = cX_{salt} \tag{5.127}$$

and

$$c_m(X_{cation}^*) = c_m(X_{anion}^*) = c_m(X_{salt})_m \tag{5.128}$$

$$K^{\pm} = K^2 \tag{5.129}$$

For the numerical solutions of the transport equations developed above, K_i^* involved in Equation 5.116 should be rewritten using the equilibrium constant K of the salt given in Equation 5.125. The derivation is too lengthy to be presented in detail. The readers who are interested in this subject should refer to the original literature [127]. As a result of the derivation, Equation 5.116 is rewritten in terms of J_1^*, J_2^*, and J_3^*, in the following way.

Defining

$$\beta = (K_{\underline{12}}/K_{\underline{14}})^2 = \frac{(D_{\underline{12}}/D_{\underline{14}})^2 (D_{Am}K_A/\delta)_{\underline{14}}^2}{(D_{Am}K_A/\delta)_{\underline{12}}^2} \qquad (5.130)$$

and

$$\gamma = (K_{\underline{12}}/K_{\underline{32}})^2 = \frac{(D_{\underline{12}}/D_{\underline{32}})^2 (D_{Am}K_A/\delta)_{\underline{32}}^2}{(D_{Am}K_A/\delta)_{\underline{12}}^2} \qquad (5.131)$$

$$J_1^* = c \left(\frac{D_1^*}{D_{\underline{12}}}\right) \left(\frac{D_{Am}K_A}{\delta}\right)_{\underline{12}} [(X_{13}^* + (X_{1_1}^* - X_{13^*}) \alpha \times$$
$$\left(\frac{(1-\beta)(X_{2_1}^*\alpha - X_{23}^*\alpha + X_{23}^*)+}{(X_{1_1}^*\alpha - X_{13}^*\alpha + X_{13}^*)+}\right.$$
$$\left.\frac{\beta(X_{1_1}^*\alpha - X_{13}^*\alpha + X_{13}^* + X_{3_1}^*\alpha - X_{33}^*\alpha + X_{33}^*)}{\gamma(X_{3_1}^*\alpha - X_{33}^*\alpha + X_{33}^*)}\right)^{1/2}$$
$$-X_{13}^* \left(\frac{(1-\beta)X_{23}^* + \beta(X_{13}^* + X_{33}^*)}{X_{13}^* + \gamma X_{33}^*}\right)^{1/2}]$$

$$(5.132)$$

$$J_2^* = c \left(\frac{D_2^*}{D_{\underline{12}}}\right) \left(\frac{D_{Am}K_A}{\delta}\right)_{\underline{12}} [(X_{23}^* + (X_{2_1}^* - X_{23}^*) \alpha \times$$
$$\left(\frac{(X_{1_1}^*\alpha - X_{13}^*\alpha + X_{13}^*)+}{(1-\beta)(X_{2_1}^*\alpha - X_{23}^*\alpha + X_{23}^*)+}\right.$$
$$\left.\frac{\gamma(X_{3_1}^*\alpha - X_{33}^*\alpha + X_{33}^*)}{\beta(X_{1_1}^*\alpha - X_{13}^*\alpha + X_{13}^* + X_{3_1}^*\alpha - X_{33}^*\alpha + X_{33}^*)}\right)^{1/2}$$
$$-X_{23}^* \left(\frac{X_{13}^* + \gamma X_{33}^*}{(1-\beta)X_{23}^* + \beta(X_{13}^* + X_{33}^*)}\right)^{1/2}]$$

$$(5.133)$$

$$J_3^* = c \left(\frac{D_3^*}{D_{32}} \right) \left(\frac{D_{Am} K_A}{\delta} \right)_{32} \left[\left(X_{33}^* + (X_{31}^* - X_{33}^*) \alpha \right) \times \right.$$

$$\left(\frac{(1-\beta)(X_{21}^* \alpha - X_{23}^* \alpha + X_{23}^*) + }{(X_{31}^* \alpha - X_{33}^* \alpha + X_{33}^*) + } \right.$$

$$\left. \frac{\beta(X_{11}^* \alpha - X_{13}^* \alpha + X_{13}^* + X_{31}^* \alpha - X_{33}^* \alpha + X_{33}^*)}{\frac{1}{\gamma}(X_{11}^* \alpha - X_{13}^* \alpha + X_{13}^*)} \right)^{1/2}$$

$$\left. -X_{33} \left(\frac{(1-\beta)X_{23}^* + \beta(X_{13}^* + X_{33}^*)}{X_{33}^* + \frac{1}{\gamma}X_{13}^*} \right)^{1/2} \right] \qquad (5.134)$$

Special Case of a Mixed-Solute System with a Common Ion

Let the common ion be the cation so that the general mixed-solute system considered above contains ions **1**, **2**, and **4** only, for which $X_3^* = 0$ in each phase. Then Equations 5.132, 5.133, 5.134, 5.112, 5.121, and 5.124 become

$$J_1^* = c \left(\frac{D_{1^*}}{D_{12}} \right) \left(\frac{D_{Am} K_A}{\delta} \right)_{12} \left[\left(X_{13}^* + (X_{11}^* - X_{13}^*) \alpha \right) \times \right.$$

$$\left(\frac{(1-\beta)(X_{21}^* \alpha - X_{23}^* \alpha + X_{23}^*) + \beta(X_{11}^* \alpha X_{13}^* \alpha + X_{13}^*)}{(X_{11}^* \alpha - X_{13}^* \alpha + X_{13}^*)} \right)^{1/2}$$

$$\left. -X_{13}^* \left(\frac{(1-\beta)X_{23}^* + \beta X_{13}^*}{X_{13}^*} \right)^{1/2} \right] \qquad (5.135)$$

$$J_2^* = c \left(\frac{D_2^*}{D_{12}} \right) \left(\frac{D_{Am} K_A}{\delta} \right)_{12} \left[\left(X_{23}^* + (X_{21}^* - X_{23}^*) \alpha \right) \times \right.$$

$$\left(\frac{X_{11}^* \alpha - X_{13}^* \alpha + X_{13}^*}{(1-\beta)(X_{21}^* \alpha - X_{23}^* \alpha + X_{23}^*) + \beta(X_{11}^* \alpha - X_{13}^* \alpha + X_{13}^*)} \right)^{1/2}$$

$$\left. -X_{23}^* \left(\frac{X_{13}^*}{(1-\beta)X_{23}^* + \beta X_{13}^*} \right)^{1/2} \right] \qquad (5.136)$$

$$J_3^* = 0 \qquad (5.137)$$

$$J_4^* = J_1^* - J_2^* \qquad (5.138)$$

$$B_{av} = \frac{(B_{12} + B_{14})}{4} \tag{5.139}$$

$$J_B = A(p_2 - p_3) - 2AB_{av}(X_{1_1}^* - X_{1_3}^*)\alpha \tag{5.140}$$

Furthermore, it should be noted that the average mass transfer coefficient k_{av} is necessary for calculating the quantity α, using Equation 5.114. The mass transfer coefficient for an ion i can be expressed as

$$k_i^* = k_{NaCl}(D_i^*/D_{NaCl})^{2/3} \tag{5.141}$$

so that

$$k_{av} = \frac{k_1^* + k_3^* + k_2^* + k_4^*}{4} \tag{5.142}$$

$$= \frac{k_{NaCl}}{4D_{NaCl}^{2/3}} \left[(D_1^*)^{2/3} + (D_3^*)^{2/3} + (D_2^*)^{2/3} + (D_4^*)^{2/3} \right] \tag{5.143}$$

The data on diffusivity of different ions are available in the literature [133].

By using the transport equations developed above, the mole fraction of each ion involved (i.e., X_{13}^*, X_{33}^*, X_{23}^*, and X_{42}^*, and J_B) can be obtained numerically. The solute separation of an ion i can then be calculated as

$$f_i^* = \frac{X_{i1}^* - X_{i3}^*}{X_{i1}^*} \tag{5.144}$$

Example 7

Calculate the separation of Na^+ (1), Cl^- (2), and NO_3^- (4) ions from the reverse osmosis experiment involving the feed $NaCl$-$NaNO_3$ solution of 0.250 ($NaCl$) and 0.789 ($NaNO_3$) molal at the operating pressure of 10,342 kPa (gauge).

Use the following numerical values for the calculation:

$$A = 1.186 \times 10^{-7} \text{ kmol/m}^2 \text{ s kPa}$$

$$(D_{Am}K_A/\delta)_{12} = (D_{Am}K_A/\delta)_{NaCl} = 2.02 \times 10^{-7} \text{ m/s}$$

$$(D_{Am}K_A/\delta)_{14} = (D_{Am}K_A/\delta)_{NaNO_3} = 4.32 \times 10^{-7} \text{ m/s}$$

$$D_{12} = D_{NaCl} = 1.61 \times 10^{-9} \text{ m}^2/\text{s}$$

$$D_{14} = D_{NaNO_3} = 1.568 \times 10^{-9} \text{ m}^2/\text{s}$$

$$D_1^* = D_{Na^+}^* = 1.35 \times 10^{-9} \text{ m}^2/\text{s}$$

$$D_2^* = D_{Cl^-}^* = 2.03 \times 10^{-9} \text{ m}^2/\text{s}$$

$$B_{av} = 1.325 \times 10^5 \text{ kPa}$$

$$c = 55.48 \text{ kmol/m}^3$$

$$S = 7.6 \times 10^{-4} \text{ m}^2$$

Assume, $k_{av} = \infty$.

· Calculation of mole fraction:

$$X_{1_1}^* \simeq \frac{0.250 + 0.789}{(1000)/(18.02)} = k0.01873$$

$$X_{2_1}^* \simeq \frac{0.250}{(1000)/(18.02)} = 0.00451$$

· Calculation of parameters α, from Equation 5.114,

$$\alpha = \exp(J_B/k_{av}c) = 1, \text{ since } k_{av} = \infty$$

From Equation 5.130,

$$\beta = \left(\frac{1.61}{1.568}\right)^2 \left(\frac{4.32}{2.02}\right)^2 = 4.822$$

Using $\alpha = 1$, Equations 5.135, 5.136, and 5.140 become

$$J_1^* = c \left(\frac{D_1^*}{D_{12}}\right) \left(\frac{D_{Am}K_A}{\delta}\right)_{12} \left[X_{1_1}^* \left(\frac{(1-\beta)X_{2_1}^* + \beta X_{1_1}^*}{X_{1_1}^*}\right)^{1/2} - \right.$$
$$\left. X_{13}^* \left(\frac{(1-\beta)X_{23}^* + \beta X_{13}^*}{X_{13}^*}\right)^{1/2} \right] \tag{5.145}$$

$$J_2^* = c \left(\frac{D_2^*}{D_{12}}\right) \left(\frac{D_{Am}K_A}{\delta}\right)_{12} \left[X_{2_1}^* \left(\frac{X_{1_1}^*}{(1-\beta)X_{2_1}^* + \beta X_{1_1}^*}\right)^{1/2} - \right.$$
$$\left. X_{23}^* \left(\frac{X_{13}^*}{(1-\beta)X_{23}^* + \beta X_{13}^*}\right)^{1/2} \right] \tag{5.146}$$

$$J_B = A(p_2 - p_3) - 2AB_{av}(X_{1_1}^* - X_{13}^*) \tag{5.147}$$

Furthermore, the following relationships are used:

$$X_{13}^* = \frac{J_1^*}{J_B} \tag{5.148}$$

$$X_{23}^* = \frac{J_2^*}{J_B} \tag{5.149}$$

Combining Equations 5.145, 5.147, and 5.148,

$$
X_{13}^* \left[A(p_2 - p_3) - 2AB_{av} \left(X_{11}^* - X_{13}^* \right) \right] -
$$

$$
c \left(\frac{D_1^*}{D_{\underline{12}}} \right) \left(\frac{D_{Am}K_A}{\delta} \right)_{\underline{12}} \left[X_{11}^* \left(\frac{(1 - \beta)X_{21}^* + \beta X_{11}^*}{X_{11}^*} \right)^{1/2} - \right.
$$

$$
\left. X_{13}^* \left(\frac{(1 - \beta)X_{23}^* + \beta X_{13}^*}{X_{13}^*} \right)^{1/2} \right]
$$

$$
= 0 \qquad (5.150)
$$

Combining Equations 5.146, 5.147, and 5.149,

$$
X_{23}^* \left[A (p_2 - p_3) - 2AB_{av} \left(X_{11}^* - X_{13}^* \right) \right] -
$$

$$
c \left(\frac{D_2^*}{D_{\underline{12}}} \right) \left(\frac{D_{Am}K_A}{k} \delta \right)_{\underline{12}} \left[X_{21}^* \left(\frac{X_{11}^*}{(1 - \beta)X_{21}^* + \beta X_{11}^*} \right)^{1/2} - \right.
$$

$$
\left. X_{23}^* \left(\frac{X_{13}^*}{(1 - \beta)X_{23}^* + \beta X_{13}^*} \right)^{1/2} \right]
$$

$$
= \qquad (5.151)
$$

In order to simplify Equations 5.150 and 5.151, the following symbols are defined:

$$
a = A(p_2 - p_3) \qquad (5.152)
$$

$$
b = -2AB_{av}X_{11}^* \qquad (5.153)
$$

$$
c = 2AB_{av} \qquad (5.154)
$$

$$
d = -c \left(\frac{D_1^*}{D_{\underline{12}}} \right) \left(\frac{D_{Am}K_A}{\delta} \right)_{\underline{12}} X_{11}^* \left(\frac{(1 - \beta)X_{21}^* + \beta X_{11}^*}{X_{11}^*} \right)^{1/2} \qquad (5.155)
$$

$$
e = c \left(\frac{D_1^*}{D_{\underline{12}}} \right) \left(\frac{D_{Am}K_A}{\delta} \right)_{\underline{12}} \qquad (5.156)
$$

$$
f = -c \left(\frac{D_2^*}{D_{\underline{12}}} \right) \left(\frac{D_{Am}K_A}{\delta} \right)_{\underline{12}} X_{21}^* \left(\frac{X_{11}^*}{(1 - \beta)X_{21}^* + \beta X_{11}^*} \right) \qquad (5.157)
$$

$$
g = c \left(\frac{D_2^*}{D_{\underline{12}}} \right) \left(\frac{D_{Am}K_A}{\delta} \right)_{\underline{12}} \qquad (5.158)
$$

$$
x = X_{13}^* \qquad (5.159)
$$

$$y = X_{23}^* \tag{5.160}$$

Then Equations 5.150 and 5.151 can be written as

$$(a+b)x + cx^2 + d + ex\left(\frac{(1-\beta)y + \beta x}{x}\right)^{1/2} = 0 \tag{5.161}$$

$$(a+b)y + cxy + f + gy\left(\frac{x}{(1-\beta)y + \beta x}\right)^{1/2} = 0 \tag{5.162}$$

Numerical values are inserted into Equations 5.152–5.158 to calculate parameters a to g.

$$a = (1.186 \times 10^{-7})(10, 342)$$
$$= 1.227 \times 10^{-3} \text{kmol/m}^2\text{s}$$
$$b = -(2)(1.186 \times 10^{-7})(1.325 \times 10^5)(0.01873)$$
$$= -5.887 \times 10^{-4} \text{kmol/m}^2\text{s}$$
$$c = (2)(1.186 \times 10^{-7})(1.325 \times 10^5)$$
$$= 3.143 \times 10^{-2} \text{kmol/m}^2\text{s}$$
$$d = -(55.48)\left(\frac{1.35 \times 10^{-9}}{1.61 \times 10^{-9}}\right)(2.02 \times 10^{-7})(0.01873) \times$$
$$\left(\frac{(1-4.822)(0.00451) + (4.822)(0.01873)}{(0.01873)}\right)^{1/2}$$
$$= -0.3477 \times 10^6 \text{kmol/m}^2\text{s}$$
$$e = (55.48)\left(\frac{1.35 \times 10^{-9}}{1.61 \times 10^{-9}}\right)(2.02 \times 10^{-7})$$
$$= 9.397 \times 10^{-6} \text{kmol/m}^2\text{s}$$
$$f = -(55.48)\left(\frac{2.03 \times 10^{-9}}{1.61 \times 10^{-9}}\right)(2.02 \times 10^{-7})(0.00451) \times$$
$$\left(\frac{(0.01873)}{(1-4.822)(0.00451) + (4.822)(0.01873)}\right)^{1/2}$$
$$= -0.03226 \times 10^{-6} \text{kmol/m}^2\text{s}$$
$$g = (55.48)\left(\frac{2.03 \times 10^{-9}}{1.61 \times 10^{-9}}\right)(2.02 \times 10^{-7})$$
$$= 14.13 \times 10^{-6} \text{kmol/m}^2\text{s}$$

Inserting the above numerical values into Equations 5.161 and 5.162, we obtain

$$0.6383 \times 10^{-3}x + 3.143 \times 10^{-2}x^2 - 0.3477 \times 10^{-6} +$$

$$9.397 \times 10^{-6}x \left(\frac{-3.822y + 4.822x}{x} \right)^{1/2} = 0 \quad (5.163)$$

$$0.6383 \times 10^{-3}y + 3.143 \times 10^{-2}xy - 0.03226 \times 10^{-6} +$$

$$14.13 \times 10^{-6}y \left(\frac{x}{-3.822y + 4.822x} \right)^{1/2} = 0 \quad (5.164)$$

Solving Equations 5.163 and 5.164, we obtain

$$x = X_{13}^* = 0.000516 \text{ and } y = X_{23}^* = 0.0000448$$

Then

$$
\begin{aligned}
f_1^* &= f_{Na+}^* \\
&= \frac{0.01873 - 0.000516}{0.01873} \\
&= 0.972 \\
f_2^* &= f_{Cl-}^* \\
&= \frac{0.00451 - 0.0000488}{0.00451} \\
&= 0.989
\end{aligned}
$$

Furthermore, from Equations 5.108 and 5.110,

$$
\begin{aligned}
X_{41}^* &= X_{11}^* - X_{21}^* \\
&= 0.01873 - 0.00451 \\
&= 0.01422 \\
X_{43}^* &= X_{13}^* - X_2^* \\
&= 0.000516 - 0.0000488 \\
&= 0.0004672
\end{aligned}
$$

Therefore,

$$
\begin{aligned}
f_4^* &= f_{NO_3^-}^* \\
&= \frac{0.01422 - 0.0004672}{0.01422} \\
&= 0.967
\end{aligned}
$$

From Equations 5.147, 5.152, 5.153, 5.154, 5.159, and $\alpha = 1$,

$$
\begin{aligned}
J_B &= a + b + cx \\
&= (1.227 \times 10^{-3}) + (-5.887 \times 10^{-4}) + (3.143 \times 10^{-2})(5.16 \times 10^{-4}) \\
&= 0.6545 \times 10^{-3} \text{kmol/m}^2\text{s}
\end{aligned}
$$

and

$$
\begin{aligned}
(PR) &= (0.6545 \times 10^{-3})(18.02)(7.6 \times 10^{-4})(3600) \\
&= 3.227 \times 10^{-2} \text{kg/h} \\
&= 32.27 \text{g/h}
\end{aligned}
$$

where (PR) is the permeation rate defined as the amount of permeate collected by a membrane of the effective area S (m^2) in 1 h.

5.3 Membrane Gas Transport

A historic perspective on the transport of gases through synthetic polymeric membrane is given by Stannett [134]. Reports of pertinent literature by Barrer [135], Hopfenberg [136], Hwang and Kammermeyer [137], Stern [138], Stannett et al. [139], and Stern and Frisch [140], based on their extensive research, provide the necessary background material for the understanding of vapor and gas transport through membranes. From the mechanistic point of view of gas and vapor transport through "nonporous membranes," the free volume theory and the dual sorption theory have been highly successful in the analysis of sorption, unsteady, and steady-state permeation in such membrane systems.

The concept of free volume in a polymer is an extension of the ideas of Cohen and Turnbull [141], first used to describe the self-diffusion in a liquid of hard spheres. Such theories suggest that the permeant diffuses by a cooperative movement between the permeant and the polymer segments, from one "hole" to the other within the polymer. The creation of a "hole" is caused by fluctuations of local density due to thermal motion. Based on the concept of the redistribution of free volume to represent the thermodynamic diffusion coefficient [142], and the standard reference state for free volume [143], Stern and Fang [144] interpreted their permeability data for nonporous membranes, and Fang, Stern, and Frisch [145] extended the theory to include the case of permeation of gas and liquid mixtures.

The dual sorption model invokes the existence of two thermodynamically distinct populations of the penetrant gas, namely, molecules dissolved in the polymer by an ordinary dissolution mechanism (obeying Henry's law), and molecules residing in a limited number of preexisting microcavities in the polymer matrix (obeying Langmuir type of sorption isotherm), with rapid exchange between these two populations. The development and the illustration of the applicability of the dual sorption model for permeation through glassy polymers are the results of extensive investigations by Paul, Koros, and associates, Vieth and associates, and Stern and co-workers. The work of Koros et al. [146] on the sorption and transport of gases in polycarbonate, Chan et al. [147] on hydrocarbon gas sorption and transport in ethyl cellulose, and Erb and Paul [148] on CO_2, CH_4, Ar, and N_2 transport through polysulfone, and the relaxation of immobilization assumption by Petropoulos, [149] resulted in a coherent approach to the understanding of gases through glassy polymer membranes. Two extensive reviews by Vieth et al. [150] and Paul [151] furnished detailed information pertaining to different aspects of the dual sorption model. In addition to the free volume theory and the dual sorption theory, a number of molecular models have also been proposed. A brief description of the molecular models is presented by Stern and Frisch [140].

The general transport equations for gas start from the same equation of the irreversible thermodynamics used for reverse osmosis. The chemical potential gradient is considered to be the driving force, following the approach by Vieth [152].

$$J_i = -L_{ii} \nabla \mu_i \qquad (5.165)$$

In the case of transport of a single gas species, let us use the subscript g instead of i in the above equation, which becomes

$$J_g = -L_{gg} \nabla \mu_g \qquad (5.166)$$

The chemical potential of the gas molecules in the membrane can be written as

$$\nabla \mu_g = \mathbf{R}T \nabla \ln a_{gm} + v_{gm} \nabla p \qquad (5.167)$$

where a_{gm} (mol/m^3) and v_{gm} (m^3/mol) are the activity and molar volume of the gaseous species in the membrane. Equation 5.167 can also be written as

$$\nabla \mu_g = \mathbf{R}T \nabla \ln \gamma c_{gm} + v_{gm} \nabla p \qquad (5.168)$$

where γ is the activity coefficient $(-)$ and is no longer assumed to be constant. Since the molar volume of the gaseous species in the membrane is considered to be as small as that of the liquid, the second term can be ignored, as compared with the first term. Then Equation 5.168 can be approximated by

$$\nabla \mu_g = RT \nabla \ln \gamma c_{gm} \qquad (5.169)$$

Integrating Equation 5.169 from one side of the membrane to the other side of the membrane,

$$\Delta \mu_g = RT \Delta \ln \gamma c_{gm} \qquad (5.170)$$

Again, in analogies to reverse osmosis transport, the phenomenological coefficient can be defined as

$$L_{gg} = \frac{c_{gm}}{f_{gm}} \qquad (5.171)$$

Assuming a constant gradient of chemical potential across the membrane, Equation 5.166 can be given as

$$J_g = -\frac{c_{gm}}{f_{gm}} \frac{\Delta \mu_g}{\delta} \qquad (5.172)$$

$$= -\frac{c_{gm}}{f_{gm}} \frac{RT \Delta \ln \gamma c_{gm}}{\delta} \qquad (5.173)$$

Approximating

$$\Delta \ln \gamma c_{gm} \simeq \left[\frac{\partial \ln \gamma}{\partial \ln c_{gm}} + 1 \right] \frac{\Delta c_{gm}}{c_{gm}} \qquad (5.174)$$

Equation 5.173 becomes

$$J_g = -\frac{RT}{f_{gm}} \left[\frac{\partial \ln \gamma}{\partial \ln c_{gm}} + 1 \right] \frac{\Delta c_{gm}}{\delta} \qquad (5.175)$$

Defining the molecular diffusion constant D (m^2/s) by

$$J_g = -D \frac{\Delta c_{gm}}{\delta} \qquad (5.176)$$

$$D = \frac{RT}{f_{gm}} \left[\frac{\partial \ln \gamma}{\partial \ln c_{gm}} + 1 \right] \qquad (5.177)$$

The molecular diffusion coefficient D, therefore, changes with concentration in the membrane, according to Equation 5.177, even when the friction, f_{gm}, working against the movement of the gas molecule in the membrane remains constant. The concentration of the gas molecules in the membrane, c_{gm}, is not known unless the solubility of the gas in the polymer has been determined. In many cases it is customary to assume Henry's law in the form

$$c_{gm} = Sp \tag{5.178}$$

where S and p are the solubility coefficient (mol/m^3 Pa) and gas pressure (Pa), respectively. Equation 5.176 is then given as

$$J_g = -DS\frac{\Delta p}{\delta} \text{ or} \tag{5.179}$$

$$= -P\frac{\Delta p}{\delta} \tag{5.180}$$

where

$$P = DS \tag{5.181}$$

and is called the permeability coefficient (mol m/m^2 s Pa). In the above derivation, D and S were treated as if they were constant. In many cases, however, D depends on c_{gm}, and S depends on p. As a product of D and S, P is not a fundamental property, being dependent on both diffusivity and solubility characteristics. However, the permeability coefficient is often of practical usefulness, since it gives the measure of the separation potential offered by a polymeric membrane. When D is a function of c_{gm} and given as $D(c_{gm})$, Equation 5.176 should be written in its differential form, and

$$J_g = -D(c_{gm})\frac{dc_{gm}}{dx} \tag{5.182}$$

where x is the distance toward the gas flow direction from the high-pressure side of the membrane. Integration of the above equation from the high-pressure to the low-pressure side of the membrane yields

$$J_g\delta = -\int_0^\delta D(c_{gm})dc_{gm} \tag{5.183}$$

Therefore,

$$J_g = -\frac{\int_0^\delta D(c_{gm})dc_{gm}}{\Delta c_{gm}} \times \frac{\Delta c_{gm}}{\delta} \tag{5.184}$$

Analogous to Equation 5.176, the average diffusion coefficient can be derived from Equation 5.184 as

$$\overline{D} = \frac{\int_0^\delta D(c_{gm})dc_{gm}}{\Delta c_{gm}} \tag{5.185}$$

The temperature dependence of D obeys a relationship similar to the Arrhenius equation:

$$D = D_0 \exp\left[-\frac{E_D}{RT}\right] \tag{5.186}$$

where D_0 and E_D are the preexponential factor (m^2/s) and the activation energy of diffusion (J/mol), respectively. The activation energy is considered to be the energy required to create a gap between polymer segments through which the penetrant molecule can diffuse. The temperature dependence of the solubility coefficient S can also be expressed by an exponential form:

$$S = S_0 \exp\left[-\frac{\Delta H_s}{RT}\right] \tag{5.187}$$

where S_0 and ΔH_s are the preexponential factor (mol/m^3 Pa) and the enthalpy of solution (J/mol), respectively. Combining Equations 5.186 and 5.187, the temperature dependence of P is given by

$$P = D_0 S_0 \exp\left[-\frac{E_D + \Delta H_s}{RT}\right] \tag{5.188}$$

$$= P_0 \exp\left[-\frac{E_p}{RT}\right] \tag{5.189}$$

where

$$E_p = E_D + \Delta H_s \tag{5.190}$$

Note that E_p is the sum of E_D and ΔH_s. While E_D is usually positive, ΔH_s is sometimes negative when the sorption process is endothermic. Depending on their magnitudes, E_p may be either positive or negative.

The sorption-diffusion model described in the foregoing section is very general. The manifestation of the fundamental principles in the transport depends on whether the membrane polymer is in a rubbery state or in a glassy state. All polymers undergo a transition from rubbery to glassy state when the temperature is lowered below the transition temperature, T_g, that is characteristic to the polymer.

Table 7. Penetrant Solubility and Transport in Rubbery
Polymers $(T > T_g)^a$[140]

$T > T_c$	$T \leq T_c$
Henry's law obeyed:	*Henry's law generally not obeyed*
$S = S_0$, a constant	$S = S(c)$
D independent of concentration:	
$D = D_0$, a constant	$D = D(c)$
Energy of activation for diffusion,	
E_D, *is constant:*	
$-\text{R}d \ln D_0/dT = E_D = $ constant	$E_D = f(c, T)$
$\overline{P}$ *constant or a weak function of* Δp:	$\overline{P}$ *can be a strong function of* Δp:
$d \ln \overline{P}/d\Delta p = m \leq 0$	$d \ln \overline{P}/d\Delta p = m(\Delta p) > 0$
where $\Delta p = p_2 - p_3$	

^{a} Single-mode solution (sorption), Fickian diffusion.

5.3.1 Gas Transport in Rubbery Polymers

When the temperature is above T_g, the polymer is in the rubbery state. In the rubbery state the segmental motion of the polymeric molecules is as rapid as that of liquid molecules. The dependence of the permeability coefficient, P, and flux, J_g, on gas pressure and on temperature will depend on how D and S are affected by the above operating variables. The behavior of D, S, and P depends on whether the temperature is above or below the critical temperature, T_c, of the permeant gas, and can be summarized as shown in Table 7. For many simple gases, such as H_2, He, Ne, Ar, O_2, N_2, CH_4, etc., $T > T_c$ at ambient temperature, and D, S, and P behave as shown in Table 7 for $T > T_c$. On the other hand, for a number of organic vapors, $T < T_c$ at ambient temperature. Depending on the reduced temperature, T/T_c, many permeant-membrane systems exhibit intermediate behavior. A most generalized picture of the dependency of the permeability coefficient on temperature and pressure (or the pressure difference on both sides of the membrane) is illustrated in Figure 5.2. Note that the permeability coefficient is almost constant when the temperature is far above the critical temperature of gas, whereas it is strongly pressure dependent when the temperature is far below the critical temperature. It should be recalled that the energy of activation of the permeability coefficient is the sum of the activation energy of diffusion and the enthalpy of solution: $E_p = E_D + \Delta H_s$.

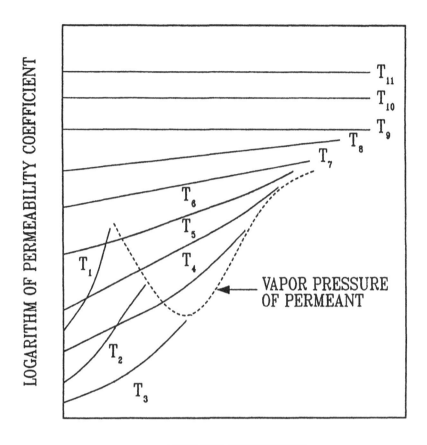

Figure 5.2. Generalized picture of the dependency of permeability coefficient on temperature and pressure at $T > T_g$. (Reproduced from [140] with permission.)

It was also pointed out that E_D is always positive, and ΔH_s is often negative. In Figure 5.2, the permeability coefficient, P, increases with temperature when the temperature is much higher than the critical temperature, which means $E_p > 0$. This is so because E_D is the dominant factor for the temperature effect on P, due to low solubility of gas at high temperature. When the temperature is far lower than the critical temperature, on the other hand, the permeability coefficient, P, decreases with an increase in temperature, which means $E_p < 0$. This occurs because the solubility is dominant in this temperature range and ΔH_s is highly negative, which means the gas solution in the membrane is an endothermic phenomena. Thus, the transport of a given gas through a membrane made of a rubbery polymer is diffusion controlled when the temperature is high and solubility controlled when the temperature is low. For a specific system of gas permeant-membrane, all the features of gas permeation depicted in Figure 5.2 are not necessarily exhibited, because the polymer may undergo the phase transition if the temperature range covered by Figure 5.2 is too broad.

5.3.2   Gas Transport in Glassy Polymers

When the temperature is lower than the glass transition temperature of the polymer, the polymer is at the glassy state. The motion of the polymer chain is not sufficiently rapid to homogenize the polymer structure. This allows the heterogeneity of the polymer structure and the localization of the permeant gas. Below glass transition temperature, free segmental rotation of the polymer chains is restricted in the glassy states, resulting in the presence of fixed microvoids throughout the polymer. These microvoids in the glassy polymer network act to immobilize a portion of the permeant gas molecules by entrapment and by binding at high energy sites at their molecular peripheries. As a result the sorption of the permeant gas molecule is split into two types: one that is the ordinary sorption in which the sorbed gas molecules reside in the polymer matrix, and the other where the gas molecules are trapped in the microvoids. While the first type of sorption is considered to be described by Henry's law, the second type is described by a Langmuir isotherm. This dual mode of sorption is currently widely accepted for the sorption of small permeant gases such as carbon dioxide, methane, argon, and nitrogen, in glassy polymers.

The dual sorption model is given by

$$c_{gm} = c_{gm,D} + c_{gm,H} \tag{5.191}$$

$$= k_D p + \frac{c_H' b p}{1 + b p} \tag{5.192}$$

where $c_{gm,D}$ and $c_{gm,H}$ are the concentration of gas molecules sorbed in the polymer matrix by an ordinary Henry's law isotherm and that of gas molecules sorbed in the microvoids of polymer by a Langmuir sorption isotherm, respec-

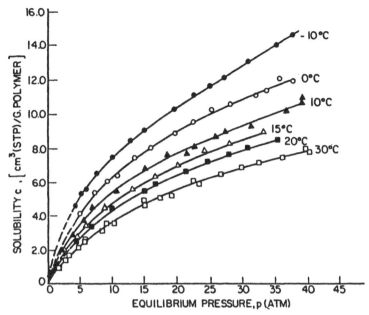

Figure 5.3. Some experimental data for typical examples of dual sorption. (From Stern, S.A. and Kulkarni, S.S., *J. Membrane Sci.*, 1982, 10, 235, with permission.)

tively. k_D, c_H', and b are called the Henry's law dissolution constant (mol/m^3 Pa), the hole saturation constant (mol/m^3), and the hole affinity constant (1/Pa), respectively. When p is small, the above equation is approximated by

$$c_{gm} = [k_D + c_H' b] p \qquad (5.193)$$

and when p is large, Equation 5.192 is approximated by

$$c_{gm} = k_D p + c_H' \qquad (5.194)$$

Thus, the dual sorption model predicts that the isothermal plot of c_{gm} vs. p will consist of a low-pressure linear region and a high-pressure linear region that are connected by a nonlinear region. Some experimental results are illustrated in Figure 5.3, for typical examples of dual sorption.

The membrane transport of the gas molecule was formulated by the dual sorption model with the following basic assumptions:

1. Two modes of sorption, Henry's law sorption and Langmuir sorption, take place simultaneously.

2. Between the above two modes of sorption, local equilibrium is maintained in the membrane.

3. The gas molecules sorbed by the Langmuir mode are partially immobilized.

4. All gas molecules sorbed by the Henry mode are mobile.

In this model the transport of gas molecules in a glassy polymer is described by

$$J_g = J_{g,D} + J_{g,H} \tag{5.195}$$

$$= -D_D \frac{dc_{g,D}}{dx} - D_H \frac{dc_{g,H}}{dx} \tag{5.196}$$

Since

$$c_{g,D} = k_D p \tag{5.197}$$

$$c_{g,H} = \frac{c_H' b p}{1 + bp} \tag{5.198}$$

Then

$$J_g = -D_D \frac{dk_D p}{dx} - D_H \frac{d\left[c_H' bp/(1 + bp)\right]}{dx} \tag{5.199}$$

and

$$J_g dx = -D_D k_D dp - D_H d\left(\frac{c_H' bp}{1 + bp}\right) \tag{5.200}$$

Integration of both sides of the above equation from one side of the membrane with gas pressure p_2 to the other side of the membrane with zero gas pressure yields

$$J_g \delta = -D_D k_D \Delta p - \left(0 - D_H \frac{c_H' b p_2}{1 + b p_2}\right) \tag{5.201}$$

$$= -D_D k_D \left[1 + \left(\frac{D_H}{D_D} \frac{c_H' b}{k_D}\Big/ 1 + b p_2\right)\right] \Delta p \tag{5.202}$$

since the low-pressure side is maintained at zero pressure and $\Delta p = 0 - p_2$. Then

$$J_g = -D_D k_D \left[1 + \left(FK/1 + b p_2\right)\right] \frac{\Delta p}{\delta} \tag{5.203}$$

where $F = D_H/D_D$ and $K = c_H' b/k_D$. In analogy to Equation 5.180,

$$P = D_D k_D \left[1 + \left(FK/1 + b p_2\right)\right] \tag{5.204}$$

According to the above equation, the permeability coefficient should decrease with the feed gas pressure, since an increased amount of gas is sorbed to the microvoids and partially immobilized.

When the temperature is lower than the critical temperature, T_c, of gas, the solubility of the permeant gas in the membrane may become so high that the diffusion constant no longer remains constant. A modification was done in the above dual transport model, with partial immobilization of the Langmuir sorption mode to include the concentration dependency of the diffusion coefficient [153]. Instead of Equation 5.204, the following equation expresses the permeability coefficient:

$$P = \left(\frac{D_0}{\beta p_2}\right)\left\{\exp\left[\beta k_D p_2\left(1 + \frac{FK}{1 + bp_2}\right)\right] - 1\right\} \qquad (5.205)$$

where D_0 is the value of D_D when the concentration of gas in the membrane, c_{gm}, is nearly equal to zero, and β is the constant that characterizes the concentration dependence of D_D. When β is nearly equal to 0 and $K \neq 0$, meaning D_D does not depend on the concentration of gas in the membrane and a significant amount of gas is sorbed by the Langmuir mode, Equation 5.205 reduces to Equation 5.204. When $\beta > 0$ and K is nearly equal to zero, meaning D_D depends strongly on the concentration of gas in the membrane, but the sorption mode is almost entirely by the Henry mode, Equation 5.205 reduces to the form describing the pressure-dependent transport through rubbery polymer. When both β and K are greater than zero, the dependence of P on the applied pressure becomes as illustrated in Figure 5.4, and a minimum appears in the permeability coefficient vs. applied pressure relationship. The minimum is caused by two opposing effects of the pressure on the permeability coefficient. One effect is the increase in the amount of gas molecules trapped and immobilized in the microvoids with an increase in the operating pressure, leading to a decrease in the permeability coefficient. The other effect is an increase in diffusivity in D_D, with an increase in the amount of sorbed gas molecules in the membrane.

5.4 Pervaporation

Pervaporation is a membrane separation process where the upstream side of the membrane is in contact with feed liquid while vacuum is applied on the downstream side of the membrane. The permeant vaporizes somewhere between the upstream and the downstream side of the membrane; therefore, the permeate is obtained as vapor. The potential of the pervaporation process in commercial applications was first pointed out by Binning and co-workers [154], [155]. Lee attempted to compare the transport of pervaporation with those of reverse

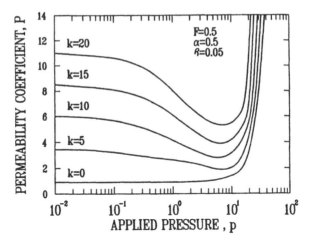

Figure 5.4. Dependence of permeability coefficient on pressure at $T < T_g$. ($\alpha = b/k_D$.) (Reproduced from [140] with permission.)

osmosis and gas permeation based on the solution-diffusion model under an isothermal condition [156]. In his approach the gradient of the chemical potential across the membrane is considered to be the driving force for the mass transfer. The chemical potential gradient for species A and B can be written as

$$\nabla \mu_A = RT \nabla \ln a_{Am} + v_A \nabla p \qquad (5.206)$$

$$\nabla \mu_B = RT \nabla \ln a_{Bm} + v_B \nabla p \qquad (5.207)$$

Before integrating the above two equations, an assumption was made that the pressure, p, remains constant, the same as that of the feed solution throughout the membrane cross-section. The pressure falls from the upstream pressure to the downstream pressure discontinuously at the permeate side of the membrane, as illustrated in Figure 5.5. Therefore, the second term of the right-hand side of Equations 5.206 and 5.207 is ignored, and the flux equations become exactly the same as that of solute species (the second species) in reverse osmosis.

$$J_A = -D_{Am} \frac{\Delta c_{Am}}{\delta} \qquad (5.208)$$

$$J_B = -D_{Bm} \frac{\Delta c_{Bm}}{\delta} \qquad (5.209)$$

Thermodynamic equilibrium should be established at both sides of the membrane, and

$$\mu_{A2} = \mu_{Am2} \qquad (5.210)$$

Membrane

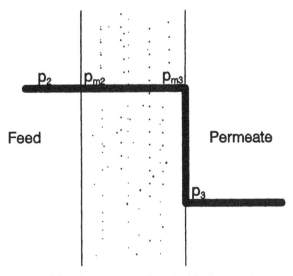

Figure 5.5. Pressure gradient inside the membrane.

$$\mu_{A3} = \mu_{Am3} \tag{5.211}$$

$$\mu_{B2} = \mu_{Bm2} \tag{5.212}$$

$$\mu_{B3} = \mu_{Bm3} \tag{5.213}$$

The subscripts 2 and 3 in the above equations indicate the upstream side of the membrane, facing the feed liquid, and the downstream side, facing the permeate vapor, respectively. Since

$$\mu = \mu_0 + RT \ln a + \int_{p_{ref}}^{p} v \, dp \tag{5.214}$$

where μ_0 is the chemical potential of pure permeant at $p = p_{ref}$, thermodynamic equilibrium at both sides of the membrane for the species A and B can be written as

$$a_{Am2} = a_{A2} \exp[-v_A(p_{m2} - p_2)/RT] \tag{5.215}$$

$$a_{Am3} = a_{A3} \exp[-v_A(p_{m3} - p_3)/RT] \tag{5.216}$$

$$a_{Bm2} = a_{B2} \exp[-v_B(p_{m2} - p_2)/RT] \tag{5.217}$$

$$a_{Bm3} = a_{B3} \exp[-v_B(p_{m3} - p_3)/RT] \tag{5.218}$$

With the assumptions that the pressure applied to the feed solution, p_2, prevails across the membrane, and the pressure decreases to that of the permeate at the membrane/permeate boundary, the above equations can be rewritten as

$$a_{Am2} = a_{A2} \tag{5.219}$$

$$a_{Am3} = a_{A3} \exp[-v_A(p_2 - p_3)/RT] \tag{5.220}$$

$$a_{Bm2} = a_{B2} \tag{5.221}$$

$$a_{Bm3} = a_{B3} \exp[-v_B(p_2 - p_3)/RT] \tag{5.222}$$

Activities are equal to the product of the activity coefficient and the concentration; hence,

$$c_{Am2} = \frac{\gamma_{A2}}{\gamma_{Am2}} c_{A2} \tag{5.223}$$

$$c_{Am3} = \frac{\gamma_{A3}}{\gamma_{Am3}} c_{A3} \exp[-v_A(p_2 - p_3)/RT] \tag{5.224}$$

$$c_{Bm2} = \frac{\gamma_{B2}}{\gamma_{Bm2}} c_{B2} \tag{5.225}$$

$$c_{Bm3} = \frac{\gamma_{B3}}{\gamma_{Bm3}} c_{B3} \exp[-v_B(p_2 - p_3)/RT] \tag{5.226}$$

· Defining the partition coefficient, K, as the ratio of the activity coefficients, the above equations become

$$c_{Am2} = K_{A2} c_{A2} \tag{5.227}$$

$$c_{Am3} = K_{A3} c_{A3} \exp[-v_A(p_2 - p_3)/RT] \tag{5.228}$$

$$c_{Bm2} = K_{B2} c_{B2} \tag{5.229}$$

$$c_{Bm3} = K_{B3} c_{B3} \exp[-v_B(p_2 - p_3)/RT] \tag{5.230}$$

Insertion of the above equations into Equations 5.208 and 5.209 yields

$$J_A = \frac{D_{Am}}{\delta} \{ K_{A2} c_{A2} - K_{A3} c_{A3} \exp[-v_A(p_2 - p_3)/RT] \} \tag{5.231}$$

$$J_B = \frac{D_{Bm}}{\delta} \{ K_{B2} c_{B2} - K_{B3} c_{B3} \exp[-v_B(p_2 - p_3)/RT] \} \tag{5.232}$$

Further,

$$J_A = \frac{P_A}{\delta} (c_{A2} - \alpha_A c_{A3} \exp[-v_A(p_2 - p_3)/RT]) \tag{5.233}$$

$$J_B = \frac{P_B}{\delta} (c_{B2} - \alpha_B c_{B3} \exp[-v_B(p_2 - p_3)/RT]) \tag{5.234}$$

where

$$P_A = D_{Am}K_{A2} \tag{5.235}$$

$$P_B = D_{Bm}K_{B2} \tag{5.236}$$

and

$$\alpha_A = \frac{K_{A3}}{K_{A2}} \tag{5.237}$$

$$\alpha_B = \frac{K_{B3}}{K_{B2}} \tag{5.238}$$

or

$$\alpha_A = \frac{\gamma_{A3}}{\gamma_{A2}} \tag{5.239}$$

$$\alpha_B = \frac{\gamma_{B3}}{\gamma_{B2}} \tag{5.240}$$

assuming that the activity coefficient is constant in the membrane.

In the case of pervaporation, the feed is in liquid phase, whereas the permeate is in the vapor phase. Rearranging Equation 5.233 and 5.234,

$$J_A = \frac{P_A c_{A2}}{\delta}\left(1 - \frac{\gamma_{A3}c_{A3}}{\gamma_{A2}c_{A2}}\exp[-v_A(p_2 - p_3)/\mathbf{R}T]\right) \tag{5.241}$$

$$J_B = \frac{P_B c_{B2}}{\delta}\left(1 - \frac{\gamma_{B3}c_{B3}}{\gamma_{B2}c_{B2}}\exp[-v_B(p_2 - p_3)/\mathbf{R}T]\right) \tag{5.242}$$

since

$$\gamma_A c_A = p_A/p_{A*} \tag{5.243}$$

$$\gamma_B c_B = p_B/p_{B*} \tag{5.244}$$

where p_A, p_B, p_{A*}, and p_{B*} are partial vapor pressures (Pa) of component A and component B, and saturation vapor pressure (Pa) of component A and component B, respectively. Equations 5.243 and 5.244 should be valid at both sides of the membrane. Then Equations 5.241 and 5.242 can be written as

$$J_A = \frac{P_A c_{A2}}{\delta}\left(1 - \frac{p_{A3}}{p_{A2}}\exp[-v_A(p_2 - p_3)/\mathbf{R}T]\right) \tag{5.245}$$

$$J_B = \frac{P_B c_{B2}}{\delta}\left(1 - \frac{p_{B3}}{p_{B2}}\exp[-v_B(p_2 - p_3)/\mathbf{R}T]\right) \tag{5.246}$$

A note regarding the partial vapor pressures is in order. In Equation 5.245, p_{A3} is the real partial vapor pressure of component 3 on the permeate side of the membrane, since the permeant is in the vapor state. On the other hand, p_{A2} is the partial vapor pressure of component A, which is in equilibrium with the feed solution. The partial vapor pressure is that of an imaginary vapor phase that satisfies Equation 5.243. The molar volume, v_A, is considered to be close to that of liquid, since the permeant is more like the condensed phase in the membrane. Then $-v_A(p_2 - p_3)/RT$ becomes very small because of a very small value of v_A, and the exponential term becomes nearly equal to unity. Applying the above approximation to both component A and component B, we obtain

$$J_A = \frac{P_A c_{A2}}{\delta}\left(1 - \frac{p_{A3}}{p_{A2}}\right) \tag{5.247}$$

$$J_B = \frac{P_B c_{B2}}{\delta}\left(1 - \frac{p_{B3}}{p_{B2}}\right) \tag{5.248}$$

When separation factor SF_B^A is defined by

$$SF_B^A = \frac{X_{A3}/X_{A2}}{X_{B3}/X_{B2}} \tag{5.249}$$

and since

$$X_{A3} = \frac{J_A}{J_A + J_B} \tag{5.250}$$

$$X_{B3} = \frac{J_B}{J_A + J_B} \tag{5.251}$$

Equation 5.249 can be written as

$$SF_B^A = J_A c_{A2}/J_B c_{B2} \tag{5.252}$$

Then inserting Equations 5.247 and 5.248,

$$SF_B^A = \frac{P_A \left(1 - (p_{A3}/p_{A2})\right)}{P_B \left(1 - (p_{B3}/p_{B2})\right)} \tag{5.253}$$

Furthermore, when the pressure on the permeate side is nearly equal to zero, p_{3A} and p_{3B} are nearly equal to zero, and therefore,

$$J_A = \frac{P_A c_{A2}}{\delta} \tag{5.254}$$

$$J_B = \frac{P_B c_{B2}}{\delta} \qquad (5.255)$$

and then

$$SF_B^A = \frac{P_A}{P_B} \qquad (5.256)$$

or

$$SF_A^B = \frac{P_B}{P_A} \qquad (5.257)$$

Let us now review reverse osmosis transport. Both feed and permeates are in the liquid phase, and we do not need to convert the concentration to partial vapor pressure to calculate the chemical potential difference as the driving force. Instead, Equations 5.241 and 5.242 are used as given. Let us define component B as the solvent and component A as the solute. Then Equations 5.242 and 5.241 can be written as

$$J_B = \frac{P_B c_{B2}}{\delta}\left(1 - \frac{c_{B3}\gamma_{B3}}{c_{B2}\gamma_{B2}}\exp[-v_B(p_2 - p_3)/RT]\right) \qquad (5.258)$$

$$J_A = \frac{P_A c_{A2}}{\delta}\left(1 - \frac{c_{A3}}{c_{A2}}\exp[-v_A(p_2 - p_3)/RT]\right) \qquad (5.259)$$

In Equation 5.259 an assumption was made that the activity coefficients of component A (solute) on both sides of the membrane are the same; i.e., $\alpha_A = 1$. For component B (solvent) this assumption was not necessary. Since

$$a_B = c_B \gamma_B \qquad (5.260)$$

Equation 5.258 can be written as

$$\begin{aligned} J_B &= \frac{P_B c_{B2}}{\delta}\left(1 - \frac{c_{B3}\gamma_{B3}}{c_{B2}\gamma_{B2}}\exp[-v_B(p_2 - p_3)/RT]\right) \\ &= \frac{P_B c_{B2}}{\delta}\left(1 - \frac{a_{B3}}{a_{B2}}\exp[-v_B(p_2 - p_3)/RT]\right) \end{aligned} \qquad (5.261)$$

/Since the osmotic pressure is defined as

$$\pi = \frac{-RT}{v_B}\ln a_B \qquad (5.262)$$

Equation 5.261 can be written as

$$J_B = \frac{P_B c_{B2}}{\delta}(1 - \exp[-v_B(p_2 - p_3 - \pi_2 + \pi_3)/RT]) \qquad (5.263)$$

The total molar flux of reverse osmosis can be approximated by that of the solvent. In analogy to Equation 5.252, the separation factor SF_A^B becomes

$$SF_A^B = J_B c_{B2}/J_A c_{A2} \tag{5.264}$$

Then

$$SF_A^B = \frac{P_B(1 - \exp[-v_B(p_2 - p_3 - \pi_2 + \pi_3)/RT])}{P_A(1 - \frac{c_{A1}}{c_{A2}} \exp[-v_A(p_2 - p_3)])} \tag{5.265}$$

Furthermore, the mole fraction of the solvent is nearly equal to unity on both sides of the membrane, and SF_A^B can be approximated by

$$SF_A^B = \frac{X_{A2}}{X_{A3}} \tag{5.266}$$

$$= \frac{c_{A2}}{c_{A3}} \tag{5.267}$$

Combining Equations 5.265 and 5.267, we can obtain

$$SF_A^B = \exp[-v_A(p_2 - p_3)/RT] + \frac{P_B}{P_A}(1 - \exp[-v_B(p_2 - p_3 - \pi_2 + \pi_3)/RT]) \tag{5.268}$$

This is a general equation for the separation factor of reverse osmosis transport.

As mentioned earlier, the molar volumes v_B and v_A are so small that the exponential terms in Equations 5.263 and 5.268 can be approximated by

$$\exp x \simeq 1 + x \tag{5.269}$$

Then Equations 5.263 and 5.268 become

$$J_B = \frac{P_B c_{B2} v_B}{\delta RT}(p_2 - p_3 - \pi_2 + \pi_3) \tag{5.270}$$

$$SF_A^B = 1 + \frac{P_B}{P_A}\left[\frac{(p_2 - p_3 - \pi_2 + \pi_3)v_B}{RT}\right] \tag{5.271}$$

Equations 5.270 and 5.271 are the equations for reverse osmosis. It should be recalled that the solvent concentration in the membrane was assumed constant and is equal to c_{Bm2} (or c_{Bm3}) in Lonsdale's derivation, whereas the pressure was assumed to be constant and equal to p_2 across the membrane in Lee's approach. Let us now compare the separation f' derived by Lonsdale and Lee.

From Equation 5.26 and Equation 5.267,

$$f' = 1 - \frac{1}{SF_A^B} \tag{5.272}$$

Combining Equations 5.235, 5.236, 5.271, and 5.272,

$$f' = 1 - \frac{1}{1 + \frac{D_{Bm}K_{B2}}{D_{Am}K_{A2}}\left[\frac{(p_2 - p_3 - \pi_2 + \pi_3)v_B}{RT}\right]} \tag{5.273}$$

$$= \frac{1}{1 + \frac{D_{Am}K_{A2}RT}{D_{Bm}K_{B2}(p_2 - p_3 - \pi_2 + \pi_3)v_B}} \tag{5.274}$$

Equation 5.274 looks different from Equation 5.30 derived by Lonsdale. However, when the distribution coefficients, K_A and K_B, and the concentration of the component B, c_{Bm}, are the same throughout the membrane, K_{B2} can be written as $\frac{c_{Bm}}{c_{B3}}$, and the above equation becomes

$$f' = \frac{1}{.1 + \frac{D_{Am}K_A RTc_{B3}}{D_{Bm}c_{Bm}(p_2 - p_3 - \pi_2 + \pi_3)v_B}} \tag{5.275}$$

which is exactly the same as Equation 5.30.

Let us examine an imaginary case where the feed pressure approaches infinity in a reverse osmosis operation. Then from Equations 5.258 and 5.268,

$$J_B = \frac{P_B c_{B2}}{\delta} \tag{5.276}$$

$$SP_A^B = \frac{P_B}{P_A} \tag{5.277}$$

The above equations are exactly the same as those for pervaporation. Therefore, according to the above approach, the flux and the separation factor of a reverse osmosis experiment should approach those of pervaporation (with zero permeate pressure) when the operating pressure approaches infinity. This is because the second term in the flux equation (contribution of the permeant activity to the driving force) becomes negligible when the pressure on the feed side becomes infinity. As for pervaporation, the second term in the flux equation becomes zero when the pressure on the permeate side becomes zero. It should be noted that this conclusion is valid only on the basis of the assumption; i.e., the pressure is equal to the feed pressure across the membrane.

There are a number of other papers in which the pervaporation phenomena was discussed by the solution-diffusion model [157]–[169].

Example 8

Calculate f' and J_B under the same operating conditions as those of Example 1, using Lee's equation.

Combining Equations 5.272 and 5.268,

$$f' = 1 - \cfrac{1}{\exp[-v_A(p_2 - p_3)/RT] + \frac{P_B}{P_A}(1 - \exp[-v_B(p_2 - p_3 - \pi_2 + \pi_3)/RT])} \tag{5.278}$$

where

$$P_B = D_{Bm}K_B$$
$$= D_{Bm}c_{Bm}/c_{B3}$$
$$P_A = D_{Am}K_A$$

From Example 1,

$$
\begin{aligned}
D_{Bm}c_{Bm} &= 2.7 \times 10^{-8} \text{ kg/m s} \\
D_{Am}K_A &= 4.2 \times 10^{-14} \text{ m}^2\text{/s} \\
v_B &= 18.02 \times 10^{-6} \text{ m}^3\text{/mol} \\
v_A &= 26.98 \times 10^{-6} \text{ m}^3\text{/mol} \\
p_2 - p_3 &= 4.134 \times 10^6 \text{ Pa} \\
RT &= 2.479 \times 10^3 \text{ J/mol} \\
c_{B3} &= 10^3 \text{ kg/mol}
\end{aligned}
$$

An approximation is made again that $\pi_2 - \pi_3 = 0.461 \times 10^6$ Pa. Inserting all numerical values into the above equation,

$$
\begin{aligned}
f' = 1 - & \\
1 \Big/ \Big\{ &\exp\left(-\frac{(26.98 \times 10^{-6})(4.134 \times 10^6)}{2.479 \times 10^3}\right) + \frac{2.7 \times 10^{-8}}{(10^3)(4.2 \times 10^{-14})} \\
\times & \left[1 - \exp\left(-\frac{(18.02 \times 10^{-6})(4.134 \times 10^6 - 0.461 \times 10^6)}{2.479 \times 10^3}\right)\right] \Big\} \\
= 0.944 & \tag{5.279}
\end{aligned}
$$

The permeate molality is $0.1 \times (1 - 0.944) = 0.0056$. The permeate mole fraction is $(0.0056)/(0.0056 + \frac{1000}{18.02}) = 10.09 \times 10^{-5}$. The permeate osmotic pressure is $(2.5645 \times 10^8)(10.09 \times 10^{-5}) = 0.0259 \times 10^6$ Pa. Then, the osmotic pressure difference becomes

$$\pi_2 - \pi_3 = 0.461 \times 10^6 - 0.0259 \times 10^6 = 0.4351 \times 10^6 \, \text{Pa}.$$

Using the above osmotic pressure difference,

$$f' = 1 - 1 \Big/ \left\{ \exp\left(-\frac{(26.98 \times 10^{-6})(4.134 \times 10^6)}{2.479 \times 10^3} \right) + \frac{2.7 \times 10^{-8}}{(10^3)(4.2 \times 10^{-14})} \right.$$
$$\left. \times \left[1 - \exp\left(-\frac{(18.02 \times 10^{-6})(4.134 \times 10^6 - 0.4351 \times 10^6)}{2.479 \times 10^3} \right) \right] \right\}$$
$$= 0.944$$

Therefore, $f' = 0.944$ is a sufficiently accurate answer. Since

$$J_B = \frac{D_{Bm} C_{Bm}}{\delta} \left[1 - \exp\left(-\frac{v_B(p_2 - p_3 - \pi_2 + \pi_3)}{RT} \right) \right]$$

inserting numerical values, the flux (kg/m² s) is

$$J_B = \frac{2.7 \times 10^{-8}}{10^{-7}} \times$$
$$\left[1 - \exp\left(-\frac{(18.02 \times 10^{-6})(4.134 \times 10^6 - 0.4351 \times 10^6)}{2.479 \times 10^3} \right) \right]$$
$$= 71.63 \times 10^{-4}$$

When there is no sodium chloride in the solution, there is no osmotic pressure effect. Therefore,

$$J_B = \frac{2.7 \times 10^{-8}}{10^{-7}} \times \left[1 - \exp\left(-\frac{(18.02 \times 10^{-6})(4.134 \times 10^6)}{2.479 \times 10^3} \right) \right]$$
$$= 79.93 \times 10^{-4}$$

Furthermore, when $p_2 - p_3$ approaches infinity, f' approaches $1 - P_A/P_B = 1 - D_{Am} K_A C_{B3}/D_{Bm} C_{Bm} = 1 - (4.2 \times 10^{-14})(10^3)/2.7 \times 10^{-8} = 0.998$; J_B approaches $D_{Bm} C_{Bm}/\delta = 2.7 \times 10^{-8}/10^{-7} = 2.7 \times 10^{-1}$ kg/m² s. J_B in the absence of sodium chloride should also approach the latter value. According to Lee, J_B at $p_2 - p_3 = \infty$ should be equal to the pervaporation flux. In other words, the pervaporation flux should be 34 times higher than the reverse osmosis flux corresponding to the operating pressure of 4.134×10^6 Pa (gauge).

Example 9

Greenlaw et al. discussed pervaporation of a single solvent [158]. In Lee's derivation, Equations 5.208 and 5.209 were obtained by integrating Equations 5.206 and 5.207, with the assumptions that the pressure is constant across the membrane and the diffusivities are constant. It is known that the diffusivity is highly dependent on the concentration; therefore, Equations 5.208 and 5.209 can be written only in differential form. Let us now drop subscripts A and B in the equations, since we are dealing with only one component. Then Equation 5.208 can be written as

$$ J dx = -D_m dc_m \tag{5.280} $$

where x is the distance toward the permeant flow direction from the upstream side of the membrane.

The diffusion coefficient, D_m (m^2/s), is given by Rogers et al. [170] as

$$ D_m = D_{m0}(1 + \alpha c_m^n) \tag{5.281} $$

where D_{m0}, α, and n are constants. The equilibrium sorption, on the other hand, is given by the same authors as

$$ c_m = \sigma(p/p_*) + \tau(p/p_*)^m \tag{5.282} $$

where σ, τ, and m are constants, and p and p_* are the vapor pressure (Pa) of the permeant and the saturation vapor pressure (Pa), respectively. Since the ratio p/p_* is equal to the activity of the vapor, Equation 5.282 may be written as

$$ c_m = \sigma a + \tau a^m \tag{5.283} $$

Substituting Equation 5.281 for D_m of Equation 5.280 and integrating, we obtain

$$ \int_0^\delta J\, dx = -\int_{c_{2m}}^{c_{3m}} D_{m0}(1 + \alpha c^n)\, dc_m \tag{5.284} $$

Integration of the above equation yields

$$ J\delta = D_{m0}(c_{2m} - c_{3m}) + D_{m0}\frac{\alpha}{n+1}\left(c_{2m}^{n+1} - c_{3m}^{n+1}\right) \tag{5.285} $$

The concentrations c_{2m} and c_{3m} will be given as the function of upstream and downstream pressures in the following.

1. For the pure liquid stream at pressure p_2, assuming constant molar volume v, the chemical potential becomes

$$\mu_2 = \mu^* + v(p_2 - p_*) \qquad (5.286)$$

where μ^* is the chemical potential of the pure liquid at the saturation vapor pressure.

2. For the dissolved permeant at the upstream face,

$$\mu_2 = \mu^* + v(p_2 - p_*) + RT \ln a_2 \qquad (5.287)$$

assuming that the molar volume of the permeant in the membrane is the same as that of the liquid.

3. For the dissolved permeant at the downstream face,

$$\mu_3 = \mu^* + v(p_2 - p_*) + RT \ln a_3 \qquad (5.288)$$

since an assumption was made that the pressure remains constant and is p_2 across the membrane.

4. For pure liquid in contact with the downstream face,

$$\mu_3 = \mu^* + v(p_3 - p_*) \qquad (5.289)$$

5. For pure vapor in contact with the downstream face, assuming perfect gas behavior,

$$\mu_3 = \mu^* + RT \ln(p_3/p_*) \qquad (5.290)$$

Equating Equations 5.286 and 5.287, we find that the upstream activity must be unity; therefore,

$$a_2 = 1 \qquad (5.291)$$

Equating Equations 5.288 and 5.289, the downstream activity a_3 is written as

$$a_3 = \exp\left[-\frac{v}{RT}(p_2 - p_3)\right] \qquad (5.292)$$

192   MEMBRANE TRANSPORT/SOLUTION-DIFFUSION MODEL

when $p_3 > p_*$ and the permeate is in liquid phase. Equating Equations 5.288 and 5.290, the downstream activity a_3 is written as

$$a_3 = \frac{p_3}{p_*} \exp\left[-\frac{v}{RT}(p_2 - p_*)\right] \qquad (5.293)$$

when $p_3 < p_*$ and the permeate is vapor. Using Equation 5.283, Equation 5.285, Equation 5.291, Equation 5.292, or Equation 5.293, we can calculate the permeate flux J.

Example 10

For the permeation of hexane through polyethylene film, the following numerical parameters are known.

$$
\begin{aligned}
D_{m0} &= 24.2 \times 10^{-13} \text{ m}^2/\text{s} \\
\alpha &= 0.001788 \\
n &= 1.33 \\
m &= 3.4 \\
\tau &= 545.3 \text{ mol/m}^3 \\
\sigma &= 137.9 \text{ mol/m}^3 \\
p_* &= 24{,}918 \text{ Pa} \\
v &= 130.58 \times 10^{-6} \text{ m}^3/\text{mol}
\end{aligned}
$$

Calculate the permeation flux of hexane when the film thickness is 2.54×10^{-5} m. The upstream pressure is 101,325 Pa (atmospheric pressure), and the downstream pressure is 40,000 Pa.

From Equation 5.291, $a_2 = 1$; therefore, from Equation 5.283,

$$c_{2m} = 137.9 \times 1 + 545.3 \times 1 = 683.2 \text{ mol/m}^3$$

From Equation 5.292,

$$
\begin{aligned}
a_3 &= \exp\left[-\frac{130.58 \times 10^{-6}}{2.479 \times 10^3}(1.01325 \times 10^5 - 0.4 \times 10^5)\right] = 0.9968 \\
c_{3m} &= 137.9 \times 0.9968 + 545.3 \times 0.9968^{3.4} = 676.9
\end{aligned}
$$

Then from Equation 5.285, the flux (mol/m² s) becomes

$$
\begin{aligned}
J = &\left\{(24.2 \times 10^{-13})(683.2 - 676.9) + \right.\\
&\left. \frac{(24.2 \times 10^{-13})(0.001788)}{2.33}(683.2^{2.33} - 676.9^{2.33})\right\} \Big/ 2.54 \times 10^{-5}
\end{aligned}
$$

$$= 6.322 \times 10^{-6}$$

Example 11

Calculate the flux when the downstream pressure is 0 Pa and 13,332 Pa.

At the downstream pressure of 0 Pa, $a_3 = 0$ and $c_3 = 0$; therefore,

$$J = \frac{(24.2 \times 10^{-13})(683.2) + \frac{(24.2 \times 10^{-13})(0.001788)}{2.33}(683.2^{2.33})}{2.54 \times 10^{-5}}$$

$$= 35.92 \times 10^{-5}$$

At the downstream pressure of 13,332 Pa, using Equation 5.293,

$$a_3 = \frac{13332}{24918} \exp\left(-\frac{130.58 \times 10^{-6}}{2.479 \times 10^3}(10,1325 - 24,918)\right) = 0.5329$$

$$c_{3m} = 137.9 \times 0.5329 + 545.3 \times 0.5329^{3.4} = 137.6$$

Therefore,

$$J = \frac{24.2 \times 10^{-13}(683.2 - 137.6)}{2.54 \times 10^{-5}} +$$

$$\frac{\{[(24.2 \times 10^{-13})(0.001788)]/(2.33)\}(683.2^{2.33} - 137.6^{2.33})}{2.54 \times 10^{-5}}$$

$$= 33.91 \times 10^{-5}$$

Example 12

Calculate the pervaporation flux when the upstream pressure is 1,205,768 Pa and the downstream pressure is 40,000, 0, and 13,332 Pa.

At the downstream pressure of 40,000 Pa,

$$a_3 = \exp\left[-\frac{130.58 \times 10^{-6}}{2.479 \times 10^3} / (1,205,800 - 40,000)\right] = 0.9404$$

$$c_{3m} = 137.9 \times 0.9404 + 545.3 \times 0.9404^{3.4} = 572.2$$

Therefore,

$$J = \frac{24.2 \times 10^{-13}(683.2 - 572.2) +}{2.54 \times 10^{-5}}$$

$$\frac{[(24.2 \times 10^{-13})0.001788]/2.33(683.2^{2.33} - 572.2^{2.33})}{2.54 \times 10^{-5}}$$

$$= 11.01 \times 10^{-5}$$

At the downstream pressure of 0 Pa,

$$J = 35.92 \times 10^{-5}$$

At the downstream pressure of 13,332 Pa,

$$a_3 = \tfrac{13332}{24918} \exp(-\tfrac{130.58 \times 10^{-6}}{2.479 \times 10^3}(1,205,800 - 24,918)) = 0.5028$$
$$c_{3m} = 137.9 \times 0.5028 + 545.3 \times 0.5028^{3.4} = 122.0$$

Therefore,

$$J = \frac{24.2 \times 10^{-13}(683.2 - 122.0)}{2.54 \times 10^{-5}} +$$
$$\frac{\frac{(24.2 \times 10^{-13})(0.001788)}{2.33}(683.2^{2.33} - 122.0^{2.33})}{2.54 \times 10^{-5}}$$
$$= 34.23 \times 10^{-5}$$

Example 13
Calculate the permeation flux when the upstream pressure is infinity and the downstream pressure is 40,000 Pa.

Since $p_2 = \infty$ in Equation 5.292, $a_3 = 0$ and $c_{3m} = 0$. Then

$$J = 35.92 \times 10^{-5} \text{ mol/m}^2 \text{ s}$$

Example 14
Develop a transport equation for pervaporation of a single component, considering the change in the chemical potential of an imaginary pure liquid (or pure vapor) phase of the permeant, which is in equilibrium with the permeant in the membrane.

The flux through a membrane segment of thickness dx is written in analogy to Equation 5.16. The equation is written in a differential form, and subscript B is dropped, since we are dealing with pervaporation of a single-component system. Then

$$J = -\frac{c_m}{f_m}\frac{d\mu}{dx} \tag{5.294}$$

where c_m and f_m are the permeant concentration in the membrane (mol/m^3) and the friction (J s/m^2 mol) to the movement of the permeant in the membrane.

At the upstream face of the membrane, the permeant in the membrane is in equilibrium with the feed liquid under the upstream pressure, p_2. Therefore,

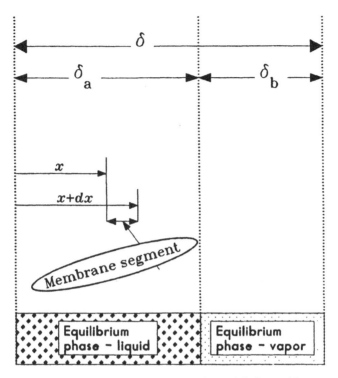

Figure 5.6. Imaginary liquid or vapor phase in equilibrium with permeant in the membrane phase.

the chemical potential of the permeant in the membrane is equal to that of the feed liquid, which can be written as

$$\mu = \mu^* + v(p_2 - p_*) \qquad (5.295)$$

where μ^* is the chemical potential of liquid at the saturation vapor pressure.

At a distance x from the upstream interface, a membrane segment whose thickness is dx is considered (see Figure 5.6). It would be possible to assume the presence of an imaginary pure liquid phase that is in equilibrium with the permeant in the membrane segment when x is not large. The pressure of the imaginary liquid phase changes from p to $p + dp$ as the distance increases from x to dx. Of course, dp should have a negative sign. Since the molar volume of the permeant is constant, the change in the chemical potential of the permeant from distance x to $x + dx$ should be

$$d\mu = vdp \qquad (5.296)$$

in the imaginary liquid phase, and so it is also in the membrane phase. The pressure of the imaginary liquid phase continues to decrease as the distance x increases; at a distance δ_a, the pressure becomes the saturation vapor pressure. The pressure in the membrane at a distance x is assumed to be the same as that of the imaginary phase at a distance x.

Assuming the concentration of the permeant in the membrane, c_m, to be constant from $x = 0$ to $x = \delta_a$ (Lonsdale made the same assumption when he developed his reverse osmosis transport equations), Equation 5.294 can be integrated into

$$J \int_0^{\delta_a} dx = -\frac{c_m}{f_m} v \int_{p_2}^{p_*} dp \qquad (5.297)$$

Therefore,

$$J = \frac{c_m}{f_m} v \frac{(p_2 - p_*)}{\delta_a} \qquad (5.298)$$

An assumption was made in the above integration that f_m is also unchanged throughout the integration. Using Equation 5.283 in the foregoing illustration,

$$c_m = \sigma + \tau \qquad (5.299)$$

since the activity $a_2 = 1$ when the permeate in the membrane is in equilibrium with the imaginary liquid phase of a single component and at the same pressure as in the imaginary phase. Then

$$D_m = D_{m0}[1 + \alpha(\sigma + \tau)^n] \qquad (5.300)$$

Since

$$f_m = \frac{RT}{D_m} \qquad (5.301)$$

the equation for the pervaporation flux becomes

$$J = \frac{(D_{m0}/RT)(\sigma + \tau)[1 + \alpha(\sigma + \tau)^n]v}{\delta_a}(p_2 - p_*) \qquad (5.302)$$

From the distance $x = \delta_a$ to $x = \delta$, the imaginary phase is considered to be in the vapor phase (see Figure 5.6), and the pressure changes from p_* to p_3. The change in the chemical potential in a small segment dx can be written as

$$d\mu = \frac{RT}{p} dp \qquad (5.303)$$

assuming the ideal gas law. Then the permeation flux becomes

$$J = -\frac{c_m}{f_m}\frac{d\mu}{dx} \tag{5.304}$$

$$= -\frac{c_m}{f_m}\frac{RT}{p}\frac{dp}{dx} \tag{5.305}$$

Rearranging,

$$J dx = -\frac{c_m RT}{f_m}\frac{dp}{p} \tag{5.306}$$

When the permeate concentration, c_m, is given as a function of vapor pressure by Equation 5.283, and Equation 5.301 is substituted for f_m,

$$J dx = -\frac{[\sigma(p/p_*) + \tau(p/p_*)^m]RT}{[RT/D_{m0}(1 + \alpha(\sigma(p/p_*) + \tau(p/p_*)^m)^n)]}\frac{d(p/p_*)}{(p/p_*)} \tag{5.307}$$

Rearranging,

$$J dx = -\frac{D_{m0}[\sigma(p/p_*) + \tau(p/p_*)^m][1 + \alpha(\sigma(p/p_*) + \tau(p/p_*)^m)^n]}{(p/p_*)}d(p/p_*) \tag{5.308}$$

Setting $\zeta = p/p_*$,

$$J dx = -\frac{D_{m0}(\sigma\zeta + \tau\zeta^m)[1 + \alpha(\sigma\zeta + \tau\zeta^m)^n]}{\zeta}d\zeta \tag{5.309}$$

Integrating from $x = \delta_a$ to $x = \delta$,

$$J\int_{\delta_a}^{\delta} dx = -\int_{p_*/p_*}^{\zeta} \frac{D_{m0}(\sigma\zeta + \tau\zeta^m)[1 + \alpha(\sigma\zeta + \tau\zeta^m)^n]}{\zeta}d\zeta \tag{5.310}$$

and

$$J = \frac{\int_{\zeta}^{1} \frac{D_{m0}(\sigma\zeta + \tau\zeta^m)[1 + \alpha(\sigma\zeta + \tau\zeta^m)^n]}{\zeta}d\zeta}{\delta_b} \tag{5.311}$$

where $\delta_b = \delta - \delta_a$.

Furthermore, from Equation 5.302,

$$\delta_a = \frac{(D_{m0}/RT)(\sigma + \tau)[1 + \alpha(\sigma + \tau)^n]v}{J}(p_2 - p_*) \tag{5.312}$$

and from Equation 5.311,

$$\delta_b = \frac{\displaystyle\int_\zeta^1 \frac{D_{m0}(\sigma\zeta + \tau\zeta^m)[1 + \alpha(\sigma\zeta + \tau\zeta^m)^n]}{\zeta}\,d\zeta}{J} \tag{5.313}$$

Since $\delta_a + \delta_b = \delta$,

$$\delta = \frac{(D_{m0}/\mathbf{R}T)(\sigma + \tau)[1 + \alpha(\sigma + \tau)^n]v}{J}(p_2 - p_*) + \frac{\displaystyle\int_\zeta^1 \frac{D_{m0}(\sigma\zeta + \tau\zeta^m)[1 + \alpha(\sigma\zeta + \tau\zeta^m)^n]}{\zeta}\,d\zeta}{J} \tag{5.314}$$

Therefore,

$$J = \frac{(D_{m0}/\mathbf{R}T)(\sigma + \tau)[1 + \alpha(\sigma + \tau)^n]v}{\delta}(p_2 - p_*) + \frac{\displaystyle\int_\zeta^1 \frac{D_{m0}(\sigma\zeta + \tau\zeta^m)[1 + \alpha(\sigma\zeta + \tau\zeta^m)^n]}{\zeta}\,d\zeta}{\delta} \tag{5.315}$$

The above equation is valid when $p_3 < p_*$. When $p_3 > p_*$, the imaginary equilibrium phase is liquid across the membrane, and the flux should be written as

$$J = \frac{(D_{m0}/\mathbf{R}T)(\sigma + \tau)[1 + \alpha(\sigma + \tau)^n]v}{\delta}(p_2 - p_3) \tag{5.316}$$

Equations 5.315 and 5.316 explicitly give the effect of the upstream pressure and the downstream pressure on the pervaporation rate. Note that this approach is closely related to Lonsdales's approach for reverse osmosis transport, and allows for the pressure change across the membrane.

Example 15
Use the same numerical values as those given in Example 10 and calculate the pervaporation flux when the upstream pressure is 101,325 Pa and the downstream pressure is 0, 12,466, 19,934, and 40,000 Pa.

For the downstream pressure of 0 Pa, Equation 5.315 is used. Since $0 < 24,918$, the flux (mol/m^2 s) becomes

$$
\begin{aligned}
J = \{&(24.2 \times 10^{-13}/2.479 \times 10^3)(137.9 + 545.3) \\
&[1 + 0.001788 \times (137.9 + 545.3)^{1.33}]\,(130.58 \times 10^{-6})\}\,/2.54 \times 10^{-5} \\
&\times (101,325 - 24,918) +
\end{aligned}
$$

$$\left(\frac{24.2 \times 10^{-13}}{2.54 \times 10^{-5}}\right) \cdot$$

$$\int_0^\zeta \frac{(137.9\zeta + 545.3\zeta^{3.4})[1 + (0.001788)(137.9\zeta + 545.3\zeta^{3.4})^{1.33}]}{\zeta} d\zeta$$

$$= 13.843 \times 10^{-5}$$

Similarly, for the downstream pressure of 12,466 and 19,934 Pa, the pervaporation fluxes are 12.753×10^{-5} and 9.295×10^{-5} mol/m^2 s, respectively.

When the downstream pressure is 40,000 Pa, $40,000 > 24,918$. Therefore, from Equation 5.316,

$$J = \{(24.2 \times 10^{-13}/2.479 \times 10^3)(137.9 + 545.3)$$
$$[1 + 0.001788 \times (137.9 + 545.3)^{1.33}](130.58 \times 10^{-6})\}/2.54 \times 10^{-5}$$
$$\times(101,325 - 40,000)$$
$$= 0.243 \times 10^{-5}$$

Example 16
Calculate pervaporation fluxes when the upstream pressure is 1.2058×10^6 Pa for the downstream pressures of 0, 12,466, 19,934 and 40,000 Pa.
The answers are

Downstream pressure, Pa	Flux, mol/m^2 s
0	18.226×10^{-5}
12,466	17.136×10^{-5}
19,934	13.678×10^{-5}
40,000	4.626×10^{-5}

Note that the pervaporation flux at zero downstream pressure increases with an increase in the upstream pressure, whereas by using the model of Greenlaw et al., the upstream pressure has no effect.

5.5   Design of Composite Membranes for Gas Separation/Resistance Theory

The composite membrane is defined as a membrane that consists of several barrier layers of distinct nature stacked together. There is a clear discontinuity at the boundary of two neighboring barrier layers, either in the chemical structure or in the morphology of the material of which the barrier layer consists. There are several different methods to prepare composite membranes, as already discussed in Chapter 3.

Henis and Tripodi developed composite membranes for gas separation, using the resistance model as a theoretical guiding principle [171]. According to the model, the resistance from each component barrier layer to the permeant flow is combined mathematically, following the law of electrical circuit [171]. This enables us to obtain the overall resistance of a composite membrane against the flow of a permeant through the membrane. In order to show the analogy between the membrane permeation and the electrical circuit, we adopt the sorption-diffusion model, since this model relates the permeation rate to the driving force, which may be either the concentration difference or the pressure difference across the membrane, in linear fashion.

The permeation rate for a component i through a membrane can be written as

$$Q_i = -P_i A \Delta p_i / \delta \qquad (5.317)$$

where Q_i is the permeation rate of the component i (mol/s), P_i the permeability coefficient of the membrane (mol·m/m²·s·Pa), A the surface area of the membrane (m²), and δ the thickness of the membrane (m). Considering an electric circuit, the following Ohm's law describes the current, I, flowing through a resistance, R, under an electrical potential difference, E,

$$I = E/R \qquad (5.318)$$

Conceptually, the current, I, is analogous to the permeation rate, Q_i, and the electrical potential difference, E, to the driving force, $-\Delta p$. Then equating Equations 5.317 and 5.318, we can define resistance for the membrane permeation as

$$R_i = \delta / P_i A \qquad (5.319)$$

Accordingly, Equation 5.317 can be written as

$$Q_i = -\Delta p_i / R_i \qquad (5.320)$$

As long as Equation 5.317 is valid, the permeation rate can be given by Equation 5.320, and all the theories developed for the electric circuit become valid. It should be noted that Equation 5.320 is not always valid. As will be shown later, the gas permeation rate is not always related to the pressure difference in a linear fashion. Similarly, the permeation rate of a component of a liquid mixture is not necessarily linear to the concentration difference. Therefore, the validity of the resistance model suffers severe restrictions under those circumstances. Nevertheless, the model offers a practical guideline to the design of the composite membrane. It should also be noted that the model is valid when the permeation rate of the fluid component i at every barrier layer is given in a form

$$Q_i = -\Delta f(p_i)/R_i \qquad (5.321)$$

where $f(p_i)$ is some functional form of an independent variable p_i and common to every barrier layer. The linear relationship between Q_i and p_i is one of many possible functional relationships.

Case 1: Two Resistances Connected in Series

Imagine two layers with resistances R_1 and R_2 combined in series, 1 on the top of 2. Denoting two different permeants by subscripts a and b, the overall resistance R for each permeant is written as

$$(R)_a = (R_1)_a + (R_2)_a \qquad (5.322)$$
$$(R)_b = (R_1)_b + (R_2)_b \qquad (5.323)$$

Defining the ratio of the resistances for permeants a and b as

$$\alpha_i = \frac{(R_i)_b}{(R_i)_a}$$
$$= \frac{(P_i)_a}{(P_i)_b} \qquad (5.324)$$

where $i = 1$ and 2. The ratio of the overall resistance for permeants a and b, designated as α, becomes

$$\alpha = \frac{(R)_b}{(R)_a} \qquad (5.325)$$
$$= \frac{\alpha_1(R_1)_a + \alpha_2(R_2)_a}{(R_1)_a + (R_2)_a} \qquad (5.326)$$
$$= \frac{\alpha_1 + \alpha_2[(R_2)_a/(R_1)_a]}{1 + [(R_2)_a/(R_1)_a]} \qquad (5.327)$$

When $[(R_2)_a/(R_1)_a] \simeq 0$,

$$\alpha \simeq \alpha_1 \qquad (5.328)$$

The above equation indicates that the selectivity of the composite membrane is controlled by a barrier component whose resistance is far greater than the other. This is the principle underlying the development of the composite reverse osmosis membrane where an active surface layer is supported by a porous support layer (Figure 5.7). The resistance to the permeant flow is contributed almost

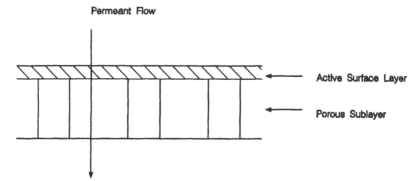

Figure 5.7. Resistances connected in series in composite membranes.

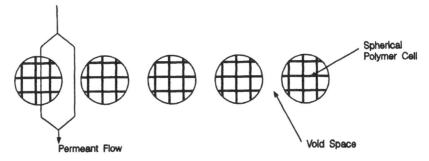

Figure 5.8. Resistances connected in parallel in a porous skin layer.

completely from the top active layer because of its high polymer density, even though this layer is much thinner than the porous support layer. Therefore, the selectivity is controlled by the active surface layer. The above theory teaches us, however, that both layers will equally control the selectivity when the resistances from both layers are nearly equal. Therefore, even when an extremely selective top surface layer is prepared, the overall selectivity could be lowered if the support layer is thick and its pore size is relatively small.

Case 2: Two Resistances Connected in Parallel

Imagine a membrane where two distinctive components of the membrane are combined parallel to each other. The gas flow through a membrane with a structure described in Figure 5.8 is an example. The network of the polymeric molecule in a spherical polymer cell is one component, and the void space generated between the spheres is the other; they are combined in parallel. Designating the resistance to the flow through the network of the polymer molecule as R_2, the resistance to the flow through the void space as R_3, and the two components of a gas mixture to be separated as a and b, the overall resistances for a and b can be written as

$$R_a = \frac{(R_2)_a \times (R_3)_a}{(R_2)_a + (R_3)_a} \tag{5.329}$$

$$R_b = \alpha R_a = \frac{\alpha_2(R_2)_a \times \alpha_3(R_3)_a}{\alpha_2(R_2)_a + \alpha_3(R_3)_a} \tag{5.330}$$

Division of Equation 5.330 by Equation 5.329 yields

$$\alpha = \alpha_2\alpha_3 \frac{(R_2)_a + (R_3)_a}{\alpha_2(R_2)_a + \alpha_3(R_3)_a} \tag{5.331}$$

$$= \alpha_2\alpha_3 \frac{1 + \left[(R_3)_a/(R_2)_a\right]}{\alpha_2 + \alpha_3\left[(R_3)_a/(R_2)_a\right]} \tag{5.332}$$

Since R_2 represents the resistance of the network component and R_3 represents the resistance from the void space component of the membrane, $R_3/R_2 \simeq 0$. Then

$$\alpha \simeq \alpha_3 \tag{5.333}$$

The overall selectivity is therefore controlled by the membrane component with a smaller resistance. The gas transport of a membrane with a structure described in Figure 5.8 is mostly through the void space between spherical polymer cells, and therefore the selectivity should be extremely low. It should be noted, however, that the resistance given by Equation 5.319 is inversely proportional to the area, A. Therefore, if the area occupied by the void space is very small (in other words, if the void spaces are eliminated by the deformation of the spherical polymer cells, as shown by an electron micrographic picture in Figure 4.7 of Chapter 4), the selectivity may become high, since it is governed by the network component of the membrane.

Case 3: Parallel Combination of Two Resistance Arms
Imagine a homogeneous film of relatively high permeability laminated on the top of a porous sublayer. The structure of such bilayer membranes can be described by Figure 5.9. The difference between Figure 5.7 and Figure 5.9 is that in the former membrane the top layer is very selective and its resistance is much higher than that of the porous sublayer. In the latter membrane, on the other hand, the selectivity of the top layer is not high and its resistance is lower than that of the porous sublayer. It is also assumed that the gas flow is strictly vertical to the surface of the membrane. In other words, the gas flow through a portion of the homogeneous film that covers the polymer network should pass through the polymer network, and the gas flow through the portion

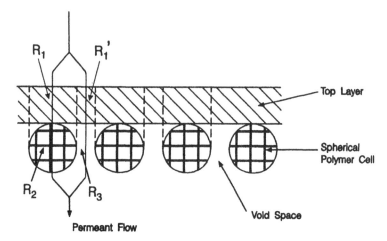

Figure 5.9. Homogeneous membrane laminated on the top of a porous substrate membrane. (Two series resistances are connected in parallel.)

of the homogeneous film that covers the void space should pass through the void space. No mixing is allowed between these two flows. The overall resistance for the component gas a and b becomes

$$R_a = \frac{[(R_1)_a + (R_2)_a] \times [(R'_1)_a + (R_3)_a]}{[(R_1)_a + (R_2)_a + (R'_1)_a + (R_3)_a]} \tag{5.334}$$

and

$$
\begin{aligned}
R_b &= \alpha R_a \\
&= \frac{[\alpha_1(R_1)_a + \alpha_2(R_2)_a] \times [\alpha_1(R'_1)_a + \alpha_3(R_3)_a]}{[\alpha_1(R_1)_a + \alpha_2(R_2)_a + \alpha_1(R'_1)_a + \alpha_3(R_3)_a]}
\end{aligned} \tag{5.335}
$$

The meanings of resistances R_1, R'_1, R_2, and R_3 should be clear from Figure 5.9. Comparing two arms of the parallel resistances, $R_1 + R_2$ and $R'_1 + R_3$, the former arm will govern the overall resistance when

$$R'_1 + R_3 \gg R_1 + R_2 \tag{5.336}$$

In the arm $R_1 + R_2$, R_2 will govern the overall resistance of this series resistance when

$$R_2 \gg R_1 \tag{5.337}$$

In other words, the resistance R_2, which corresponds to the resistance of the polymer network in the spherical polymer cell, can govern the overall resis-

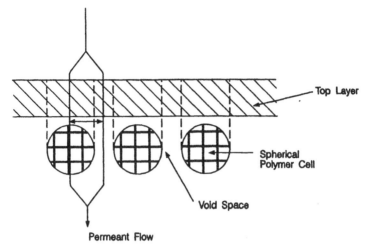

Figure 5.10.   Two arms of parallel resistances connected by a cross-flow.

tance of the composite membrane consisting of two laminated layers when Equations 5.336 and 5.337 are satisfied simultaneously. This means that the gas leakage through the void space between the spherical polymer cells can be stopped by the lamination of the top layer. That Equation 5.337 can be satisfied is quite obvious. Equation 5.336 can also be satisfied when the area occupied by the void space is very small.

Case 4: Wheatstone Bridge Model — Introduction of Gas Mixing Between Two Resistance Arms [172]

When a flow parallel to the membrane surface occurs, the gas flows through two arms of the resistance described by Figure 5.9 will be connected (see Figure 5.10). The electric circuit analog is then given by Figure 5.11, which is exactly the same as that of the Wheatstone bridge. The overall resistance of such an electric circuit is

$$R = \frac{R_x \times (R_1 + R_2)(R'_1 + R_3) + R_1 R'_1(R_2 + R_3) + R_2 R_3(R_1 + R'_1)}{R_x \times (R_1 + R_2 + R'_1 + R_3) + (R_1 + R'_1)(R_2 + R_3)} \quad (5.338)$$

where R_x is the resistance corresponding to the gas flow connecting two resistance arms. When R_x approaches infinity, the Wheatstone bridge model becomes equal to the parallel series model, and Equation 5.338 becomes Equation 5.334. It has to be recalled that Equation 5.338 was written for the gas component a.

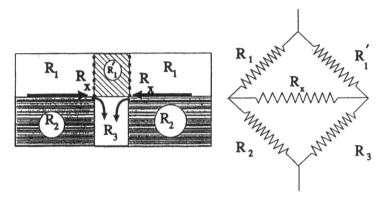

Figure 5.11. Wheatstone bridge model. (Reproduced from [172] with permission.)

When R_x becomes zero, the overall resistance is given by

$$R = \frac{R_1 \times R_1'}{(R_1 + R_1')} + \frac{R_2 \times R_3}{(R_2 + R_3)} \tag{5.339}$$

Since the resistances R_1 and R_1' are two parts of the same homogeneous film, one on the top of the polymer network and the other on the top of the void space connected in parallel, the overall resistance for the homogeneous film, R_{hom}, can be written as

$$R_{hom} = \frac{R_1 \times R_1'}{(R_1 + R_1')} \tag{5.340}$$

Then Equation 5.339 becomes

$$R = R_{hom} + \frac{R_2 \times R_3}{(R_2 + R_3)} \tag{5.341}$$

The above equation is considered to be a circuit formed by three component resistances, as illustrated in Figure 5.12. In other words, a homogeneous film is connected in series to a parallel resistance circuit of R_2 and R_3. This model is exactly the same as that proposed by Henis and Tripodi when their Prism membrane was developed [173], [174]. Let us now examine how the overall resistance and the selectivity of the Wheatstone bridge changes as the component resistance for mixing, R_x, changes. Figure 5.13 shows that both the overall resistance and selectivity decrease with a decrease in R_x. When the contribution of the cross-flow between two arms increases with a decrease in R_x, more gas flows into the void space, lowering the overall resistance. The selectivity is also

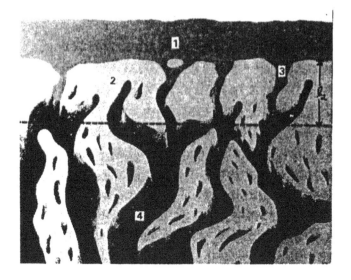

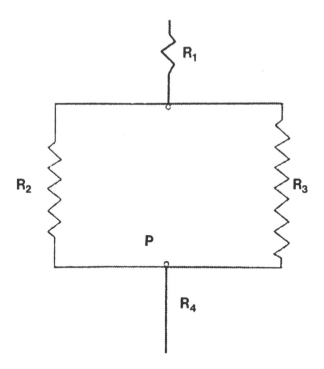

Figure 5.12.   Three component resistances involved in Hennis-Tripodi's model. (Reproduced from [174] with permission)

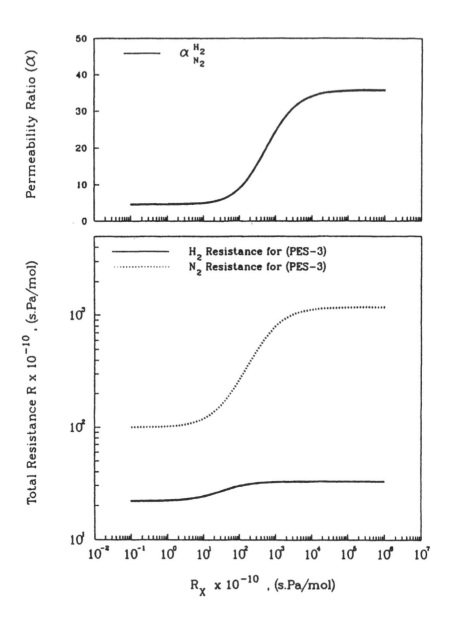

Figure 5.13. Overall resistance of Wheatstone bridge as a function of the cross-flow resistance. (Reproduced from [172] with permission.)

lowered, since the selectivity of the void space is low. The resistance model has been further developed in the literature [31],[175],[176].

Since the Henis and Tripodi's resistance model led to the development of the Prism hollow fiber, their model is examined in detail in the following example.

Example 17

Suppose there is a porous substrate with an area ratio of the void space and the polymer matrix, A_3/A_2, of 1.9×10^{-6} and an effective film thickness of 10^{-7} m (100 nm). The material of the membrane is a polysulfone polymer with permeability coefficients for H_2 and CO of 4.019×10^{-13} mol m/m^2 s Pa $(1.2 \times 10^{-9}$ cc[STP]·cm/cm^2·s·cmHg) and 10.047×10^{-15} mol m/m^2 s Pa $(3.0 \times 10^{-11}$ cc[STP]·cm/cm^2·s·cmHg), respectively. This substrate is coated with a silicone rubber polymer that exhibits permeability coefficients for H_2 and CO of 17.415×10^{-12} mol m/m^2 s Pa $(5.2 \times 10^{-8}$ cc[STP]·cm/cm^2·s·cmHg) and 8.373 $\times 10^{-12}$ mol m/m^2 s Pa $(2.5 \times 10^{-8}$ cc[STP]·cm/cm^2·s·cmHg), respectively. The coating thickness is 10^{-6} m (1 μm). Let us calculate the overall resistance of the composite membrane for each component of the gaseous mixture. Assume that the void space is filled with silicone rubber after the coating is applied. According to Equation 5.319,

$$(R_{hom})_{H_2} = (10^{-6})/(17.415 \times 10^{-12})A_{hom}$$
$$= 0.05742 \times 10^6/A_{hom} \qquad (5.342)$$
$$(R_{hom})_{CO} = (10^{-6})/(8.373 \times 10^{-12})A_{hom}$$
$$= 0.1194 \times 10^6/A_{hom} \qquad (5.343)$$

$$(R_2)_{H_2} = (10^{-7})/(4.019 \times 10^{-13})A_2$$
$$= 0.2488 \times 10^6/A_2 \qquad (5.344)$$
$$(R_2)_{CO} = (10^{-7})/(10.047 \times 10^{-15})A_2$$
$$= 0.09953 \times 10^8/A_2 \qquad (5.345)$$

$$(R_3)_{H_2} = (10^{-7})/(17.415 \times 10^{-12})A_3$$
$$= 0.05742 \times 10^5/A_3 \qquad (5.346)$$
$$(R_3)_{CO} = (10^{-7})/(8.373 \times 10^{-12})A_3$$
$$= 0.1194 \times 10^5/A_3 \qquad (5.347)$$

From Equation 5.341,

$$(R)_{H_2} = (0.05742 \times 10^6/A_{hom}) + \frac{(0.2488 \times 10^6/A_2)(0.05742 \times 10^5/A_3)}{(0.2488 \times 10^6/A_2) + (0.05742 \times 10^5/A_3)} \tag{5.348}$$

$$(R)_{CO} = (0.1194 \times 10^6/A_{hom}) + \frac{(0.09953 \times 10^8/A_2)(0.1194 \times 10^5/A_3)}{(0.09953 \times 10^8/A_2) + (0.1194 \times 10^5/A_3)} \tag{5.349}$$

Assuming $A_{hom} \simeq A_2$, and since $A_3 = 1.9 \times 10^{-6}A_2$, the permeability ratio α, defined as H_2 permeability/CO permeability, becomes

$$1/\alpha = \frac{\left[(0.05742 \times 10^6) + \frac{(0.2488 \times 10^6)(0.05742 \times 10^5/1.9 \times 10^{-6})}{(0.2488 \times 10^6) + (0.05742 \times 10^5/1.9 \times 10^{-6})}\right]}{\left[(0.1194 \times 10^6) + \frac{(0.09953 \times 10^8)(0.1194 \times 10^6/1.9 \times 10^{-6})}{(0.09953 \times 10^8) + (0.1194 \times 10^6/1.9 \times 10^{-6})}\right]} \tag{5.350}$$

$$1/\alpha = 0.03041 \tag{5.351}$$

Therefore,

$$\alpha = 33 \tag{5.352}$$

Note that the ratio of the permeability coefficient of H_2 and CO by the polysulfone polymer matrix is $4.019 \times 10^{-13}/10.047 \times 10^{-15} = 40$; therefore, the selectivity of the composite membrane is close to that of the polysulfone polymer matrix. It should be also noted that two important assumptions were made in the above calculation. They are (1) the permeability coefficient of the void space was assumed to be the same as that of silicone rubber, and (2) the effective thickness of the void space was assumed to be the same as that of the polymer network. This implies that the void space was completely filled with silicone rubber, and its effective thickness is the same as that of the polymer network. Furthermore, we can calculate the loss in permeability by coating. The resistance without coating is that of the porous substrate, and therefore, it is designated as R_{sub}. With respect to hydrogen gas, it becomes

$$(R_{sub})_{H_2} = \frac{(0.2488 \times 10^6/A_2)(0.05742 \times 10^5/A_3)}{(0.2488 \times 10^6/A_2) + (0.05742 \times 10^5/A_3)} \tag{5.353}$$

The flux ratio of the coated membrane to that of the substrate membrane, which is equal to $(R_{sub})_{H_2}/(R)_{H_2}$, then becomes, from Equations 5.348 and 5.353,

$$(R_{sub})_{H_2}/(R)_{H_2} = 0.8125 \tag{5.354}$$

In other words, about 20% of the hydrogen flux was sacrificed by silicone coating. Thus, coating the surface of the porous sublayer and blocking the void space with silicone material enables us to bring the selectivity of the composite membrane close to the value expected from the intrinsic selectivity of the polymer network, without sacrificing much of the hydrogen flux. When the coating thickness is increased, however, the resistance of the silicone layer increases, and consequently its contribution to the overall resistance increases. Since the silicone layer exhibits a low selectivity, an increase in its contribution in the series circuit (described in Figure 5.12) means a decrease in the overall selectivity. The effect of the silicone layer thickness was discussed by Henis and Tripodi and was experimentally confirmed [31]. Figure 5.14 shows that the initial increase in the selectivity with the number of coatings is followed by a decrease in the selectivity. The selectivity idecrease is due to a greater contribution of the silicone rubber layer to the overall selectivity.

Example 18

The following quantities are known from the gas permeation experiments with silicone rubber film, porous polyethersulfone substrate membrane, and silicone rubber/polyethersulfone laminated membrane:

· $(P_1/\delta)_{H_2}$: Hydrogen permeability of silicone rubber film, 80.18×10^{-10} mol/m^2 s Pa

· $(P_{sub}/\delta)_{H_2}$: Hydrogen permeability of substrate membrane, 112.2×10^{-10} mol/m^2 s Pa

· $(P/\delta)_{H_2}$: Hydrogen permeability of laminated membrane, 29.2×10^{-10} mol/m^2 s Pa

· α_1: H$_2$/N$_2$ permeability ratio of silicone rubber film, 2.2

· α_2: H$_2$/N$_2$ permeability ratio of polysulfone polymer matrix, 66.67

· α_{sub} : H$_2$/N$_2$ permeability ratio of the porous substrate membrane, 6.29

· α: H$_2$/N$_2$ permeability ratio of the laminated membrane, 36.9

The total membrane area is 10.18×10^{-4} m^2. Calculate the resistance of each component.

From Equation 5.319,

$$R_i = \frac{1}{(P_i/\delta)A}$$

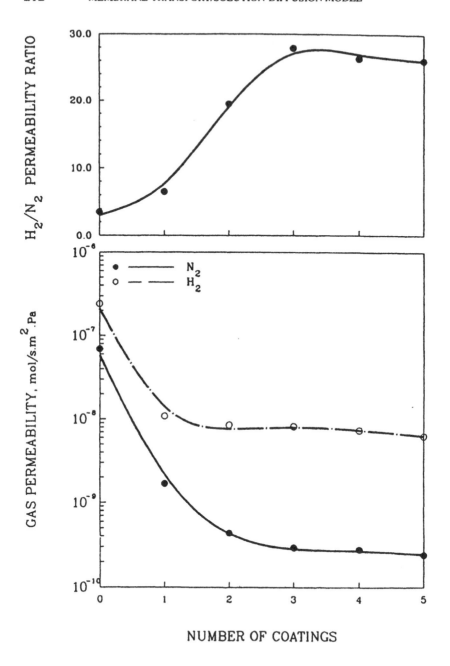

Figure 5.14. Change in the membrane selectivity with an increase in the number of coatings. (Reproduced from [31] with permission.)

Therefore,

$$(R_1)_{H_2} = \frac{1}{(80.18\times 10^{-10})(10.18\times 10^{-4})} = 12.19 \times 10^{10} \text{ s Pa/mol}$$

$$(R_{sub})_{H_2} = \frac{1}{(112.2\times 10^{-10})(10.18\times 10^{-4})} = 8.755 \times 10^{10} \text{ s Pa/mol}$$

$$(R)_{H_2} = \frac{1}{(29.2\times 10^{-10})(10.18\times 10^{-4})} = 33.64 \times 10^{10} \text{ s Pa/mol}$$

The following symbols are used for the unknown resistances: $x = (R_2)_{H_2}$, $y = (R_3)_{H_2}$, and $z = (R'_1)_{H_2}$.

Equations 5.329 and 5.330 are applicable for the overall resistance of the porous substrate membrane. Component a and b are hydrogen and nitrogen, respectively, in the example. Then inserting all numerical values,

$$8.755 \times 10^{10} = \frac{xy}{x+y} \tag{5.355}$$

$$6.29 \times (8.755 \times 10^{10}) = \frac{(66.67x)(\alpha_3 y)}{(66.67x) + (\alpha_3 y)} \tag{5.356}$$

Equations 5.334 and 5.335 are applicable, on the other hand, for the laminated membrane. Inserting all the numerical values,

$$33.64 \times 10^{10} = \frac{[(12.19 \times 10^{10}) + x](z + y)}{(12.19 \times 10^{10}) + x + z + y} \tag{5.357}$$

$$36.9 \times (33.64 \times 10^{10}) =$$
$$\frac{[(2.2 \times 12.19 \times 10^{10}) + 66.67x](2.2z + \alpha_3 y)}{(2.2 \times 12.19 \times 10^{10}) + 66.67x + 2.2z + \alpha_3 y} \tag{5.358}$$

Equations 5.355, 5.356, 5.357, and 5.358 are solved for four unknowns: x, y, z, and α_3. Division of Equation 5.356 by Equation 5.355 yields

$$6.29 = 66.67\alpha_3 \frac{(x+y)}{66.67x + \alpha_3 y} \tag{5.359}$$

First we assume $\alpha_3 = 3$. Then Equation 5.359 becomes

$$6.29 = (66.67 \times 3)\frac{(x+y)}{66.67x + 3y} \tag{5.360}$$

Unknowns x and y can be obtained by solving Equations 5.355 and 5.360 simultaneously. The results are $x = 15.98 \times 10^{10}$ and $y = 19.35 \times 10^{10}$. Then z can be solved from Equation 5.357. The result is -192.59×10^{10}. Obviously, a negative value is unrealistic, meaning that the assumption of $\alpha_3 = 3$ is inappropriate.

Assuming $\alpha_3 = 4$, we obtain $x = 22.61 \times 10^{10}$, $y = 14.29 \times 10^{10}$, and $z = 994.9 \times 10^{10}$. Then inserting all numerical values into the right side of Equation 5.358,

$$
\begin{aligned}
&\{[(2.2 \times 12.19 \times 10^{10}) + (66.67 \times 22.61 \times 10^{10})] \\
&\quad \times [(2.2 \times 994.9 \times 10^{10}) + (4 \times 14.29 \times 10^{10})]\} / \\
&\{(2.2 \times 12.19 \times 10^{10}) + (66.67 \times 22.61 \times 10^{10}) + \\
&\quad (2.2 \times 994.9 \times 10^{10}) + (4 \times 14.29 \times 10^{10})\} \\
&= 911.54 \times 10^{10}
\end{aligned}
$$

The left side of equation, on the other hand, is $36.9 \times 33.64 \times 10^{10} = 1241.32 \times 10^{10}$. The agreement of the right and the left sides of Equation 5.358 is not satisfactory. Assuming $\alpha_3 = 3.9$, we obtain $x = 21.69 \times 10^{10}$, $y = 14.67 \times 10^{10}$, and $z = 4733.8 \times 10^{10}$, respectively. The right side of Equation 5.358 is 1291.25×10^{10}. This is a great improvement, but yet not sufficiently close to the left side.

Assuming $\alpha_3 = 3.91$, we obtain $x = 21.78 \times 10^{10}$, $y = 14.64 \times 10^{10}$, and $z = 3448.2 \times 10^{10}$, respectively. The right side of Equation 5.358 is 1239×10^{10}. This is considered to be sufficiently close to 1241.32×10^{10}. Therefore, the answers are

$$
\begin{aligned}
(R_2)_{H_2} &= 21.78 \times 10^{10} \text{ s Pa/mol} \\
(R_3)_{H_2} &= 14.64 \times 10^{10} \text{ s Pa/mol} \\
(R_1')_{H_2} &= 3448.2 \times 10^{10} \text{ s Pa/mol} \\
(R_2)_{N_2} &= 66.67 \times (R_2)_{H_2} = 1452 \times 10^{10} \text{ s Pa/mol} \\
(R_3)_{N_2} &= 3.91 \times (R_3)_{H_2} = 57.24 \times 10^{10} \text{ s Pa/mol} \\
(R_1')_{N_2} &= 2.2 \times (R_1')_{H_2} = 7586 \times 10^{10} \text{ s Pa/mol}
\end{aligned}
$$

As for H_2 gas,

$$
\begin{aligned}
R_1' + R_3 &= 3448.2 \times 10^{10} + 14.64 \times 10^{10} \\
&= 3462.84 \times 10^{10} \text{ s Pa/mol}
\end{aligned}
\tag{5.361}
$$

$$
\begin{aligned}
R_1 + R_2 &= 12.19 \times 10^{10} + 21.78 \times 10^{10} \\
&= 33.97 \times 10^{10} \text{ s Pa/mol}
\end{aligned}
\tag{5.362}
$$

$R_1' + R_3$ is two orders of magnitude higher than $R_1 + R_2$.

Nomenclature for Chapter 5

5.1 Reverse Osmosis

a = activity, mol/m^3

c = concentration, mol/m^3

D = diffusion coefficient, m^2/s

f' = solute separation defined by Equation 5.26

if = solute separation defined by Equation 5.46

f = friction working between unit mol of permeant and the membrane material, J s/m^2 mol

J = flux

K_A = distribution constant of solute, —

k = mass transfer coefficient in the concentrated boundary layer, m/s

L = phenomenological coefficient

p = pressure, Pa

$\mathbf{R}$ = gas constant, 8.314 J/mol K

T = absolute temperature, K

v = molar volume, m^3/mol, or permeation velocity, m/s

X = force that causes the flux

y = distance toward the flow direction from the interface between the feed bulk solution phase and the concentrated boundary solution phase, m

μ = chemical potential, J/mol

π = osmotic pressure, Pa

Δ = defined as quantities at the low-pressure side of the membrane minus quantities at the high-pressure side of the membrane

δ = membrane thickness, m

δ_{bl} = thickness of the concentrated boundary layer, m

Subscripts

A = quantities concerning solute

B = quantities concerning solvent

i = quantities concerning ith component

j = quantities concerning jth component

m = quantities concerning inside the membrane

1 = quantities concerning the bulk feed solution

<table>
<tr><td>2</td><td>=</td><td>quantities concerning the concentrated boundary layer that is in contact with the high-pressure side of the membrane</td></tr>
<tr><td>3</td><td>=</td><td>quantities concerning the low-pressure side of the membrane</td></tr>
</table>

5.2 Prediction of Reverse Osmosis Membrane Performance

A	=	pure-water permeability constant, mol/m^2 s Pa
B	=	proportionality constant for osmotic pressure, Pa
B_{av}	=	average value of B, Pa
$\ln C^*$	=	constant representing the porous structure of the membrane surface, —
$\ln C^*_{NaCl}$	=	constant representing the porous structure of the membrane surface, —
c	=	total molar concentration of the solution, mol/m^3
c_A	=	solute concentration, mol/m^3
D	=	diffusion coefficient, m^2/s
ΣE_s	=	Taft's steric parameter for the substituent group in the organic molecule, —
f	=	solute separation, —
$\Delta\Delta G/RT$	=	free energy required to bring an ion from the bulk water to the interfacial water, —
J	=	molar flux, mol/m^2 s
K	=	equilibrium constant for the salt, —
$K^{\pm}$	=	Donnan equilibrium constant written for neutral membrane, —
k	=	mass transfer coefficient, m/s
k_{av}	=	average mass transfer coefficient, m/s
n_a	=	number of moles of anion in 1 mol of ionized solute, —
n_c	=	number of moles of cation in 1 mol of ionized solute, —
n_{hy}	=	number of moles of hydrolyzed species resulting from hydrolysis, —
n_{ipa}	=	number of moles of anion involved in 1 mol of the ion pair, —
n_{ipc}	=	number of moles of cation involved in 1 mol of the ion pair, —
PR	=	product permeation rate per effective membrane area, kg/h

p	=	pressure, Pa
R	=	gas constant, 8.314 J/mol K
S	=	effective membrane area, m^2
T	=	absolute temperature, K
v	=	molar volume, m^3/mol, or permeation velocity, m/s
X	=	mole fraction, —
z	=	ionic valence, —
α_D	=	degree of dissociation, —
α_H	=	degree of hydrolysis, —
δ	=	effective thickness of the membrane, m
δ^*	=	constant associated with Taft's steric parameter, —
π	=	osmotic pressure, Pa
ρ^*	=	constant associated with Taft's polar parameter,
$\Sigma\sigma^*$	=	Taft's polar parameter, —

Superscript

$*$	=	ionic properties

Subscripts

1	=	bulk feed solution
2	=	concentrated boundary solution
3	=	permeate solution
1	=	ion 1
2	=	ion 2
3	=	ion 3
4	=	ion 4
12	=	salt **12**
14	=	salt **14**
32	=	salt **32**
34	=	salt **34**
A	=	solute
B	=	solvent
i	=	ionic species
m	=	membrane phase

5.3 Membrane Gas Transport   5.4 Pervaporation

$$
\begin{aligned}
a &= \text{activity, mol/m}^3 \\
b &= \text{hole affinity constant, 1/Pa} \\
c &= \text{concentration, mol/m}^3 \\
c'_H &= \text{hole saturation constant, mol/m}^3 \\
D &= \text{diffusion coefficient, m}^2/\text{s} \\
D_0 &= \text{preexponential factor of diffusion coefficient, m}^2/\text{s} \\
\bar{D} &= \text{average diffusion coefficient, m}^2/\text{s} \\
E_D &= \text{activation energy of diffusion, J/mol} \\
E_p &= \text{activation energy of permeability coefficient, J/mol} \\
f' &= \text{solute rejection defined by Equation 5.272} \\
f &= \text{friction working between unit mol of permeant and the} \\
  &\quad \text{membrane material, J s/m}^2\text{ mol} \\
\Delta H_s &= \text{enthalpy of solution, J/mol} \\
J &= \text{flux} \\
K &= \text{partition coefficient, ---} \\
K_A &= \text{distribution constant of solute, ---} \\
K_B &= \text{distribution constant of solvent, ---} \\
k_D &= \text{Henry's law dissociation constant, mol/m}^3\text{ Pa} \\
L &= \text{phenomenological coefficient} \\
m &= \text{constant defined by Equation 5.282} \\
n &= \text{constant defined by Equation 5.281} \\
P &= \text{permeability coefficient, mol m/m}^2\text{ s Pa} \\
p &= \text{pressure, Pa} \\
p_{\text{ref}} &= \text{reference pressure, Pa} \\
\mathbf{R} &= \text{gas constant, 8.314 J/mol K} \\
S &= \text{solubility coefficient, mol/m}^3\text{ Pa} \\
S_0 &= \text{preexponential factor of solubility coefficient, mol/m}^3 \\
  &\quad \text{Pa} \\
SF &= \text{separation factor} \\
T &= \text{absolute temperature, K} \\
v &= \text{molar volume, m}^3/\text{mol} \\
X &= \text{mole fraction, ---} \\
\alpha &= \text{constant defined by Equation 5.281} \\
\alpha_A, \alpha_B &= \text{quantities defined by Equations 5.237 and 5.238, or} \\
  &\quad \text{5.239 and 5.240} \\
\beta &= \text{constant that characterizes the concentration depen-} \\
  &\quad \text{dence of } D_D \\
\gamma &= \text{activity coefficient} \\
\zeta &= p/p_*
\end{aligned}
$$

μ = chemical potential, J/mol

μ_0 = chemical potential of pure liquid at $p = p_{ref}$, J/mol

μ^* = chemical potential of pure liquid at the saturation vapor pressure, J/mol

π = osmotic pressure, Pa

Δ = defined as quantities at the downstream side of the membrane minus quantities at the upstream side of the membrane

δ = membrane thickness, m

δ_a = thickness of the imaginary liquid phase, m

δ_b = thickness of the imaginary vapor phase, m

σ = constant defined by Equation 5.282

τ = constant defined by Equation 5.282

Subscripts

A = quantities concerning solute

B = quantities concerning solvent

D = quantities concerning Henry's law isotherm

g = quantities concerning gas transport

H = quantities concerning Langmuir sorption isotherm

i = quantities concerning ith component

m = quantities concerning inside the membrane

1 = quantities concerning the bulk feed solution

2 = quantities concerning the concentrated boundary layer that is in contact with the upstream side of the membrane

3 = quantities concerning the downstream side of the membrane

$*$ = quantities concerning the saturation vapor pressure

5.5 Design of Composite Membranes for Gas Separation Resistance Theory

A = surface area of the membrane, m^2

E = electric potential difference

I = electric current

J = permeant flux, mol/m^2 s

P = membrane permeability coefficient, mol m/m^2 s Pa

R = resistance to the permeant flow, s Pa/mol

R_1 = resistance of component 1, s Pa/mol

R_1' = resistance of component 1$'$, s Pa/mol

R_2 = resistance of component 2, s Pa/mol
R_3 = resistance of component 3, s Pa/mol
R_{hom} = resistance of the homogeneous film, s Pa/mol
R_x = resistance to the gas flow connecting two resistance arms, s Pa/mol
R_{sub} = resistance of the substrate membrane, s Pa/mol
α = ratio of resistances for gaseous component b and a, —
δ = membrane thickness, m

Subscripts

a = gas component a
b = gas component b
i = ith gas component
n = nth resistance

6

Membrane Transport/Pore Model

6.1   Reverse Osmosis

The solution-diffusion model was outlined in Chapter 5 as a transport model for reverse osmosis. Models other than the solution-diffusion model exist, the most typical ones being the irreversible thermodynamic model [177]–[181] and the fine-pore model [182]–[187]. The fine-pore model was developed assuming the presence of open micropores on the active surface layer of the membrane through which the mass transport occurs. In contrast, the presence of pores is not assumed in the solution-diffusion model, and the mass transport is considered to occur through a film without heterogeneous structure. In the preferential sorption-capillary flow model proposed by Sourirajan [51], the influence of the interaction force between the solute, solvent, and membrane material on the flow of mass through the open capillaries is considered. Because the radii of the micropores that are present in the reverse osmosis membrane are so small that the flowing mass cannot escape from the influence of the interaction force exerted from the membrane material, Sourirajan's model seems to be the closest to reality. Later a quantitative expression was given to the above model by various authors, in developing the surface force-pore flow model [188]-[207].

The differences between the surface force-pore flow model and the solution-diffusion model are (1) the microscopic structure of the membrane is incorporated explicitly into the transport equations as the pore radius in the surface force-pore flow model; (2) the interaction force working between the permeant and the membrane is also incorporated into the transport equations as an interaction force parameter in the surface force-pore flow model; (3) as mentioned in Chapter 5, the solution-diffusion model describes the transport of permeants through the membrane, as an uncoupled diffusive flow. Mass transfer by the diffusive flow is expressed by a set of transport parameters that are intrinsic to the polymeric material. Any flow other than the above intrinsic diffusive flow is

221

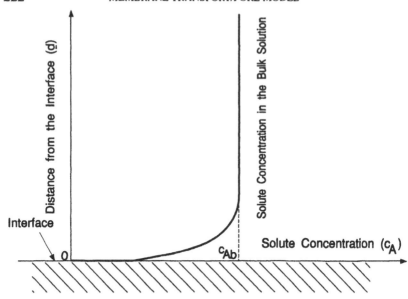

Figure 6.1. Solute concentration profile at the interface for negative solute adsorption.

due to the presence of pores that are considered as defects to the polymer matrix. There is a sharp distinction between the flow through the membrane matrix and that through the pore. There is no mass flow intrinsic to a given material in the surface force-pore flow model. Hence, there is no distinction between the flow intrinsic to the polymeric material and the flow-through defects.

The preferential sorption-capillary flow model starts from the consideration of the solid-liquid interface. For example, aqueous sodium chloride solution is in contact with a solid surface. Sodium chloride solution represents the reverse osmosis system where the separation of solute (sodium chloride) from solvent (water) occurs. This system also represents one of the most important applications of reverse osmosis, i.e., seawater desalination. A concentration gradient should inevitably appear at the solution-solid interface, as shown in Figure 6.1. The Gibbs adsorption isotherm

$$\Gamma = -\frac{1}{RT}\frac{\partial \sigma}{\partial \ln a} \qquad (6.1)$$

is known to be the most appropriate to describe such a concentration profile, where Γ is the surface excess of the solute (mol/m^2), σ is the interfacial tension at the solution-solid interface (N/m), and a is the activity of solute. The definition

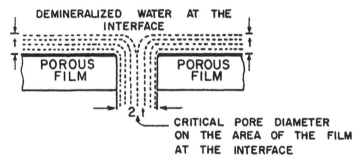

Figure 6.2. Preferential sorption-capillary flow model. (Reproduced from [51] with permission.)

of the surface excess, Γ, is given by

$$\Gamma = \int_0^\infty (c_{Ai} - c_{Ab})d(\underline{d}) \tag{6.2}$$

where c_{Ai} is the molar solute concentration (mol/m^3) at distance $\underline{d}$ (m) from the interface, and c_{Ab} is the molar solute concentration (mol/m^3) of the bulk solution at infinite distance. Since the interfacial tension σ of sodium chloride solution increases with an increase in the solute concentration, with an increase in the solute activity, a, Γ should be negative, according to Equation 6.1. This means that the solute concentration in the vicinity of the interface is less than that in the bulk and approaches zero with a decrease in the distance $\underline{d}$. The concentration profile is described schematically in Figure 6.1. According to a calculation based on the surface tension data of aqueous sodium chloride solution, the sodium chloride concentration is zero when the distance from the interface is below 3 to 6×10^{-10} m (see Example 1), indicating the presence of a thin layer of pure water at the air-solution interface. Similarly, the formation of a pure-water layer is expected at the solid-solution interface. Suppose the thickness of the pure-water layer is t_i, and the radius of the pore across a solid film is t_i. If water flows through the pore under a pressure drop across the solid film, as illustrated schematically in Figure 6.2, it is obvious that from the feed sodium chloride solution, only pure water can be recovered in the permeate. If the solid film is a thin polymeric film, the above system represents desalination by a membrane. However, the solid material is not restricted to polymers. It can also be an inorganic material, as long as a thin water layer can be formed at the solution-solid material interface. According to the mechanism described above, two requirements should be satisfied to enable the separation of sodium chloride and water. One is the presence of a steep concentration gradient at the solution-solid interface, by which the pure-water layer is formed, and the

Table 1. Some Physicochemical Data Pertinent to
Sodium Chloride Solution

Molality, m	Activity coefficient, α	Density, ρ ($\times 10^{-3}$, kg/m^3)	Surface tension, γ ($\times 10^3$, J/m^2)
0.0000	—	—	72.80
0.2010	0.751	1.00675	73.17
0.5030	0.688	1.01876	73.71
1.0204	0.650	1.0385	74.515
2.0988	0.614	1.06984	76.27
3.1920	0.714	1.1152	78.08
4.3628	0.790	1.1507	80.02
4.9730	0.848	1.1679	81.09
5.5410	0.874	1.1947	82.17

second is the presence of the micropore, the radius of which is either the same as or smaller than the pure-water layer thickness.

Example 1

The activity coefficient, density, and surface tension of aqueous sodium chloride solutions at 20°C are given for different molalities in Table 1. Calculate the thickness of pure water by using the data listed in Table 1.

A modification of the Gibbs adsorption isotherm Equation 6.1 is necessary. For the solution of symmetric electrolytes,

$$a_{\pm} = a^{1/2} \tag{6.3}$$

Combining Equations 6.1 and 6.3,

$$\Gamma = -\frac{1}{2RT}\left(\frac{\partial \sigma}{\partial \ln a_{\pm}}\right)_{T,A} \tag{6.4}$$

$$= -\frac{1}{2RT}\left(\frac{\partial \sigma}{\partial \ln(\alpha m_{\pm})}\right)_{T,A} \tag{6.5}$$

$$= -\frac{\alpha m_{\pm}}{2RT}\left(\frac{\partial \sigma}{\partial(\alpha m_{\pm})}\right)_{T,A} \tag{6.6}$$

$$= -\frac{\alpha m}{2RT}\left(\frac{\partial \sigma}{\partial(\alpha m)}\right)_{T,A} \tag{6.7}$$

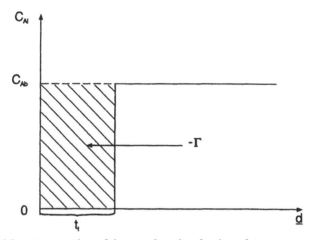

Figure 6.3. Assumption of the step function for the solute concentration profile.

Since

$$c_{Ab} = \frac{1000m}{1000 + 58.54m}\rho \tag{6.8}$$

where c_{Ab} is molar concentration of sodium chloride (mol/m^3),

$$-\frac{\Gamma}{c_{Ab}} = \frac{\alpha(1000 + 58.54m)}{2RT\rho \times 1000}\left(\frac{\partial\sigma}{\partial(\alpha m)}\right)_{T,A} \tag{6.9}$$

Assuming a stepwise concentration profile at the interface, as illustrated in Figure 6.3, and considering that $-\Gamma$ is equal to the shadowed area in the figure, it is obvious that a quantity, $-\frac{\Gamma}{c_{Ab}}$, is the thickness of the layer where sodium chloride concentration is equal to zero. Therefore,

$$t_i = -\frac{\Gamma}{c_{Ab}} \tag{6.10}$$

Table 2 was made based on the data given in Table 1. The last column of Table 2 shows the thickness of the pure-water layer at the air-solution interface. It should be noted that the calculation of the pure-water layer thickness was possible only by assuming a step function for the sodium chloride concentration profile.

Let us now consider why the surface excess (or negative surface excess in the case of sodium chloride) arises at the interface. Suppose two materials of different dielectric constants (F/m), D, and D', are in contact with each other, forming a plane interface. For example, the dielectric constant of water is D, and that of the polymeric material is D'. Suppose there is a charged particle in a material of dielectric constant D at a distance $\underline{d}$ vertical to the interfacial

Table 2. Physicochemical Data of Sodium Chloride Solution
Based on the Data Given in Table 1

αm (mol/kg)	$\gamma \times 10^3$ (J/m^2)	$d\gamma/d(\alpha m)$ $\times 10^3$	α	$\rho \times 10^{-3}$ (kg/m^3)	m (mol/kg)	$t_i \times 10^{10}$ (m)
0	72.80	2.74[a]	1.0	1.0	0	5.62
0.5	74.16	2.70	0.669	1.024	0.747	3.78
1.0	75.50	2.52	0.624	1.056	1.603	3.35
1.5	76.68	2.15	0.616	1.081	2.435	2.87
2.0	77.65	1.82	0.640	1.103	3.125	2.57
2.5	78.50	1.67	0.685	1.122	3.650	2.54
3.0	79.32	1.62	0.745	1.139	4.027	2.68
3.5	80.12	1.58	0.795	1.152	4.403	2.82
4.0	80.90	1.49	0.833	1.164	4.802	2.79
4.5	81.61	1.35[b]	0.861	1.179	5.226	2.64

[a] Calculated as $-3 \times 72.80 + 4 \times 74.16 - 75.50$.
[b] Calculated as $80.12 - 4 \times 80.9 + 3 \times 81.61$.

plane (point A in Figure 6.4), and the particle carries an electric charge of $Z\varepsilon$ (C). Then a repulsive force works from the interface to the charged particle as if a particle carrying the same electric charge $Z\varepsilon$ were existing in a material of dielectric constant D' at a position of mirror image (point B in Figure 6.4). Therefore, the energy level at point A of Figure 6.5 is higher than at point C of Figure 6.5 where there is no force working on the charged particle. Obviously, point C should be an infinite distance from the interface in the material of dielectric constant D. Assuming the energy level of the particle in vacuum as a reference, the energy level at point A, designated as W (J), is equivalent to the work required to bring the particle from vacuum to point A. Similarly, the energy level of point C, designated as W_0, is equivalent to the work required to bring the particle from vacuum to point C. The work required to move the charged particle from point C to point A, designated as ΔW, is then

$$\Delta W = W - W_0 \tag{6.11}$$

According to Onsager and Samaras [208], the above energy is

$$\Delta W = \frac{NZ^2\varepsilon^2}{16\pi D\underline{d}} \frac{D - D'}{D + D'} \tag{6.12}$$

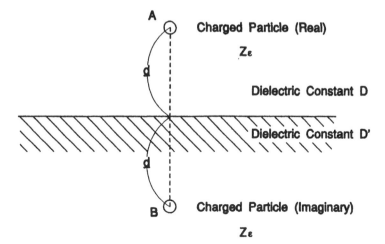

Figure 6.4. Force working on a charged particle at an interface of materials with different dielectric constants.

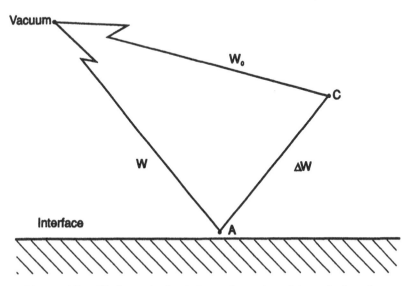

Figure 6.5. Work required to bring a charged particle to the interface.

where N is the Avogadro's number. Setting the potential at infinite distance equal to zero, the potential at distance $\underline{d}$, designated as $\phi(\underline{d})$, is

$$\phi(\underline{d}) = \Delta W \tag{6.13}$$

Furthermore, defining

$$A = \frac{NZ^2\varepsilon^2}{16\pi DRT} \frac{D - D'}{D + D'} \tag{6.14}$$

Equation 6.13 becomes

$$\phi(\underline{d}) = ART/\underline{d} \tag{6.15}$$

So far the electric charge has been considered as a point. In reality, however, the charged particle has a size. If the charged particle is a sphere with a radius D, one end of the particle touches the plane interface when the center of the particle is at a distance D from the interface. Representing the position of the particle by that of the center of the sphere, the repulsive force working between the particle and the interface becomes extremely strong due to the steric hindrance when the distance (from the center of the sphere to the interface) is between zero and D. The center of the particle cannot approach the interface any closer than D. In other words, the potential is infinity when the distance, $\underline{d}$, is between zero and D. For a distance greater than D, Equation 6.15 should become valid. Then

$$\phi(\underline{d}) \;=\; \infty \text{ when } 0 < \underline{d} \le D \tag{6.16}$$

$$=\; \frac{ART}{\underline{d}} \text{ when } D < \underline{d} \tag{6.17}$$

Applying the Maxwell-Boltzmann equation, the interfacial concentration, c_{Ai}, at a distance $\underline{d}$ from the interface can be given as

$$c_{Ai} = c_{Ab} \exp(-\phi/RT) \tag{6.18}$$

where c_{Ab} is the solute concentration in the bulk solution (mol/m^3) corresponding to the distance $\underline{d} = \infty$, and ϕ is given by Equations 6.16 and 6.17.

Example 2

The dielectric constant of water at 25°C is 78.54. Assume that the dielectric constant of cellulose acetate material is one tenth that of water. Calculate the parameter A. Use the following numerical values:

$$\text{Dielectric constant of vacuum} = 8.854 \times 10^{-12}\text{Fm}^{-1}$$
$$\text{Electric charge} = 1.602 \times 10^{-19}\text{C}$$
$$\text{Avogadro number} = 6.023 \times 10^{23}\text{mol}^{-1}$$
$$\text{Gas constant} = 8.314\text{JK}^{-1}\text{mol}^{-1}$$

Inserting all numerical values into Equation 6.14, $A = 1.459 \times 10^{-10}\text{m}$.

Example 3

The Stokes law radii of Na^+ and Cl^- ions are $1.83 \times 10^{-10}\text{m}$ and $1.20 \times 10^{-10}\text{m}$, respectively. Assuming that the average value of the above Stokes radii ($1.52 \times 10^{-10}\text{m}$) is D_{NaCl}, calculate the surface excess of NaCl.

Using the numerical value obtained in Example 2, in Equations 6.16 and 6.17,

$$\frac{\phi(\underline{d})}{RT} = \infty \text{ when } \underline{d} \le 1.52 \times 10^{-10} \tag{6.19}$$

$$= \frac{1.46 \times 10^{-10}}{\underline{d}} \text{ when } \underline{d} > 1.52 \times 10^{-10} \tag{6.20}$$

ϕ/RT, $\exp(-\phi/RT)$, and $\exp(-\phi/RT) - 1$ were calculated for various values of $\underline{d}$, and the results listed in Table 3.

Combining Equations 6.2 and 6.18 yields

$$\frac{\Gamma}{c_{Ab}} = \int_{\mathbf{D}_{water}}^{\infty} [\exp(-\phi/\mathbf{R}T) - 1]d(\underline{d}) \tag{6.21}$$

where $\mathbf{D}_{water}$ is the Stokes law radius of a water molecule. $\mathbf{D}_{water}$ is used as the lower integration limit, instead of zero, because the center of the water molecule cannot approach the polymer-solution interface any closer than $\mathbf{D}_{water}$ due to steric hindrance. Therefore, there are no solvent water molecules from the distance zero to $\mathbf{D}_{water}$. Numerical integration was carried out using data in Table 3.

Table 3. Φ, $e^{-\Phi}$, $e^{-\Phi} - 1$ as Functions of $\underline{d}$

$\underline{d} \times 10^{10}$, m	Φ^a	$e^{-\Phi}$	$e^{-\Phi} - 1$
0.87 – 1.52	∞	0	–1.0
1.52	0.961	0.3825	–0.6175
2	0.730	0.4820	–0.5180
3	0.487	0.6145	–0.3855
4	0.365	0.6942	–0.3058
5	0.292	0.7468	–0.2532
6	0.243	0.7843	–0.2157
7	0.209	0.8114	–0.1886
8	0.183	0.8328	–0.1672
9	0.162	0.8504	–0.1496
10	0.146	0.8642	–0.1358
20	0.073	0.9296	–0.0704
30	0.0487	0.9525	–0.0475
40	0.0365	0.9642	–0.0358
50	0.0292	0.9712	–0.0288
60	0.0243	0.9760	–0.0240
70	0.0209	0.9793	–0.0207
80	0.0183	0.9819	–0.0181
90	0.0162	0.9838	–0.0161
100	0.0146	0.9855	–0.0145
110	0.0133	0.9868	–0.0132

$^a\Phi = \phi/\mathbf{R}T$.

The upper integration limit was set at $\underline{d} = 110 \times 10^{-10}$ m, the distance where $\exp(-\phi/RT)$ becomes nearly equal to 0.99. According to Equation 6.18, the interfacial concentration reaches 99% of the bulk concentration at this distance. The result of integration is -6.324×10^{-10}m, indicating that there is a negative surface excess that is similar to the value calculated based on Gibbs' adsorption isotherm.

Suppose we assume a step function for the interfacial concentration profile, as described in Figure 6.3. The concentration remains $\bar{c}_{Ai}$ from the interface to a distance t_i and then abruptly changes to c_{Ab}. The quantity $\bar{c}_{Ai}$ is therefore considered to be the average interfacial concentration in the interfacial region that extends from distance zero to t_i. The surface excess is written according to the above concentration profile, as

$$\Gamma = (\bar{c}_{Ai} - c_{Ab}) \times t_i \qquad (6.22)$$

or

$$\frac{\Gamma}{c_{Ab}} = \left(\frac{\bar{c}_{Ai}}{c_{Ab}} - 1\right) \times t_i \qquad (6.23)$$

Going back to the chromatography theory developed in Chapter 2, the partition coefficient K in Equation 6.18 was defined as the solute concentration in the stationary phase/solute concentration in the mobile phase. Furthermore, in Chapter 2, the mobile phase of chromatography was considered equal to the bulk water, whereas the stationary phase was considered equal to the interfacial bound water. Since $\bar{c}_{Ai}$ is the average concentration in the interfacial water, and c_{Ab} is the concentration in the bulk water, $K = \frac{\bar{c}_{Ai}}{c_{Ab}}$. Similarly, V_s, the volume of the stationary phase, is equal to that of the interfacial bound water. Then the thickness of the interfacial bound water can be given as $t_i = V_s/A_t$, where A_t is the total area of the interface. Combining Equations 17 and 18 of Chapter 2 and Equation 6.23 above yields

$$\frac{\Gamma}{c_{Ab}} = \left(\frac{V_{R'} - [V_{R'}]_{min}}{[V_{R'}]_{water} - [V_{R'}]_{min}} - 1\right) \times \frac{[V_{R'}]_{water} - [V_{R'}]_{min}}{A_t} \qquad (6.24)$$

$$= \frac{V_{R'} - [V_{R'}]_{min} - [V_{R'}]_{water} + [V_{R'}]_{min}}{[V_{R'}]_{water} - [V_{R'}]_{min}} \times \frac{[V_{R'}]_{water} - [V_{R'}]_{min}}{A_t} \qquad (6.25)$$

$$= \frac{V_{R'} - [V_{R'}]_{water}}{A_t} \qquad (6.26)$$

Example 4

For the system cellulose acetate (E-398-3)/sodium chloride/water, the retention volume data of $[V_{R'}] = 3.578 \times 10^{-6} \text{m}^3$ and $[V_{R'}]_{water} = 4.050 \times 10^{-6} \text{m}^3$ were obtained from liquid chromatography experiments. The total surface area of the cellulose acetate powder packed in the chromatography column is known to be 708.7 m². Calculate Γ/c_{Ab} and compare the result with the value obtained in Example 3.

From Equation 6.26,

$$\Gamma/c_{Ab} = \frac{(3.578 - 4.050) \times 10^{-6}}{708.7} = -6.66 \times 10^{-10} \text{m}$$

The above examples indicate that negative surface excess arises at the air-water or polymer-water interfaces when the solute is sodium chloride. The case will be the same when the solute is an electrolyte or a charged particle. In other words, water is preferentially sorbed at the membrane polymer-aqueous solution interface when the solute is an electrolyte, and an electrostatic repulsive force is exerted on charged ions from the interface due to the discontinuity of the dielectric constant. As for the nonionized organic solute and the surface of the polymer material, the best way to express the interaction force is to adopt the Lennard–Jones type potential function, which can be written as

$$\phi(\underline{d}) = -\frac{BRT}{\underline{d}^6} + \frac{CRT}{\underline{d}^{12}} \tag{6.27}$$

where the first and the second term of the right side of the equation correspond to an attractive force due to the van der Waals force and the short-range repulsive force arising from the overlap of the electron clouds of interacting atoms, respectively. The parameters B and C are force constants applicable to the attractive and the repulsive force, respectively. Both parameters were multiplied by the gas constant and the absolute temperature, RT, in Equation 6.27 in order to simplify the dimensionless form of the potential function to be used in the latter part of the section. The above equation is applicable to the interaction force working between two points. When the force is working between a point and a surface, the better expression would be

$$\phi(\underline{d}) = -\frac{BRT}{\underline{d}^3} + \frac{CRT}{\underline{d}^{12}} \tag{6.28}$$

An approximation of the Lennard-Jones equation was made by assuming a potential barrier with a thickness of D, to express the repulsive force. Equation 6.28 then becomes

$$\phi(\underline{d}) = \infty \text{ when } \underline{d} \leq D \tag{6.29}$$

$$= -\frac{BRT}{\underline{d}^3} \text{ when } \underline{d} > D \tag{6.30}$$

The potential functions given by Equations 6.16, 6.17, 6.29, and 6.30 are illustrated schematically in Figures 6.6a and b. Equations 6.29 and 6.30 are known as the Sutherland potential function.

The electrostatic repulsive force working between the surface of a polymeric material without electric charge and an electrolyte, and the van der Waals interaction force working between the polymer surface and a nonionized organic solute, were expressed by potential function Equations 6.16, 6.17, 6.29, and 6.30. There are also cases where both interaction forces are working. In

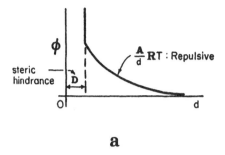

a

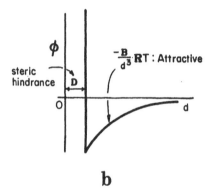

b

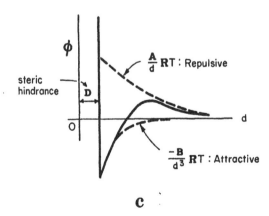

c

Figure 6.6. Potential functions expressing the interfacial forces working on the solute molecule from the polymer surface. (Reproduced from [5] with permission.)

such a case,

$$\phi(\underline{d}) \; = \; \infty \text{ when } \underline{d} \leq D \tag{6.31}$$

$$= \; \frac{ART}{\underline{d}} - \frac{BRT}{\underline{d}^3} \text{ when } \underline{d} > D \tag{6.32}$$

Equations 6.31 and 6.32 are also illustrated schematically in Figure 6.6. The physicochemical significance of parameters A, B, and D, involved in the potential functions, are summarized. All three parameters depend on the chemical nature of the solute, solvent, and polymeric material and their mutual interactions. A is a measure of the resultant electrostatic force; it is always positive; and its dimension is length. B is a measure of the resultant short-range van der Waals force; its signs can be positive or negative; and its dimension is length³. D is a measure of the steric hindrance for the solute at the interface; its sign is always positive; and its dimension is length. Steric hindrance is always associated with the effective size of the solute molecule. As indicated schematically in Figure 6.7, D may be regarded as the radius of the hydration sphere formed around ions at the interface (Figure 6.7a), the distance from the center of a molecule to the point where the molecule touches the interface (Figure 6.7b,c,d), or the radius of an agglomerate of several solute molecules (Figure 6.7e). For a given solute, the value of D depends on the chemical nature of the solvent and the material of the surface; consequently, D is not the size of the solute molecule in the bulk solution, such as that represented by the Stokes radius. However, as a first approximation one may equate the numerical value of D to the Stokes radius, for the purpose of analysis.

Example 5

For the system cellulose acetate (E-398-3)/methyl ethyl ketone/water, B and D values are known to be $35.9 \times 10^{-30} m^3$ and $1.92 \times 10^{-10} m$, respectively. Calculate the surface excess Γ/c_{Ab} of methyl ethyl ketone at the cellulose acetate/water interface.

Inserting numerical values into Equations 6.29 and 6.30,

$$\frac{\phi(\underline{d})}{RT} \; = \; \infty \text{ when } D_{water} < \underline{d} \leq 1.92 \times 10^{-10}$$

$$= \; -\frac{35.9 \times 10^{-30}}{\underline{d}} \text{ when } \underline{d} > 1.92 \times 10^{-10}$$

From Equation 6.21,

$$\frac{\Gamma}{c_{Ab}} \; = \; \int_{0.87 \times 10^{-10}}^{\infty} \left[\exp \left(-\frac{\phi}{RT} \right) - 1 \right] d(\underline{d})$$

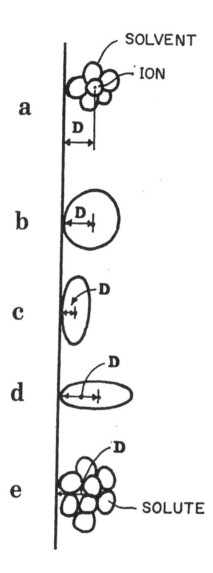

Figure 6.7. Different forms of steric hindrance. (Reproduced from [5] with permission.)

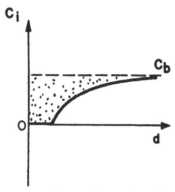

Figure 6.8. Solute concentration profile at the polymer-solution interface; repulsive force working on the electrolyte solute. (Reproduced from [209] with permission.)

$$= \int_{0.87 \times 10^{-10}}^{1.92 \times 10^{-10}} (-1)d(\underline{d})$$

$$+ \int_{1.92 \times 10^{-10}}^{\infty} \left[\exp\left(\frac{35.9 \times 10^{-30}}{\underline{d}^3} \right) - 1 \right] d(\underline{d})$$

$$= 33.40 \times 10^{-10} \text{m}$$

Example 6

For the system cellulose acetate (E-398-3)/methyl ethyl ketone/water, the retention volume data of $[V_{R'}]_{MEK} = 3.33 \times 10^{-6}$ m^3 and $[V_{R'}]_{water} = 2.04 \times 10^{-6}$ m^3 were obtained from liquid chromatography experiments. The total surface area of cellulose acetate powder packed in the column is known to be 354.4 m^2. Calculate the Γ/c_{Ab} value.

Inserting the numerical values in Equation 6.26,

$$\Gamma/c_{Ab} = \frac{(3.33 - 2.04) \times 10^{-6}}{354.4}$$

$$= 36.4 \times 10^{-10} \text{m}$$

When the Maxwell-Boltzmann equation (Equation 6.18) is applied to the above potential functions, the concentration profile at the interface can be obtained as a function of the distance from the interface. Figure 6.8 corresponds to the repulsive force working from the interface to an electrolyte solute. Starting from the interface, from the distance zero to D_{water}, even the solvent water molecule cannot exist, due to steric hindrance. (One end of the water molecule touches the surface when the center is at the distance D_{water}.) From distance D_{water} to D, the solute molecule cannot exist, because of the steric hindrance working

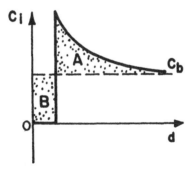

Figure 6.9. Solute concentration profile at the polymer-solution interface; attractive force working on the organic solute. (Reproduced from [239] with permission.)

on the solute molecule; therefore, the interfacial solute concentration is zero in this region. According to Equations 6.18 and 6.16, $c_{Ai} = c_{Ab} \exp(-\infty) = 0$ when $\underline{d} \leq D$. For the distance larger than D, the profile of the interfacial concentration follows the equation $c_{Ai} = c_{Ab} \exp(-\frac{A}{\underline{d}})$. According to the above equation, c_{Ai} is always smaller than c_{Ab} and gradually increases with an increase in $\underline{d}$, approaching c_{Ab} as $\underline{d}$ approaches infinity. In other words, the interfacial concentration becomes equal to the bulk concentration at the distance infinity. The concentration profile illustrated in Figure 6.8 is not restricted to the electrostatic repulsive force. It may be due to the negative van der Waals force when the parameter B in Equation 6.30 is negative. On the other hand, the concentration profile at the interface becomes as illustrated in Figure 6.9, when the interaction force is attractive. The solute concentration is equal to zero when the distance from the interface is less than D, because of the steric hindrance. At a distance D, however, the solute concentration increases discontinuously to $c_{Ab} \exp(B/D)$, which is much higher than c_{Ab} if the interaction constant, B, has a large positive value. The interfacial concentration, c_{Ai}, decreases gradually from the distance D, approaching the bulk concentration, c_{Ab}, as the distance approaches infinity. It should be noted that the surface excess, Γ, corresponds to the shadowed area. It is negative for Figure 6.8, since c_{Ai} is always less than c_{Ab}. It may become positive in Figure 6.9 when the area above c_{Ab} is greater than the area below c_{Ab}, which is usually the case. The main features of the reverse osmosis phenomena can be discussed using the interfacial concentration profiles illustrated in Figures 6.8 and 6.9 [209]. For this purpose let us first consider an extreme case where the solvent flow velocity in the pore is nearly equal to zero. This situation arises when an extremely low pressure is applied at the feed side of the membrane. Then the permeation velocity of solute due to diffusion surpasses the solvent flow velocity, and the solute concentration in the permeate increases gradually. The increase in the solute concentration stops, however, when the solute concentration in the permeate becomes equal to that in the feed. In other words, the solute concentration on both sides of the membrane are equal, and the solute separation, defined as Equation 46, Chapter 5, becomes zero when a steady state is reached.

When the pressure on the feed side is increased and, consequently, the solvent flow velocity is increased, the effect of the interfacial concentration profile becomes more explicit in the permeate concentration. Let us first consider the radial concentration profile in the pore when a repulsive force works between the solute and the membrane material. It is obvious that the pore wall is the interface formed between the polymer material and the solution in the pore. The concentration profile at the pore wall is illustrated in Figure 6.10, and it is, as expected, analogous to the one given in Figure 6.8. Note that the solute concentration is zero when the distance from the pore wall is smaller than D, and it starts to increase as the distance becomes greater than D. If the pore radius is large enough, the concentration can reach the bulk concentration in the central region of the pore. The bulk concentration is equal to the solute concentration outside the pore where there is no solute-membrane material interaction. The average concentration in the pore is lower than the bulk concentration. Therefore, there is a deficiency of the solute in the pore. As for the permeate concentration, it is known that the permeate concentration is exactly the same as the feed concentration when the applied gauge pressure on the feed side is zero, as already discussed. It is also known that the permeate concentration approaches that in the pore when the applied gauge pressure (and also the solvent velocity) approaches infinity. The permeate concentration changes from that of the feed to that in the pore monotonically as the applied gauge pressure increases. Thus, the solute separation defined by Equation 46, Chapter 5, increases gradually from zero to an asymptotic value as the pressure approaches infinity. The concentration profile in the pore becomes as shown in Figure 6.10b, when the pore size is smaller. The concentration of the bulk solution can no longer be reached even at the center of the pore. The average concentration in the pore is lower than in the pore of a larger radius (Figure 6.10a), and therefore the solute separation becomes greater for a given operating pressure. When the pore radius is even smaller and less than D (Figure 6.10c), no solute can exist in the pore, because of the steric hindrance, and the solute concentration in the pore is zero. The solute separation is 100%. The effect of the operating pressure and the pore size on the solute separation is summarized in Figure 6.11.

When the solute is preferentially sorbed, the concentration profile in the pore becomes as illustrated in Figure 6.12a. The solute concentration from the pore wall to the distance D is zero. At distance D the solute concentration increases discontinuously and surpasses the bulk concentration, c_{Ab}. As the dis-

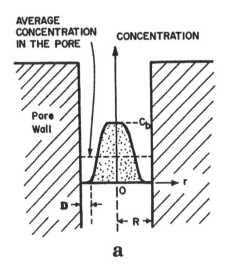

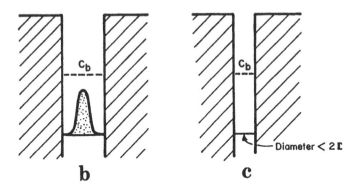

Figure 6.10. Solute concentration profile in a membrane pore; repulsive force working on the electrolyte solute. (Reproduced from [209] with permission.)

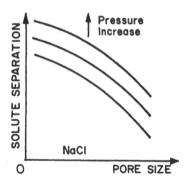

Figure 6.11. Effect of the operating pressure and pore size on the solute separation; repulsive force working on the electrolyte solute. (Reproduced from [209] with permission.)

tance from the pore wall increases, the concentration decreases, and if the pore size is large enough, it can reach the bulk concentration in the center of the pore. The average concentration in the pore is larger than the bulk concentration. Since the permeate concentration approaches the concentration in the pore when the operating pressure approaches infinity, the solute separation defined by Equation 46, Chapter 5, becomes negative. As discussed earlier, the solute concentration is zero at zero gauge operating pressure. The solute separation is negative, but higher than that corresponding to the operating pressure of infinity, at the intermediate operating pressure. The effect of the pore size is as follows. When the pore size becomes smaller, the solute concentration is higher than the bulk concentration, even at the center of the pore (Figure 6.12b). Since the high solute concentration near the pore wall dominates, the average concentration in the pore is higher than that in a pore of a larger size (Figure 6.12a). The solute separation is therefore negative and is lower than that which corresponds to Figure 6.12a. When the pore size becomes even smaller, the central region of the pore, where the solute concentration is higher than the bulk concentration, becomes narrower, and the region of zero solute concentration starts to dominate. Therefore, the average solute concentration in the pore decreases, and the solute separation starts to increase. When the pore size is further decreased and the radius becomes smaller than D, the solute cannot enter the pore, due to steric hindrance, and the solute separation becomes 100%. The effect of the operating pressure and the pore size on the solute separation is summarized in Figure 6.13 for the case where solute is preferentially sorbed. The main features of the solute separation illustrated in Figure 6.11 and 6.13 are typical for the preferential sorption of solvent and of solute, respectively. Experimental data for the reverse osmosis separation of sodium chloride represent the preferential sorption of solvent, due to the electrostatic repulsive force, and those of p-

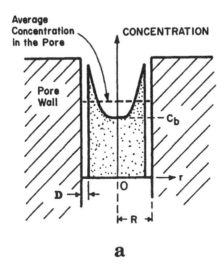

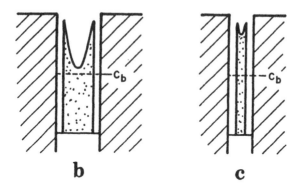

Figure 6.12. Solute concentration profile in a membrane pore; attractive force working on the organic solute. (Reproduced from [209] with permission.)

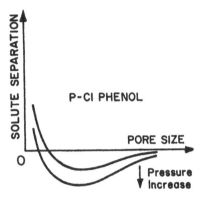

Figure 6.13. Effect of the operating pressure and pore size on the solute separation; attractive force working on the organic solute. (Reproduced from [209] with permission.)

chlorophenol represent the solute preferential sorption due to the van der Waals force. Those data are illustrated in Figures 6.14 and 6.15, respectively. The main features of the solute separation predicted theoretically are well represented by the experimental data.

Based on the concept outlined above, a set of transport equations were derived, assuming a bundle of cylindrical pores of radius R_b and effective pore length δ (see Figure 6.16), to solve for solute separations, f, and the ratio of pure-water permeation rate (PWP) and the product permeation rate (PR). PWP and PR are defined as the permeation rate (kg/h) in the absence of solute in the feed solution and that in the presence of the solute, respectively, per effective membrane area. Using the cylindrical coordinate system described in Figure 6.16, the solute concentration profile and the velocity profile of the solution inside the pore were analyzed in differential segments set in the cylindrical pore, under the steady-state operating condition of the process. The derivations involved are given in detail elsewhere [188].

Since the size of the permeating molecule is generally comparable to that of the pore, a distinction is made between the radius of the pore and that of the fluid channel. As explained previously, the position of a molecule is defined as that of its center; this means that in the region of the pore where the center of the molecule cannot exist, the molecule as an entity cannot exist. Then denoting D_w as the radius of the water molecule (m), the radius of the water channel, R_a (m), can be given as

$$R_a = R_b - D_w \tag{6.33}$$

where R_b is the pore radius (m) (see Figure 6.17). Furthermore, noting that r

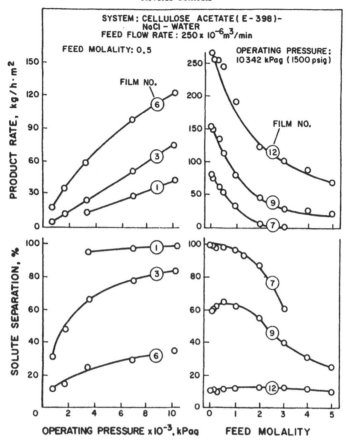

Figure 6.14. Experimental data for reverse osmosis separation of sodium chloride. (Pore size: Film 1 < Film 3 < Film 6.) (Reproduced from [5] with permission.)

represents the radial distance, and $\underline{d}$ represents the distance from the pore wall,

$$\underline{d} = R_b - r \tag{6.34}$$

Relations such as those given by Equations 6.33 and 6.34 were taken into consideration when the following transport equations were developed.

For simplification of expression and analysis, the following dimensionless quantities are defined:

· Dimensionless radial distance:

$$\rho = r/R_a \tag{6.35}$$

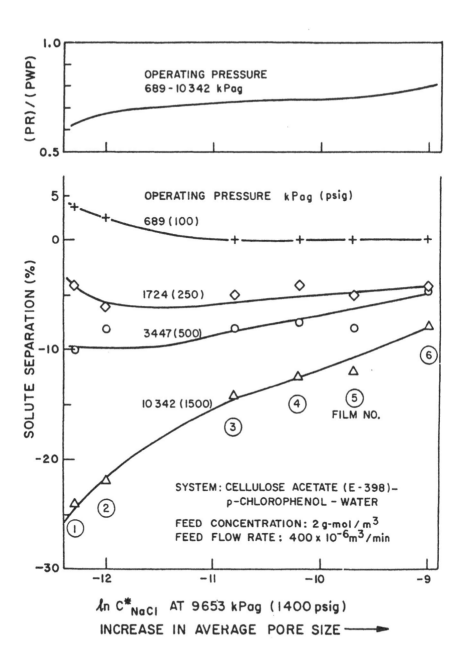

Figure 6.15. Experimental data for reverse osmosis separation of p-chlorophenol. (Reproduced from [5] with permission.)

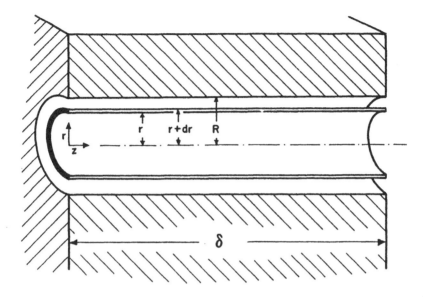

Figure 6.16. Cylindrical coordinates in a membrane pore. (Reproduced from [5] with permission.)

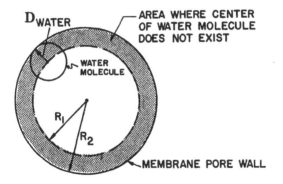

Figure 6.17. Regions where water molecules can and cannot exist in a membrane pore. (Reproduced from [5] with permission.)

· Dimensionless solute concentration at the pore outlet:

$$C_A = c_{A3}/c_{A2} \qquad (6.36)$$

· Dimensionless solution velocity in the pore:

$$\alpha(\rho) = u_B(r)\delta\chi_{AB}/RT \qquad (6.37)$$

where χ_{AB} is the solute-solvent friction function and can be obtained by

$$\chi_{AB} = RT/D_{AB} \qquad (6.38)$$

· Dimensionless solution viscosity:

$$\beta_1 = \eta/\chi_{AB}R_a^2 c_{A2} \qquad (6.39)$$

· Dimensionless operating pressure:

$$\beta_2 = (p_2 - p_3)/RTc_{A2} \qquad (6.40)$$

· Dimensionless potential function:

$$\Phi(\rho) = \phi(r)/RT \qquad (6.41)$$

· Dimensionless friction function:

$$b(\rho) = [\chi_{AB} + \chi_{AM}]/\chi_{AB} \qquad (6.42)$$

All the symbols involved in the above equations are defined in the end of the chapter.

As already discussed, the concentration in the bulk of the feed, c_{A1}, and in the region at the membrane/solution boundary, c_{A2}, are different because of the concentration polarization phenomena. (This boundary region and the interfacial region formed by the preferential sorption are different. While the thickness of the interfacial region is a multiple of 1×10^{-10}m, the thickness of the boundary region is a multiple of 1×10^{-6}m.) Two different solute separations are defined; i.e.,

$$f' = (c_{A2} - c_{A3})/c_{A2} \qquad (6.43)$$

and

$$f = (c_{A1} - c_{A3})/c_{A1} \qquad (6.44)$$

It is known from the theory of the concentration polarization that there is a relationship between these two solute separations [188].

$$f = \frac{f'}{\left(f' + \left[(1 - f') \exp(v_s/k)\right]\right)} \tag{6.45}$$

where v_s and k are the permeation velocity (m/s) and the mass transfer coefficient of the solute in the boundary region (m/s), respectively. While f' is the true solute separation by the pore, f is the apparent solute separation defined on the basis of the feed solute concentration.

Using the dimensionless quantities defined by Equations 6.35 to 6.42, the solute separation, f', can be calculated by the equation

$$f' = 1 - \frac{\int_0^1 \left[\exp\left[\alpha(\rho)\right]/1 + \left[b/e^{-\Phi(\rho)}\right]\left(\exp\left[\alpha(\rho)\right] - 1\right)\right]\alpha(\rho)\rho d\rho}{\int_0^1 \alpha(\rho)\rho d\rho} \tag{6.46}$$

The radial velocity profile involved in the above equation can be calculated by solving the differential equation

$$\frac{d^2\alpha(\rho)}{d\rho^2} + \frac{1}{\rho}\frac{d\alpha(\rho)}{d\rho} + \frac{\beta_2}{\beta_1} + \frac{1}{\beta_1}\left(1 - e^{-\Phi(\rho)}\right)\left[C_A(\rho) - 1\right]$$
$$- \left[b(\rho) - 1\right]\alpha(\rho)C_A(\rho)/\beta_1 = 0 \tag{6.47}$$

where

$$C_A(\rho) = \frac{\exp\left[\alpha(\rho)\right]}{1 + \left[b/e^{-\Phi(\rho)}\right]\left(\exp\left[\alpha(\rho)\right] - 1\right)} \tag{6.48}$$

The boundary conditions for solving Equation 6.47 are

$$\frac{d\alpha(\rho)}{d\rho} = 0 \text{ at } \rho = 0 \tag{6.49}$$

$$\alpha(\rho) = 0 \text{ at } \rho = 1 \tag{6.50}$$

The ratio PR/PWP is given by

$$PR/PWP = \frac{\left[2\int_0^1 \alpha(\rho)\rho d\rho\right]}{\left[\beta_2/8\beta_1\right]} \tag{6.51}$$

Some explanation of Equation 6.47 is in order. This equation was derived from the differential equation

$$\frac{d^2 u_B(r)}{dr^2} + \frac{1}{r}\frac{d u_B(r)}{dr} + \frac{1}{\eta}\frac{(p_2 - p_3)}{\delta} +$$

$$\frac{1}{\eta}\frac{RT}{\delta}\left[c_{A3}(r) - c_{A2}\right]\left(1 - e^{-\phi(r)/RT}\right)$$

$$- \frac{(b-1)\chi_{AB}c_{A3}(r)u_B(r)}{\eta} = 0 \qquad (6.52)$$

with boundary conditions

$$\frac{d u_B(r)}{dr} = 0 \text{ at } r = 0 \qquad (6.53)$$

$$u_B = 0 \text{ at } r = R_A \qquad (6.54)$$

by using the dimensionless quantities defined from Equation 6.35 to Equation 6.42. For the reverse osmosis of pure water, all terms that include solute concentration c_A become zero; hence, Equation 6.52 becomes

$$\frac{d^2 u_B(r)}{dr^2} + \frac{1}{r}\frac{d u_B(r)}{dr} + \frac{1}{\eta}\frac{(p_2 - p_3)}{\delta} = 0 \qquad (6.55)$$

The same boundary conditions as Equations 6.53 and 6.54 apply. Solving the differential equation and calculating the molar flow rate of the solvent water for one single pore, by

$$q_w = \frac{g}{M_B}\int_0^{R_a} u_B(r)2\pi r dr \qquad (6.56)$$

we obtain

$$q_w = \frac{g}{M_B}\pi R_a^4 (p_2 - p_3)/8\eta\delta \qquad (6.57)$$

This is the same equation as the well-known Poiseuille equation.

When there is a distribution in the pore size, the Gaussian normal distribution with an average pore radius $\overline{R}_b$ (m) and a standard deviation σ(m) is

$$Y(R_b) = \frac{1}{\sigma\sqrt{2\pi}}\exp\left[-\frac{\left(R_b - \overline{R}_b\right)^2}{2\sigma^2}\right] \qquad (6.58)$$

The solute separation then becomes

$$f' = 1 - \frac{\int_0^\infty Y(R_b) \left[\int_0^1 C_A \alpha(\rho)\rho d\rho\right]_{\text{at } R_b = R_b} dR_b}{\int_0^\infty Y(R_b) \left[\int_0^1 \alpha(\rho)\rho d\rho\right]_{\text{at } R_b = R_b} dR_b} \tag{6.59}$$

and the ratio PR/PWP becomes

$$\frac{PR}{PWP} = \frac{\int_0^\infty Y(R_b) \left[2\int_0^1 \alpha(\rho)\rho d\rho\right]_{\text{at } R_b = R_b} dR_b}{\int_0^\infty Y(R_b) \left[\beta_2/8\beta_1\right]_{\text{at } R_b = R_b} dR_b} \tag{6.60}$$

When there are more than one distribution, each distribution being expressed by a Gaussian normal distribution, the ith distribution is given by

$$Y_i(R_{b,i}) = \frac{1}{\sigma_i \sqrt{2\pi}} \exp\left[-\frac{(R_{b,i} - \bar{R}_{b,i})^2}{2\sigma_i^2}\right] \tag{6.61}$$

Furthermore, defining

$$h_i = \frac{\text{number of pores in the } i\text{th normal distribution}}{\text{number of pores in the first normal distribution}} \tag{6.62}$$

the solute separation can be calculated by

$$f' = 1 - \sum_i h_i \int_\gamma^{\gamma'} Y_i(R_{b,i}) \left[\int_0^1 C_A \alpha(\rho)\rho d\rho\right] dR_{b,i}$$
$$\div \sum_i h_i \int_\gamma^{\gamma'} Y_i(R_{b,i}) \left[\int_0^1 \alpha(\rho)\rho d\rho\right] dR_{b,i} \tag{6.63}$$

and

$$\frac{PR}{PWP} = \sum_i h_i \int_\gamma^{\gamma'} Y_i(R_{b,i}) \left[2\int_0^1 \alpha(\rho)\rho d\rho\right] dR_{b,i}$$
$$\div \sum_i h_i \int_\gamma^{\gamma'} Y_i(R_{b,i}) \left[\frac{\beta_2}{8\beta_1}\right] dR_{b,i} \tag{6.64}$$

where

$$\gamma = \bar{R}_{b,i} - 3\sigma_i \tag{6.65}$$

$$\gamma' = \bar{R}_{b,i} + 3\sigma_i \tag{6.66}$$

The dimensionless potential function required for obtaining the dimensionless velocity profile $\alpha(\rho)$ from Equation 6.47 is derived from the potential function for the electrolyte solute Equations 6.16 and 6.17, and the result is

$$\Phi(\rho) = \infty \quad \text{when} \quad \left(\frac{R_b}{R_a} - \rho\right) \leq \frac{D}{R_a} \tag{6.67}$$

$$= \frac{(A/R_a)}{[(R_b/R_a) - \rho]} \quad \text{when} \quad \left(\frac{R_b}{R_a} - \rho\right) > \frac{D}{R_a} \tag{6.68}$$

Similarly, for the nonelectrolyte solute,

$$\Phi(\rho) = \infty \quad \text{when} \quad \left(\frac{R_b}{R_a} - \rho\right) \leq \frac{D}{R_a} \tag{6.69}$$

$$= -\frac{(B/R_a^3)}{[(R_b/R_a) - \rho]^3} \quad \text{when} \quad \left(\frac{R_b}{R_a} - \rho\right) > \frac{D}{R_a} \tag{6.70}$$

can be obtained starting from Equations 6.29 and 6.30.

As for the dimensionless friction constant b, which appears in Equation 6.47, the following empirical relationship was developed based on the work of Faxen [210], Satterfield [211], and Lane and Riggle [212] who have considered the problem of friction experienced by a molecule while moving in a narrow pore.

$$b = \left(1 - 2.104\lambda_f + 2.09\lambda_f^3 - 0.95\lambda_f^5\right)^{-1}$$
$$\text{when } \lambda_f \leq 0.22 \tag{6.71}$$
$$= \left(44.57 - 416.2\lambda_f + 934.9\lambda_f^2 + 302.4\lambda_f^3\right)$$
$$\text{when } 0.22 < \lambda_f < 1 \tag{6.72}$$

where

$$\lambda_f = D/R_b \tag{6.73}$$

The above functional relationship is similar to that reported elsewhere by other authors [213].

Example 7
Use the average solution velocity and the average potential in the pore and simplify Equation 6.46. Assume a dilute feed solution.

The average solution velocity is calculated in a dimensionless form as

$$\overline{\alpha(\rho)} = \frac{\int_0^1 \alpha(\rho)\rho d\rho}{\int_0^1 \rho d\rho} \tag{6.74}$$

$$= 2\int_0^1 \alpha(\rho)\rho d\rho \tag{6.75}$$

Since a dilute solution system is assumed, there is no osmotic pressure effect and $PR/PWP = 1$. Then from Equation 6.51,

$$\overline{\alpha(\rho)} = 2\int_0^1 \alpha(\rho)\rho d\rho \tag{6.76}$$

$$= \frac{\beta_2}{8\beta_1} \tag{6.77}$$

Combining Equations 6.39, 6.40 and 6.77,

$$\overline{\alpha(\rho)} = \frac{(p_2 - p_3)\chi_{AB}R_a^2 c_{A2}}{8\eta RTc_{A2}} \tag{6.78}$$

$$= \frac{(p_2 - p_3)R_a^2}{8\eta D_{AB}} \tag{6.79}$$

Similarly, the potential function is averaged in an exponential form, since it is equivalent to averaging the solute concentration in the pore according to the Maxwell-Boltzmann equation.

$$\overline{\exp\left[-\Phi(\rho)\right]} = \frac{\int_0^1 \exp\left[-\Phi(\rho)\right]\rho d\rho}{\int_0^1 \rho d\rho} \tag{6.80}$$

$$= 2\int_0^1 \exp\left[-\Phi(\rho)\right]\rho d\rho \tag{6.81}$$

Using $\overline{\alpha(\rho)}$ and $\overline{\exp\left[-\Phi(\rho)\right]}$, Equation 6.46 is approximated by

$$f' = 1 - \frac{\exp\left[\overline{\alpha(\rho)}\right]}{1 + \left(b\Big/\overline{\exp\left[-\Phi(\rho)\right]}\right)\left(\exp\left[\overline{\alpha(\rho)}\right] - 1\right)} \tag{6.82}$$

Example 8

For the cellulose acetate/isopropyl alcohol/water system, the following numerical parameters were given: $D = 2.45 \times 10^{-10}$ m and $B = 38.40 \times 10^{-30}$ m³.

Calculate the solute separation of isopropyl alcohol by reverse osmosis, using a cellulose acetate membrane of pore radius $R_b = 10 \times 10^{-10}$ m, at the operating pressure of 1700 kPa (gauge).

Assume $k = \infty$ and use the following numerical values:

Viscosity of water $= 0.8941 \times 10^{-3}$ Pa s
Diffusivity of isopropyl alcohol $= 1.08 \times 10^{-9}$ m²/s
Effective radius of a water molecule, $D_w = 0.87 \times 10^{-10}$ m

- Calculation of $\overline{\alpha(\rho)}$

Using Equation 6.33,

$$R_a = 10 \times 10^{-10} - 0.87 \times 10^{-10}$$
$$= 9.13 \times 10^{-10} m$$

From Equation 6.79,

$$\overline{\alpha(\rho)} = \frac{(1700 \times 10^3)(9.13 \times 10^{-10})^2}{(8)(0.8941 \times 10^{-3})(1.08 \times 10^{-9})}$$
$$= 0.1834$$
$$\exp\left[\overline{\alpha(\rho)}\right] = e^{0.1834}$$
$$= 1.2013$$

- Calculation of b

From Equation 6.73,

$$\lambda_f = \frac{(2.54 \times 10^{-10})}{10 \times 10^{-10}}$$
$$= 0.245$$
$$> 0.220$$

From Equation 6.72,

$$b = (44.57) - (416.2)(0.245) + (934.9)(0.245)^2 + (302.4)(0.245)^3$$
$$= 3.17$$

- Calculation of $\overline{\exp\left[-\Phi(\rho)\right]}$

From Equation 6.69,

$$\Phi(\rho) = \infty \text{ when } \frac{10.0 \times 10^{-10}}{9.13 \times 10^{-10}} - \rho \leq \frac{2.45 \times 10^{-10}}{9.13 \times 10^{-10}}$$

or

$$\Phi(\rho) = \infty \text{ when } \rho \geq \frac{10.0 \times 10^{-10} - 2.45 \times 10^{-10}}{9.13 \times 10^{-10}} = 0.827$$

The maximum value for ρ is unity; therefore,

$$\Phi(\rho) = \infty \text{ and } \exp\left[-\Phi(\rho)\right] = 0 \text{ when } 0.827 \leq \rho \leq 1.0$$

Furthermore, from Equation 6.70,

$$\Phi(\rho) = -\frac{(38.40 \times 10^{-30})/(9.13 \times 10^{-10})^3}{\left((10.0 \times 10^{-10}/9.13 \times 10^{-10}) - \rho\right)^3} \qquad (6.83)$$

$$= -\frac{0.0505}{(1.095 - \rho)^3}$$

when $0 \leq \rho < 0.827$
.   Therefore,

$$\overline{\exp\left[-\Phi(\rho)\right]} = 2 \int_0^1 \exp\left[-\Phi(\rho)\right] \rho d\rho$$

$$= 2 \int_0^{0.827} \exp\left[\frac{0.0505}{(1.095 - \rho)^3}\right] \rho d\rho + 2 \int_{0.827}^1 (0)\rho d\rho$$

$$= 2 \int_0^{0.827} \exp\left[\frac{0.0505}{(1.095 - \rho)^3}\right] \rho d\rho$$

Numerical integration yields

$$\overline{\exp\left[-\Phi(\rho)\right]} = 1.698 \qquad (6.84)$$

Since $k = \infty, f = f'$ from Equation 6.45, then

$$f = 1 - \frac{1.2013}{1 + (3.17/1.698)(1.2013 - 1)}$$

$$= 0.127$$

Example 9

Calculate the separation of isopropyl alcohol at the operating pressures of 3400, 6800 and 10,200 kPa gauge.

$\alpha(\rho)$ and $\exp[\alpha(\rho)]$ were calculated for different operating pressures from Equation 6.79. $f(= f')$ was calculated from Equation 6.82 and the results are shown in Table 4.

Table 4. $\overline{\alpha(\rho)}$, $\exp\left[\overline{\alpha(\rho)}\right]$, and f as a Function of Operating Pressure

Pressure, kPa (gauge)	$\overline{\alpha(\rho)}$	$\exp\left[\overline{\alpha(\rho)}\right]$	f
3,400	0.3368	1.4431	0.210
6,800	0.7336	2.0826	0.311
10,200	1.1004	3.0050	0.366

Example 10

Calculate the separation of isopropyl alcohol by membranes with average pore radii of 7×10^{-10}m and 12×10^{-10}m at the operating pressures of 1700, 3400, 6800, and 10,200 kPa (gauge). Assume $k = \infty$.

All the results are shown in Table 5. The results of the above calculation are approximated values since Equation 6.82 was used. They are compared with the rigorous solution of Equation 6.46 in Table 6.

Example 11

Simplify Equation 6.59, using Equation 6.82. Split the integration range from -3σ to 3σ into ten segments, each having an increment of 0.6σ.

The Gaussian normal distribution is used for the pore size distribution. Then, Equation 6.59 can be approximated by

$$
\begin{aligned}
f' = 1 - & \\
& \sum_{i=1}^{10} \int_{\gamma}^{\gamma'} Y(R_b) \left[\int_0^1 \overline{C}_A \alpha(\rho)\rho\,d\rho\right]_{\text{at } R_b = R_b} dR_b \\
\div & \sum_{i=1}^{10} \int_{\gamma}^{\gamma'} Y(R_b) \left[\int_0^1 \alpha(\rho)\rho\,d\rho\right]_{\text{at } R_b = R_b} dR_b \qquad (6.85)
\end{aligned}
$$

Table 5. Results of Calculation for Example 10

$R_b = 7 \times 10^{-10}$m				
Operating pressure, kPa (gauge)	1,700	3,400	6,800	10,200
λ_f	0.35	0.35	0.35	0.35
b	26.4	26.4	26.4	26.4
$\overline{\exp[-\Phi(\rho)]}$	1.771	1.771	1.771	1.771
$\overline{\alpha(\rho)}$	0.0827	0.1645	0.3308	0.4962
$\exp\overline{[\alpha(\rho)]}$	1.0862	1.1798	1.3919	1.6422
f	0.525	0.679	0.797	0.845
$R_b = 12 \times 10^{-10}$m				
Operating pressure, kPa (gauge)	1,700	3,400	6,800	10,200
λ_f	0.204	0.204	0.204	0.204
b	1.698	1.698	1.698	1.698
$\overline{\exp[-\Phi(\rho)]}$	1.636	1.636	1.636	1.636
$\overline{\alpha(\rho)}$	0.2726	0.5452	1.0904	1.6356
$\exp\overline{[\alpha(\rho)]}$	1.3133	1.725	2.975	5.131
f	0.009	0.016	0.025	0.030

$$f' = 1 -$$

$$\sum_{i=1}^{10} \int_\gamma^{\gamma'} Y(R_b) \left[\overline{C}_A \int_0^1 \alpha(\rho)\rho d\rho \right]_{\text{at } R_b = R_b} dR_b$$

$$\div \sum_{i=1}^{10} \int_\gamma^{\gamma'} Y(R_b) \left[\int_0^1 \alpha(\rho)\rho d\rho \right]_{\text{at } R_b = R_b} dR_b \qquad (6.86)$$

Table 6. Comparison of Results from Equations 6.82 and 6.46

$R_b \times 10^{10}$, m	7		10		12	
			Solute separation, %			
Operating pressure, kPa (gauge)	Eq. 6.83	Eq. 6.47	Eq. 6.83	Eq. 6.47	Eq. 6.83	Eq. 6.47
1,700	52.5	65.6	12.7	30.3	1.33	4.1
3,400	67.9	74.3	21.0	38.5	2.31	1.5
6,800	79.7	82.4	31.1	45.0	3.61	−4.0
10,200	84.5	85.5	36.6	48.1	4.34	−8.3

$$f' = 1 -$$

$$\sum_{i=1}^{10} \int_{\gamma}^{\gamma'} Y(R_b) \left[\overline{C}_A \left(\frac{\int_0^1 \alpha(\rho)\rho d\rho}{\int_0^1 \rho d\rho} \right) \right]_{\text{at } R_b=R_b} dR_b$$

$$\div \sum_{i=1}^{10} \int_{\gamma}^{\gamma'} Y(R_b) \left[\left(\frac{\int_0^1 \alpha(\rho)\rho d\rho}{\int_0^1 \rho d\rho} \right) \right]_{\text{at } R_b=R_b} dR_b \qquad (6.87)$$

$$f' = 1 -$$

$$\sum_{i=1}^{10} \int_{\gamma}^{\gamma'} Y(R_b) \left[\overline{C}_A \times \overline{\alpha(\rho)} \right]_{\text{at } R_b=R_b} dR_b$$

$$\div \sum_{i=1}^{10} \int_{\gamma}^{\gamma'} Y(R_b) \left[\overline{\alpha(\rho)} \right]_{\text{at } R_b=R_b} dR_b \qquad (6.88)$$

where

$$\gamma = \overline{R_b} - 3.6\sigma + 0.6i\sigma \qquad (6.89)$$

$$\gamma' = \overline{R_b} - 3.0\sigma + 0.6i\sigma \qquad (6.90)$$

and

$$\overline{C}_A = \frac{\exp\left[\overline{\alpha(\rho)}\right]}{1 + \left\{b/\exp\left[-\overline{\Phi(\rho)}\right]\right\}\left\{\exp\left[\overline{\alpha(\rho)}\right] - 1\right\}} \qquad (6.91)$$

Designating $\overline{C}_A \times \overline{\alpha(\rho)}$ and $\overline{\alpha(\rho)}$ calculated for $R_b = \overline{R_b} - 3.3\sigma + 0.6i\sigma$ as "$(1-f')_i\overline{\alpha(\rho)}_i$" and "$\overline{\alpha(\rho)}_i$," respectively, and considering the above quantities as constant from $R_b = \overline{R}_b - 3.6\sigma + 0.6i\sigma$ to $R_b = \overline{R}_b - 3.0\sigma + 0.6i\sigma$, f' can be approximated by

$$f' = 1 -$$

$$\sum_{i=1}^{10} (1 - f')_i \overline{\alpha(\rho)_i} \int_\gamma^{\gamma'} Y(R_b) dR_b$$

$$\div \sum_{i=1}^{10} \overline{\alpha(\rho)_i} \int_\gamma^{\gamma'} Y(R_b) dR_b \qquad (6.92)$$

Example 12

Suppose a membrane has a Gaussian pore size distribution of 7×10^{-10}m and 12×10^{-10}m, corresponding to $\overline{R}_b - 3\sigma$ and $\overline{R}_b + 3\sigma$, respectively. Calculate the separation of isopropyl alcohol at the operating pressure of 10,200 kPa (gauge). Assume $k = \infty$.

$$\overline{R}_b = \left(7 \times 10^{-10} + 12 \times 10^{-10}\right)/2 = 9.5 \times 10^{-10}\text{m}$$

Since 12×10^{-10}m corresponds to $\overline{R}_b + 3\sigma$ of the distribution, $\sigma = (12 \times 10^{-10} - 9.5 \times 10^{-10})/3 = 0.8333 \times 10^{-10}$m.

Using the numerical values for $\overline{R}_b$ and σ obtained above, Equation 6.79, separation data given in Table 6 (at the operating pressure of 10,200 kPag and according to Equation 6.46, and the numerical values representing the area

$$\int_{\overline{R}_b - 3.6\sigma + 0.6i\sigma}^{\overline{R}_b - 3\sigma + 0.6i\sigma} Y(R_b) dR_b$$

Table 7 was produced.

Using the quantities given in Table 7 and Equation 6.92, f' can be calculated to be 0.517.

6.2   Pore Model for Membrane Gas Transport

Several investigations of the permeation of gases through porous media are described in the literature [214]–[229]. From these investigations it is clear that any model to analyze the data on the permeation of gases through porous membranes must take into account contributions of (1) Knudsen, (2) slip, (3) viscous, and (4) surface transport. Some attempts were made to interpret the data of gas permeation through polymeric membranes using the concept of flow through porous media [230]–[236]. It is shown in this chapter how the

Table 7. Numerical Values Involved in the Gaussian Distribution

i	$(\bar{R}_b - 3.3\sigma + 0.6i\sigma)$	$\overline{\alpha(\rho)}_i$	$(1 - f')_i$	a^{a}
1	7.25	0.5373	0.840	0.0069
2	7.75	0.6348	0.788	0.0277
3	8.25	0.7189	0.729	0.0792
4	8.75	0.8196	0.670	0.1592
5	9.25	0.9270	0.600	0.2257
6	9.75	1.0409	0.521	0.2257
7	10.25	1.1614	0.425	0.1592
8	10.75	1.2885	0.306	0.0792
9	11.25	1.4222	0.176	0.0277
10	11.75	1.5625	0.010	0.0069

$$^{\text{a}}a = \int_{\bar{R}_b - 3.6\sigma + 0.6i\sigma}^{\bar{R}_b - 3\sigma + 0.6i\sigma} Y(R_b)dR_b.$$

model for gas permeation through porous membranes is related to the preferential sorption-capillary flow mechanism that was proposed as a fundamental transport mechanism underlying the reverse osmosis phenomena.

The preferential sorption-capillary flow mechanism proposed for reverse osmosis transport includes the flow of two distinct layers in the membrane pore. One is the layer of the interfacial fluid, which is under the influence of the interaction force working from the pore wall. The other is the layer of the bulk fluid, which is no longer influenced by the above-interaction force, since such an interaction force is effective in a relatively short range and its strength diminishes quickly as the distance from the pore wall increases (Figure 6.18). Although the properties of the two fluid layers may be very different, both of them are in a liquid-like condensed phase. The change from one layer to the other is believed to be continuous. In analogy to the above mechanism for reverse osmosis transport, the flow of two distinct phases can be included for the transport of gas through membranes. One is the layer of adsorbed gas molecules that are at the gas-pore wall interface and under a strong interaction force from the pore wall. The other is gas molecules that are free from the influence of the pore wall. While molecules in the former layer form a condensed phase where molecules are densely packed, those in the latter phase are in a gas phase, and they diffuse through the membrane freely under no restriction from the pore wall (Figure 6.19). In order to describe the free gas flow in a pore, three mechanisms were proposed depending on the size of the pore relative to the mean free path of the permeating gas molecules. The mean free path is defined as the distance

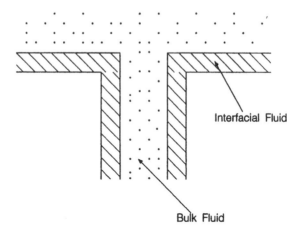

Figure 6.18.   Schematic representation of interfacial and bulk fluid layers.

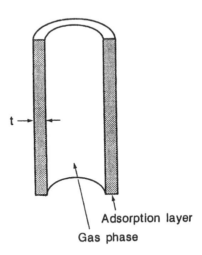

Figure 6.19.   Schematic representation of the sorbed gas layer in a membrane pore. (Reproduced from [239] with permission.)

traversed by a gas molecule from one collision to the next and is given by

$$\lambda = \frac{kT}{\sqrt{2}p\pi d^2} \qquad (6.93)$$

where k, p, and d are the Boltzmann constant (1.381×10^{-23} J/K), gas pressure (Pa), and the collision diameter of the gas (m). When the pore radius is less than 0.05λ, the gas transport in a cylindrical pore is by the Knudsen mechanism, and the molar flow rate, q_k(mol/s), through a single pore of radius R is given by [230],

$$q_k = \left(\frac{32\pi}{9MRT}\right)^{1/2} R^3 \frac{p_2 - p_3}{\delta} \qquad (6.94)$$

When the pore size is between 0.05 and 50λ, the gas transport in a cylindrical pore is by the slip flow, and the molar flux through a pore of radius R is given by [230]

$$q_{sl} = \frac{\pi R^3 (p_2 - p_3)}{M\bar{c}\delta} \qquad (6.95)$$

where $(p_2 - p_3)$ is the pressure differential across the membrane, and

$$\bar{c} = (8RT/\pi M)^{1/2} \qquad (6.96)$$

When the pore radius is larger than 50λ, the gas transport in a cylindrical pore is by the viscous flow, and the molar flux through a pore of radius R is given by [230],

$$q_v = \frac{\pi R^4 \bar{p} (p_2 - p_3)}{8\eta RT\delta} \qquad (6.97)$$

where

$$\bar{p} = \frac{p_2 + p_3}{2} \qquad (6.98)$$

Assuming a Gaussian normal distribution in the pore radius,

$$N(R_b) = \frac{N_t}{(2\pi)^{1/2}\sigma} \exp\left[-\frac{1}{2}\left(\frac{R - \bar{R}}{\sigma}\right)^2\right] \qquad (6.99)$$

where N_t is the total number of pores on the effective membrane surface area ($1/Sm^2$), $\bar{R}$ is the average pore radius (m), and σ is the standard deviation of the pore size distribution (m). The total molar flux of gas through a given membrane area, Q_g (kmol/s), can be given as the sum of Knudsen, slip, and viscous transport by integrating over the range of the pore radius applicable to each transport mechanism. Since

$$Q_g = Q_k + Q_{sl} + Q_v \tag{6.100}$$

therefore,

$$
\begin{aligned}
Q_g = \; & \frac{N_t}{\delta} \left(\frac{32\pi}{9MRT} \right)^{1/2} \frac{(p_2 - p_3)}{(2\pi)^{1/2}\sigma} \int_0^{0.05\lambda} R^3 \exp\left[-\frac{1}{2} \left(\frac{R - \overline{R}}{\sigma} \right)^2 \right] dR \\
& + \frac{N_t}{\delta} \left(\frac{\pi}{M\overline{c}} \right) \frac{(p_2 - p_3)}{(2\pi)^{1/2}\sigma} \int_{0.05\lambda}^{50\lambda} R^3 \exp\left[-\frac{1}{2} \left(\frac{R - \overline{R}}{\sigma} \right)^2 \right] dR \\
& + \frac{N_t}{\delta} \left(\frac{\pi \overline{p}}{8\eta RT} \right) \frac{(p_2 - p_3)}{(2\pi)^{1/2}\sigma} \int_{50\lambda}^{R_{\max}} R^4 \exp\left[-\frac{1}{2} \left(\frac{R - \overline{R}}{\sigma} \right)^2 \right] dR \quad (6.101)
\end{aligned}
$$

or

$$Q_g = \frac{N_t}{\delta} (G_1 I_1 + G_2 I_2 + G_3 I_3)(p_2 - p_3) \tag{6.102}$$

where

$$G_1 = \left(\frac{32\pi}{9MRT} \right)^{1/2} \tag{6.103}$$

$$G_2 = \frac{\pi}{M\overline{c}} \tag{6.104}$$

$$G_3 = \left(\frac{\pi \overline{p}}{8\eta RT} \right) \tag{6.105}$$

and

$$I_1 = \frac{1}{(2\pi)^{1/2}\sigma} \int_0^{0.05\lambda} R^3 \exp\left[-\frac{1}{2} \left(\frac{R - \overline{R}}{\sigma} \right)^2 \right] dR \tag{6.106}$$

$$I_2 = \frac{1}{(2\pi)^{1/2}\sigma} \int_{0.05\lambda}^{50\lambda} R^3 \exp\left[-\frac{1}{2} \left(\frac{R - \overline{R}}{\sigma} \right)^2 \right] dR \tag{6.107}$$

$$I_3 = \frac{1}{(2\pi)^{1/2}\sigma} \int_{50\lambda}^{R_{\max}} R^4 \exp\left[-\frac{1}{2} \left(\frac{R - \overline{R}}{\sigma} \right)^2 \right] dR \tag{6.108}$$

The contribution of different mechanisms to the total flow varies considerably depending on the value of $R_{\max}$. If $R_{\max} < 0.05\lambda$, only Knudsen flow prevails; if $R_{\max} < 50\lambda$, viscous flow is absent, and the upper limit of slip flow is $R_{\max}$; and if $R_{\max} > 50\lambda$, all mechanisms are operative simultaneously.

The flow of the condensed phase of adsorbed molecules on the pore wall is called surface flow. The expression for the surface flow was developed based on the two-dimensional force balance for an adsorbed film following Gilliland et al. [218]. The starting point for the derivation of the transport equation is that the balance is established between the spreading pressure of the adsorbed film and the friction force exerted on the film when it moves at a steady velocity on a smooth surface. The surface is considered to be the interface formed between the adsorbed film and the membrane pore wall. Assuming Henry's law for the gas adsorption, the molar flux by the surface flow through pores on a given membrane area is given by [230],

$$Q_s = \frac{RT\rho_{app}}{2\tau C_R \delta^2} k_H^2 \frac{I_4}{I_5} \bar{p}(p_2 - p_3) \tag{6.109}$$

where

$$I_4 = \frac{\pi N_t}{(2\pi)^{1/2}\sigma} \int_0^{R_{max}} R^2 \exp\left[-\frac{1}{2}\left(\frac{R - \bar{R}}{\sigma}\right)^2\right] dR \tag{6.110}$$

$$I_5 = \frac{\pi N_t}{(2\pi)^{1/2}\sigma} \int_0^{R_{max}} R \exp\left[-\frac{1}{2}\left(\frac{R - \bar{R}}{\sigma}\right)^2\right] dR \tag{6.111}$$

The total gas permeation rate, Q_t, is the sum of gas phase flow, Q_g, and the surface flow, Q_s, and therefore,

$$Q_t = Q_g + Q_s \tag{6.112}$$

$$= \frac{N_t}{\delta}(G_1 I_1 + G_2 I_2 + G_3 I_3)(p_2 - p_3) +$$

$$\frac{RT\rho_{app}}{2\tau C_R \delta^2} k_H^2 \frac{I_4}{I_5}\bar{p}(p_2 - p_3) \tag{6.113}$$

Furthermore, the permeability, A_G, is defined as

$$A_G = \frac{Q_t}{S(p_2 - p_3)} \tag{6.114}$$

and from Equation 6.113,

$$A_G = A_1(G_1 I_1 + G_2 I_2 + G_3 I_3) + A_2 \frac{I_4}{I_5}\bar{p} \tag{6.115}$$

where,

$$A_1 = \frac{N_t}{S\delta} \tag{6.116}$$

and

$$A_2 = \frac{\mathrm{R}T\rho_{app}}{2\tau C_R\delta^2 S}k_H^2 \tag{6.117}$$

Equation 6.115 has a fundamental physicochemical basis. From any set of experimental $p_2 - p_3$ vs. A_G data, the average pore radius, $\overline{R}$, on the membrane surface, its standard deviation, σ, and the quantities A_1 and A_2 can be obtained by regression analysis.

The permeation data of helium, methane, and carbon dioxide were analyzed by the method described above. The steps involved in the search of the numerical parameters, A_1, A_2, $\overline{R}$, and σ are as follows [236]:

1. The permeation data are collected at different operating pressures for each individual gas.

2. The experimental permeability, A_G, is calculated for each operating pressure.

3. The optimum set of the four gas transport parameters are obtained using either the Simplex method [234] or a grid search method. In particular, for the latter method, a set of parameters describing the pore size distribution, $\overline{R}$ and σ, is assumed corresponding to a point on a grid covering $(\overline{R}, \sigma)$ surface. An appropriate value for R_{max} is also chosen.

4. Then integrals I_1, I_2, I_3, I_4, and I_5 are calculated and substituted into Equation 6.115.

5. Parameters A_1 and A_2 are assumed, and the permeability is calculated from Equation 6.115 using the four transport parameters $(\overline{R}, \sigma, A_1,$ and $A_2)$ for different operating pressures, and the result is designated as A_G'. Note that there are n different $A_{G'}$ values calculated for n different operating pressures. Then the sum of square residuals is calculated according to

$$SS_R = \sum_{i=1}^{n} \left(A_G' - A_G\right)^2 \tag{6.118}$$

where A_G is the experimental permeability.

6. A set of parameters (A_1 and A_2) that minimizes SS_R is searched for by a nonlinear regression computer routine. Thus, we have an optimum set of A_1 and A_2, for each grid point $(\overline{R}, \sigma)$.

7. Steps 3 to 6 are repeated at all the grid points, and the parameters corresponding to the minimum value of SS_R are found.

The above calculation steps were applied to the experimental data obtained using a cellulose acetate reverse osmosis membrane dried by the solvent exchange method. The details of the condition of the membrane preparation are reported in the literature [236]. The transport parameters obtained from the above experimental data were $A_1 = 0.1277 \times 10^{21} \mathrm{m}^{-3}$, $A_2 = 0.1500 \times 10^{-7} \mathrm{kmol/m^3 sPa^2}$, $\overline{R} = 4.0 \times 10^{-10}\mathrm{m}$, and $\sigma = 1.3 \times 10^{-10}\mathrm{m}$ for CO_2 gas. Note that an average pore size $\overline{R}$ as low as $4 \times 10^{-10}\mathrm{m}$ was obtained. Furthermore, flux components, Q_k, Q_{sl}, Q_v, and Q_s were calculated and the contribution of the component flux to the total flux was determined for different gases under different operating pressures. The results are illustrated in Figure 6.20.

The graph shows a very strong contribution of the surface flow to CO_2 gas transport, and it increases with an increase in the operating pressure. This seems natural, since CO_2 adsorption to cellulose acetate material is the strongest among the gases under consideration. The amount of the adsorbed gas increases with an increase in the operating pressure. It is rather surprising that more than 20% is contributed by the surface flow to the transport of helium gas, even though helium is only weakly adsorbed onto the cellulose acetate surface. The Knudsen flow contribution is very strong for helium gas in the low operating pressure range, but it is replaced gradually by the slip flow contribution as the operating pressure increases. The Knudsen flow contribution is small for both CO_2 and CH_4 gases and becomes nearly equal to zero above a pressure of 1378 kPa gauge (200 psig). This reflects the collision diameters of these gases, which are much larger than that of He gas. CO_2 and CH_4 have smaller mean free paths than He, and the Knudsen flow contribution disappears at a relatively low operating pressure. CH_4 gas shows a very high slip flow contribution, since both Knudsen and surface flow contributions are weak for this gas.

As mentioned earlier, the viscous flow contribution occurs only when the pore radius is larger than 50λ. Some examples of mean free path, λ, are shown in Table 8 for various gases [237]. The data cover a wide range (from 30 to 200 nm) at the temperature of $0°C$.

In other words, the pore radius should be larger than 1500 or even 10,000 nm (1.5 or 10 μm) in order for viscous flow to be effective. These pores are extremely large and can be regarded as defective pinholes on the membranes that may arise when the membranes are not carefully prepared. Therefore, the gas-phase flow mechanisms in the ordinary gas separation membranes are restricted to either the slip or Knudsen flow mechanisms. Looking into the flux equations (Equations 6.95 and 6.94) applicable to the slip and Knudsen flow mechanisms, both q_{sl} and q_k are proportional to $p_2 - p_3$. Therefore, if the latter flow mechanisms govern the total flux (Q_t), the permeability A_G defined by Equation 6.114 should become independent of the operating pressure. On the other hand, according to the flux equation applicable to surface flow mecha-

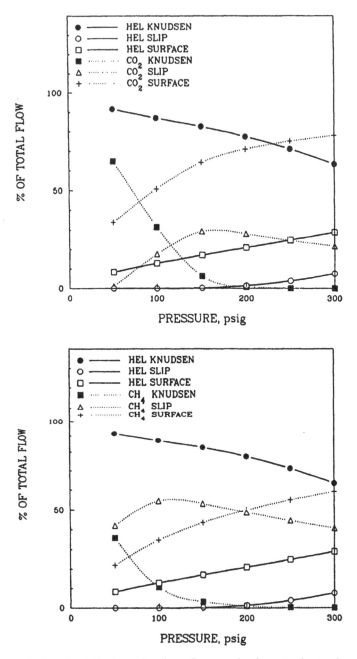

Figure 6.20. Contribution of various flow mechanisms to the total gas permeation rate. (Reproduced from [236] with permission.)

Table 8. Mean Free Path of Gases at 0°C
and 1 Atmospheric Pressure [237]

Gas	Mean free path $\times 10^{10}$(m)
Ammonia	441
Argon	635
Carbon dioxide	397
Carbon monoxide	584
Chlorine	287
Ethylene	345
Helium	1798
Hydrogen	1123
Nitrogen	600
Oxygen	647

nism (Equation 6.109), the permeation rate, Q_s, is proportional to $\bar{p}(p_2 - p_3)$. Therefore, the permeability, A_g, should be proportional to $\bar{p}$ and should increase with an increase in the feed gas pressure, p_2, when the surface flow mechanism governs the total permeation rate. Thus, the effect of the operating pressure is a criterion to judge whether the gas-phase flow mechanism (the Knudsen and slip flow) or the surface flow mechanism governs the total gas permeation through a membrane. Another important factor to distinguish the gas-phase flow that includes the slip flow and Knudsen flow mechanisms from the surface flow mechanism is the effect of the molecular weight of the permeant gas on the permeation flux. According to Equation 6.95, the permeation rate of the slip flow should be inversely proportional to the square root of the molecular weight of the permeant gas. Equation 6.94 indicates that the above-molecular weight dependency also applies to the Knudsen flow. This rule was found in the last century by Graham for gas permeation through the pore and is called Graham's law. The transport equation for the surface flow, however, does not explicitly include the molecular weight of the permeant gas. The adsorption parameter, k_H, is more important than the molecular weight, and the flux by the surface flow mechanism is intensified when the permeant adsorption to the membrane polymer becomes strong. The effect of the pore size on the gas-phase flow and the surface flow is also different in their respective transport equations. Equations 6.95 and 6.94 show that there is a third power relationship between the permeate flux and the pore radius for the slip and the Knudsen flow, whereas the ratio I_4/I_5, a term that is involved in the permeate flux by the surface flow

(Equation 6.109), implies that the flux is linearly proportional to the pore radius. Therefore, the contribution from the gas-phase flow increases rapidly as the pore radius increases.

In summary,

1. The operating pressure has no effect on the permeability defined by Equation 6.114 when the gas-phase flow (slip flow and the Knudsen flow) mechanism is dominant. The permeability should increase with an increase in the operating pressure when the surface flow is dominant.

2. The permeation rate should be inversely proportional to the square root of the molecular weight of the permeant gas when the gas phase flow mechanism is dominant. The permeation rate is affected, on the other hand, by Henry's constant when the surface flow becomes dominant.

3. As the pore radius increases, the contribution of the gas-phase flow mechanism becomes more significant.

The above tendencies expected from the transport equations for the gas-phase flow and the surface flow mechanisms can be observed experimentally in the permeation of CO_2 and CH_4 gas through cellulose acetate membranes of different pore sizes [233]. Figure 6.21 illustrates such experimental results. There are four dry cellulose acetate membranes involved in this study. They were cast under exactly the same conditions, but subjected to the post-treatment in which the membranes were heat treated at different temperatures, called shrinkage temperatures. They were then dried using methyl alcohol and isopropyl ether for solvent exchange. Apparently, different shrinkage temperatures resulted in different pore radii. Let us assume that the number of pores and the thickness of the effective skin layer of the membranes are the same for all membranes involved. Then helium permeability can be considered as the measure of the pore radius. The larger the pore radius, the higher the helium permeability. Figure 6.21 indicates that the pore radius becomes smaller as the shrinkage temperature decreases, which is contrary to the change of the pore radius of wet reverse osmosis membranes used for the reverse osmosis process. In Figure 6.21, membrane 85 shows the highest helium permeability, hence the largest pore radius. The gas-phase flow contribution for this membrane is supposed to be more dominant than for other membranes. This expectation is confirmed by the experimental results. The permeability of CH_4 gas is higher than that of CO_2

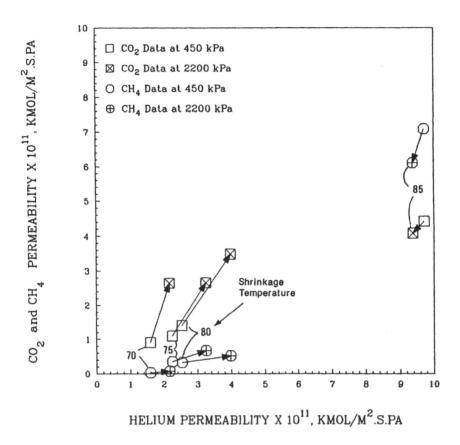

Figure 6.21. Experimental data for the permeation of CO_2 and methane gas through cellulose acetate membranes of different pore sizes. (Reproduced from [233] with permission.)

gas, since the molecular weight of CH_4 (16) is lower than that of CO_2 (44), and at higher operating pressure there is no increase in the permeability. (In fact, a slight decrease in the permeability was observed for both CH_4 and CO_2 gases.) Membrane 70 shows the lowest helium permeability, hence the smallest pore radius. The surface flow mechanism should be more dominant for this membrane than for other membranes. This expectation is also confirmed by the experimental results. Contrary to membrane 85, CH_4 permeability is lower than CO_2. This reflects the stronger adsorption of CO_2 gas onto cellulose acetate polymer than methane gas, as observed by gas chromatography experiments. Also, the CO_2 permeability increases significantly as the operating pressure increases. This result also confirms the dominant effect of the surface flow on the CO_2 transport. The effect of pressure on the methane permeability is not clear due to the extremely low permeability of methane. Thus, all the tendencies predicted theoretically by the transport equations of the gas phase flow and the surface flow were observed experimentally. Most importantly, Figure 6.21 shows that it is possible to prepare both gas-phase-flow-governing and surface-flow-governing membranes from cellulose acetate material by adjusting the membrane pore size.

Referring to Equation 6.103, this equation indicates that the permeation rate should be proportional to $\bar{p}\Delta P \left(= p_2^2 - p_3^2\right)$ when the pore size is so small that the primary contribution to the pore flow comes from the surface flow.

6.3   Pore Model for Pervaporation

Several attempts were also made to interpret the data on the pervaporation transport by the pore flow mechanism [238]–[242]. The development of pervaporation transport theory on the basis of the pore model is described most extensively by Okada and Matsuura [239]. The theory starts from the assumption that there are a bundle of straight cylindrical pores of length δ penetrating across the active surface layer of the membrane perpendicular to the membrane surface. Furthermore, it is assumed that the entire membrane is in an isothermal condition. As shown schematically in Figure 6.22, the pores are filled with liquid from the pore inlet to a distance δ_a along the cylindrical axis. The rest of the pore, of length δ_b, is filled with vapor. Therefore, there should be a liquid/vapor-phase boundary somewhere in the middle of the membrane pore. The pressure of the feed liquid and the permeate vapor are given as p_2 and p_3, respectively, and at the phase boundary the pressure should be the saturation vapor pressure of the feed liquid, which is denoted as p_*. The picture shown in Figure 6.22 has validity only when $p_3 \leq p_*$. If $p_3 > p_*$, the entire pore is filled with liquid.

Some explanation is in order for Figure 6.22. It is misleading to say that the pore is filled with "liquid" and with "vapor" and there is a clear boundary separating liquid and vapor phases in the membrane. The author's view is as follows. Suppose there is a tortuous pore in the membrane (an assumption of a straight cylindrical pore has been made in order to simplify the derivation of transport equations). Assuming many small segments in the membrane cross-section (see Figure 6.23), an increasing number, i, is given to a segment as the distance between the feed liquid-membrane interface and the segment

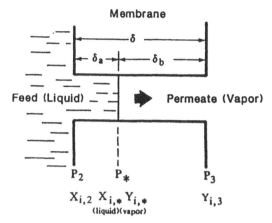

Figure 6.22. Schematic representation of pervaporation occurring in a membrane pore. (Reproduced from [239] with permission.)

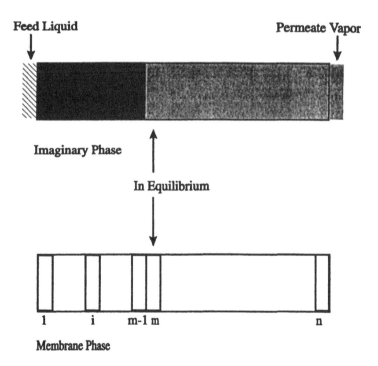

Figure 6.23. Transition from liquid to vapor transport in the membrane cross-section.

increases. Then the first segment is in sorption equilibrium with the feed liquid. Similarly, we can assume the presence of an imaginary phase that is in sorption equilibrium with the permeant in each segment. Obviously, such an imaginary phase should be liquid when i is small. As i increases, however, the pressure and the composition of the imaginary liquid phase should change gradually. When the segment number is sufficiently large, the pressure of the imaginary phase can no longer be greater than the vapor pressure of the permeant. Let us call the number of this segment, m. For $i \geq m$, the imaginary phase is vapor. Adsorption equilibrium is established between the imaginary vapor phase and the permeant in the pore. When the segment number, i, increases further and the segment eventually reaches the membrane-permeate vapor interface, the equilibrium phase is no longer imaginary, but is the permeate vapor.

As far as the imaginary phase is concerned, therefore, there exists a clear boundary between liquid and vapor, the position of which corresponds to the segment number m. Although the properties of the permeant in the pore, such as density and composition, may not show any discontinuity when the segment number increases from $m-1$ to m, the change in the density and the composition in the imaginary phase should be discontinuous due to the sharp phase change. Therefore, the liquid and vapor-filled portion of the pore, illustrated schematically in Figure 6.22, mean that a portion of the pore contains the permeant that is in equilibrium with an imaginary liquid phase, and another portion of the pore contains the permeant that is in equilibrium with a vapor phase, respectively. Hereafter, the liquid-filled portion of the pore is defined as the portion of the pore where the permeant flows by liquid transport, while the vapor-filled portion of the pore is the portion of the pore where the permeant flows by vapor-phase transport.

Let us now discuss the transport of a single-component system. Let us also assume $p_3 \leq p_*$. According to Equations 6.48 and 5.49,

$$J_B = \frac{c_{Bm} D_{Bm} v_B}{RT\delta} (p_2 - p_3) \tag{6.119}$$

in reverse osmosis. Note that there is a pressure drop from p_2 to p_3, across the membrane the thickness of which is δ. In pervaporation the pressure in the imaginary liquid phase drops from p_2 to p_* while the distance from the feed-membrane interface increases from 0 to δ_a. Then defining a liquid transport parameter as

$$A = \frac{c_{Bm} D_{Bm} v_B}{RT} \tag{6.120}$$

and changing the subscript B to liq, the molar permeation flux of liquid through the pore becomes

$$J_{\text{liq}} = \frac{A}{\delta_a} (p_2 - p_*)$$ (6.121)

where A is a proportionality constant. As for the transport of the vapor in the pore, we assume that the surface flow governs. This assumption can be justified, since the pore sizes of pervaporation membranes are usually as small as those of gas separation membranes in which the surface flow governs. (Film 70 in Figure 6.21 is a typical example.) Vapors of water and organic liquids are also known to be strongly adsorbed to the polymeric material. On the basis of the above assumption, the molar permeation flux should be proportional to the difference in the squares of pressures at both ends of the vapor-filled pores Equation 6.109. Then

$$J_{\text{vap}} = \frac{B}{\delta_b} \left(p_*^2 - p_3^2\right)$$ (6.122)

where B is another proportionality constant. At a steady state,

$$J = J_{\text{liq}} = J_{\text{vap}}$$ (6.123)

and, of course,

$$\delta = \delta_a + \delta_b$$ (6.124)

Combining Equations 6.121, 6.122, 6.123, and 6.124 and rearranging, we obtain

$$J = \frac{A}{\delta} (p_2 - p_*) + \frac{B}{\delta} \left(p_*^2 - p_3^2\right)$$ (6.125)

The flux, in terms of weight per unit area of the membrane surface, is

$$W = \frac{AM}{\delta} (p_2 - p_*) + \frac{BM}{\delta} \left(p_*^2 - p_3^2\right)$$ (6.126)

Combining Equations 6.121, 6.123, and 6.125,

$$\frac{\delta_a}{\delta} = \frac{A (p_2 - p_*)}{A (p_2 - p_*) + B \left(p_*^2 - p_3^2\right)}$$ (6.127)

Combining Equations 6.122, 6.123, and 6.125,

$$\frac{\delta_b}{\delta} = \frac{B \left(p_*^2 - p_3^2\right)}{A (p_2 - p_*) + B \left(p_*^2 - p_3^2\right)}$$ (6.128)

The above two equations indicate that the ratio δ_a/δ decreases, approaching a minimum value (but not zero) when p_3 approaches zero, while the above ratio approaches unity as p_3 approaches p_*. In other words, the pore is least filled with liquid when a high vacuum is applied on the downstream side of the membrane. When $p_3 \geq p_*$, the entire pore is filled with liquid, and only the liquid flow occurs in the pore. Then

$$J = \frac{A}{\delta}(p_2 - p_3) \tag{6.129}$$

The flux in terms of weight per unit surface area is

$$W = \frac{AM}{\delta}(p_2 - p_3) \tag{6.130}$$

and

$$\frac{\delta_a}{\delta} = 1 \tag{6.131}$$

Let us then consider a two-component system including ith and jth components. The same picture as Figure 6.22 should be applicable for this system. Furthermore, most of the pervaporation separation is carried out at a downstream pressure far lower than the saturation vapor pressure. As was pointed out earlier, the liquid-filled portion of the pore moves towards the feed-membrane interface when the downstream pressure is lowered. Let us now assume that the pore consists of the vapor-filled portion alone. In other words the position of the liquid-vapor boundary is close to the pore inlet. This assumption is considered valid when the pore radius is sufficiently small and the swelling of the membrane with feed liquid is not very strong [240]. With the above assumption, the following equations can be derived, ignoring the liquid flow contribution in Equation 6.125 and using partial pressures of each individual component instead of the total pressure in the vapor-phase transport. The saturation vapor is assumed to be in equilibrium with the feed liquid mixture.

$$J_i = \frac{B_i}{\delta}\left(p_{i,*}^2 - p_{i,3}^2\right) \tag{6.132}$$

$$J_j = \frac{B_j}{\delta}\left(p_{j,*}^2 - p_{j,3}^2\right) \tag{6.133}$$

and

$$Y_{i,3} = \frac{J_i}{J_i + J_j} \tag{6.134}$$

Therefore,

$$Y_{i,3} = \frac{\left(p_{i,*}^2 - p_{i,3}^2\right)}{\left(p_{i,*}^2 - p_{i,3}^2\right) + (B_j/B_i)\left(p_{j,*}^2 - p_{j,3}^2\right)} \tag{6.135}$$

where

$$p_{i,*} = p_* Y_{i,*} \tag{6.136}$$
$$p_{j,*} = p_* Y_{j,*} \tag{6.137}$$
$$p_{i,3} = p_3 Y_{i,3} \tag{6.138}$$
$$p_{j,3} = p_3 Y_{j,3} \tag{6.139}$$

The total molar permeation flux can be written as

$$J = \frac{B_i}{\delta}\left(p_{i,*}^2 - p_{i,3}^2\right) + \frac{B_j}{\delta}\left(p_{j,*}^2 - p_{j,3}^2\right) \tag{6.140}$$

and

$$W = \left[\frac{B_i}{\delta}\left(p_{i,*}^2 - p_{i,3}^2\right) + \frac{B_j}{\delta}\left(p_{j,*}^2 - p_{j,3}^2\right)\right](M_i Y_{i,3} + M_j Y_{j,3}) \tag{6.141}$$

The weight flux of pure jth component at $p_3 = 0$ is given as

$$W_j\left(\text{ at } X_{j,*} = 1 \text{ and } p_3 = 0\right) = \frac{B_j}{\delta}p_{j,\text{pure},*}^2 M_j \tag{6.142}$$

Defining the relative weight flux as

$$W_{\text{rel}} = W/W_j\left(\text{at } X_{j,*} = 1 \text{ and } p_3 = 0\right) \tag{6.143}$$

$$W_{\text{rel}} = \frac{\left[\frac{B_i}{\delta}\left(p_{i,*}^2 - p_{i,3}^2\right) + \frac{B_j}{\delta}\left(p_{j,*}^2 - p_{j,3}^2\right)\right](M_i Y_{i,3} + M_j Y_{j,3})}{(B_j/\delta)\,p_{j,\text{pure},*}^2 M_j} \tag{6.144}$$

and

$$W_{\text{rel}} = \frac{1}{p_{j,\text{pure},*}^2 M_j}\left[\frac{1}{(B_j/B_i)}\left(p_{i,*}^2 - p_{i,3}^2\right) + \left(p_{j,*}^2 - p_{j,3}^2\right)\right] \times (M_i Y_{i,3} + M_j Y_{j,3}) \tag{6.145}$$

As for the pervaporation of a single-component system, there are several experimental data available in the literature. An example is shown in Figure 6.24 for the pervaporation of 2-propanol by polydimethylsiloxane membrane with a thickness of 7.62×10^{-5}m [98]. The line correlating the permeate flux and the

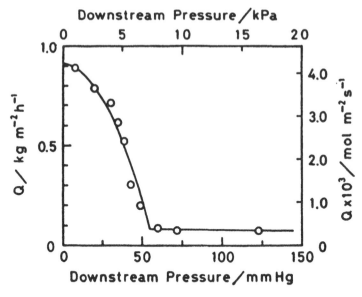

Figure 6.24. Pervaporation of 2-propanol. (Reproduced from [98] with permission.)

downstream pressure consists of two components: one the parabolic component, and the other a linear component. The best fit of the calculated permeate flux, according to the transport Equations 6.125 and 6.129, were obtained based on the parameters $A = 6.24 \times 10^{-14}$ mol m^{-1}s^{-1}Pa^{-1} and $B = 3.51 \times 10^{-15}$ mol m^{-1}s^{-1}Pa^{-2}. The calculated values are compared with experimental ones, in Figure 6.24. The agreement is satisfactory, indicating the validity of the transport equations and the associated parameters. It has to be noted that the shape of the flux curve as a function of the downstream pressure is the same as that obtained by the solution-diffusion model. According to Equation 6.125, J should increase with an increase in p_2, which was also predicted by Equation 5.315.

As for the pervaporation separation of binary mixtures, the calculation was done with respect to the binary system of water(i) / ethanol(j) mixtures. The water mole fraction in the permeate vapor, $Y_{i,3}$, and the relative weight flux, W_{rel}, were calculated from Equations 6.135 and 6.145, for different B_j/B_i values. The saturation vapor pressure and the vapor composition of water/ethanol binary mixtures for 40°C were adopted from the literature [243], and therefore the calculated values should correspond to pervaporation data for 40°C. All the results are summarized in Figure 6.25.

The main features of Figure 6.25 are as follows. When the B_j/B_i value is as small as 0.001, water is strongly enriched in the permeate. The relative permeation rate increases with an increase in the feed water mole fraction. When the B_j/B_i value is as large as 10, on the other hand, ethanol is enriched

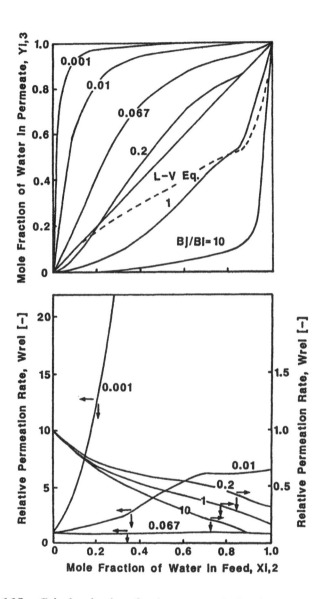

Figure 6.25. Calculated values for the water mole fraction in the permeate and the relative permeation rate for different B_j/B_i ratios. (Upstream pressure, 10,325 Pa; downstream pressure, 0 Pa; temperature, 40°C; system, water/ethanol mixtures.) (Reproduced from [240] with permission.)

in the permeate, and the relative permeation rate decreases with an increase in the feed water mole fraction. When B_j/B_i values are in the intermediate range, the tendencies to be observed in pervaporation experiments lie between those corresponding to B_j/B_i values of 0.001 and 10. It is particularly worth noting that the relative permeation rate remains almost unity, without noticeable change, while water is enriched in the permeate, when $B_j/B_i = 0.067$.

Neel et al. reported pervaporation results of water/ethanol separation at 35°C with hydrophilic membranes prepared by grafting acrylic acid on polyethylene films [244]. The membranes were preferentially permeable to water. Their results were replotted and are shown in Figure 6.26. The higher the selectivity of the membrane, the more the increase in the permeation rate with an increase in water content of the feed solution. For example, the low selectivity membrane (film 0) has shown little increase in the permeation rate, whereas the membrane of the highest selectivity (film 3) has shown a strong increase in the permeation rate with an increase in the feed water mole fraction. The results for film 0 and film 3 correspond to the calculated values for B_j/B_i ratios of 0.001 and 0.067, respectively.

Tanigaki et al. reported, on the other hand, pervaporation of water/ethanol mixtures with a silicone membrane at 25°C [245]. As shown in Figure 6.27, their membranes are preferentially permeable to ethanol, and the relative permeation rate decreases with an increase in the feed water mole fraction. Similar results were obtained with respect to the membrane with a B_j/B_i ratio of 1.0 in Figure 6.25. When we look at the data of Figures 6.27 and 6.25 closely, a more striking resemblance between the calculated data at $B_j/B_i = 1.0$ (Figure 6.25) and Tanigaki's experimental data (Figure 6.27) is found, i.e., the line correlating the permeate mole fraction to the feed mole fraction and the liquid-vapor equilibrium curve intersect at the feed mole fraction between 0.8 and 0.9 in both figures. The above agreement between the calculated and the experimental values testifies to the applicability of the transport equations.

It has to be pointed out that experimental results obtained from GFT's polyvinyl alcohol membranes, particularly the line correlating the water mole fraction in the permeate to that in the feed (see Figure 6.28) [246], cannot be reproduced by Equations 6.135 and 6.145. Presumably, the unique shape observed in the above-correlation is either due to the composite nature of the commercial PVA membrane or due to the membrane swelling occurring at the high water contents, or both. Recently, Hauser et al. explained the shape of the above-correlation curve by using the sorption-diffusion model [247].

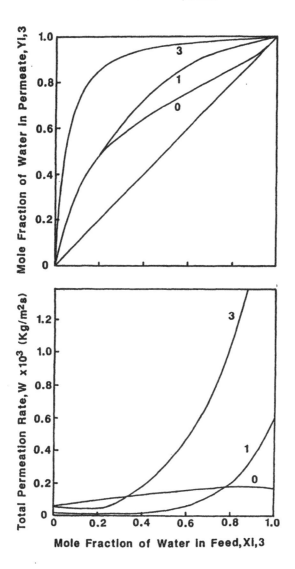

Figure 6.26. Experimental data for water mole fraction in the permeate and the permeation rate. (Temperature, 35°C; membrane, polyethylene grafted with polyacrylic acid; system, water/ethanol mixtures.) (Reproduced from [240] with permission.)

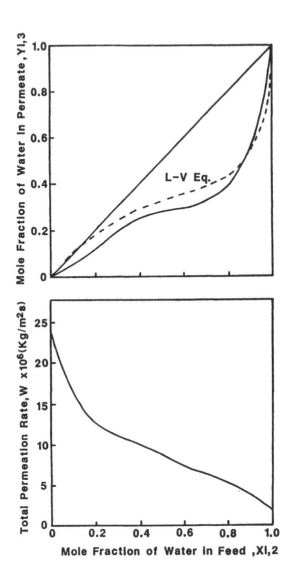

Figure 6.27.   Experimental data for water mole fraction in the permeate and the permeation rate. (Temperature, 25°C; membrane, silicone rubber; system, water/ethanol mixtures.) (Reproduced from [240] with permission.)

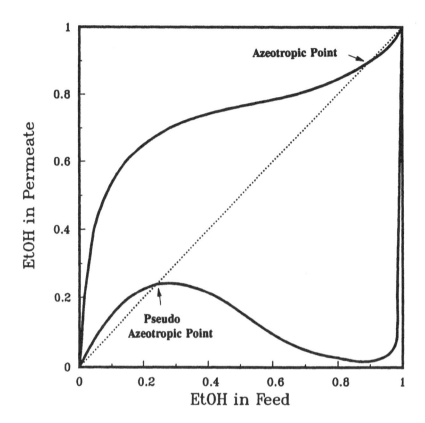

Figure 6.28. Pervaporation of water/ethanol mixtures by GFT membranes.

Equations 6.135 and 6.145 are based implicitly on the assumption that the pore size remains constant in the entire feed composition range and also in the direction of the permeant flow. The discussion on the effect of the composite structure and the membrane swelling is out of the scope of this chapter.

Figures 6.29 and 6.30 show the results of the calculation of the effect of the downstream pressure on the permeate composition and on the relative permeation rate, with respect to a water-selective membrane (Figure 6.29, $B_j/B_i = 0.001$) and also to an ethanol-selective membrane (Figure 6.30, $B_j/B_i = 1$), respectively. Figure 6.29 indicates that the downstream pressure strongly affects the membrane performance when the membrane is preferentially permeable to water. Both the water mole fraction in the permeate and the permeation rate decrease with an increase in the downstream pressure. On the other hand, in the case of the membrane that is preferentially permeable to ethanol, Figure 6.30 indicates that the effect of the downstream pressure is practically negligible. The data supporting the results shown in Figure 6.29 can be found in the work of Taketani and Minematsu [248]. The pervaporation of a water/ethanol mixture (the mole fraction of water in feed is 0.12) was attempted by cross-linked polyethyleneimine membranes at 30°C. The water mole fraction in the permeate dropped from 0.77 to 0.21 when the downstream pressure increased from 13.3 to 334 Pa. The permeation rates were 0.25 kg/m^2 h and 0.05 kg/m^2 h at downstream pressures of 13.3 and 334 Pa, respectively. A similar data was reported by Wesslein et al. for the pervaporation of a water/ethanol mixture by a polyvinyl alcohol membrane at 60°C [249].

Figure 6.31 shows some experimental data for the pervaporation of water/ethanol mixtures by a silicone rubber membrane preferentially permeable to ethanol. The experiment was conducted at 23°C for downstream pressures of 667, 1200, and 2100 Pa (5, 9, and 16 mmHg). As reported by Hoover and Hwang [250] and Tanigaki et al. [245], the silicone membrane showed preferential permeation to ethanol. Evidently, the downstream pressure has little effect on both permeate composition and permeation rate, supporting the calculated results shown in Figure 6.30. When the experimental data are closely examined, however, the relative permeation rate decreases slightly with an increase in the downstream pressure. The calculated values in Figure 6.30 show exactly the same tendency, justifying the transport model on which the calculation is based. It has to be noted, however, that the saturation vapor pressure of water and ethanol at 60°C are 1.99×10^4 and 4.69×10^4 Pa (149.4 and 351.9 mmHg), respectively. When the downstream pressure approaches the saturation vapor pressure, the assumption on which the theoretical calculation is based (i.e., the vapor permeation prevails across the membrane cross-section) becomes invalid, since liquid penetrates more deeply into the pore.

Figure 6.32 shows the results of the calculation to examine the effect of temperature on pervaporation. Two assumptions were made in the above calculation. One is that the system is isothermal, and the local temperature change can be ignored. The second assumption is that the ratio B_j/B_i involved in Equations 6.135 and 6.145 does not change with temperature. Therefore, the calcu-

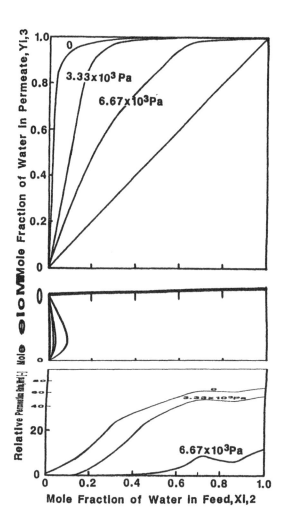

Figure 6.29. Effect of downstream pressure on pervaporation data, calculated for B_j/B_i ratio = 0.001. (Upstream pressure, 10,325 Pa; temperature, 40°C; system, water/ethanol mixtures.) (Reproduced from [240] with permission.)

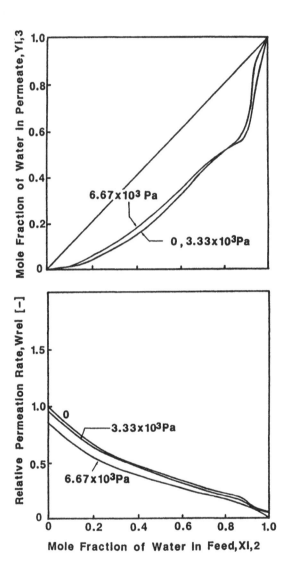

Figure 6.30. Effect of downstream pressure on pervaporation data, calculated for B_j/B_i ratio $= 1.0$. (Upstream pressure, 10,325 Pa; temperature, 40°C; system, water/ethanol mixtures.) (Reproduced from [240] with permission.)

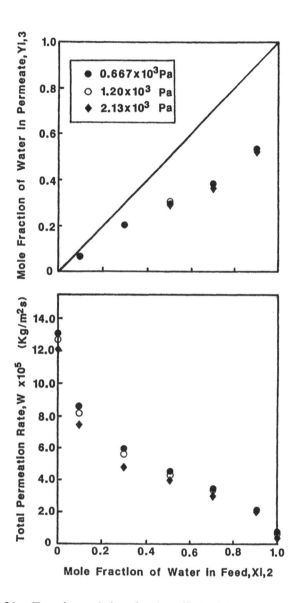

Figure 6.31. Experimental data for the effect of downstream pressure on pervaporation. (Upstream pressure, 10,325 Pa; temperature, 23°C; membrane, silicone rubber; system, water/ethanol mixtures.) (Reproduced from [240] with permission.)

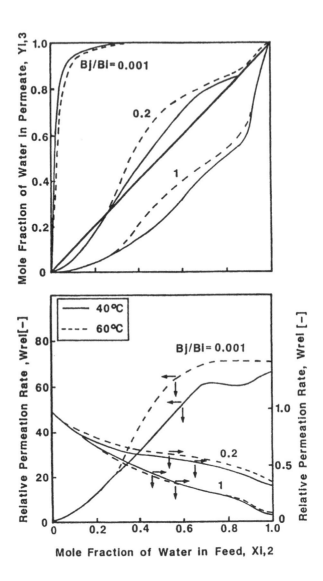

Figure 6.32. Effect of temperature on pervaporation data, calculated for B_j/B_i ratio = 0.001, 0.2 and 1. (Upstream pressure, 10,325 Pa; system, water/ethanol mixtures.) (Reproduced from [240] with permission.)

lated data depict the effect of the change in the liquid-vapor equilibrium on the pervaporation performance as a result of the temperature change. According to Figure 6.32, the effect of the temperature on the permeate composition and on the relative permeation rate is insignificant for any B_j/B_i ratios involved in the calculation within the temperature range of 40 to 60°C. Of course, the absolute permeation rate should increase with an increase in temperature, since the saturation vapor pressures of both water and ethanol increase. Brüschke et al. studied pervaporation of water/ethanol mixtures with cellulose triacetate membranes at temperatures of 30 and 80°C [246]. Their data indicate that cellulose acetate is preferentially permeable to water, and the permeate composition changes only little with temperature. Lee et al. studied pervaporation of water/ethanol mixtures with polyvinylidene fluoride membranes that are preferentially permeable to ethanol [7]. Their data also indicates only a slight decrease in the separation factor with an increase in the temperature. When there is any temperature effect that is more than expected from Figure 6.32, it is due to the change of the B_j/B_i ratio due to temperature, which was not taken into account when Figure 6.32 was prepared.

The applicability of the transport equations for pervaporation of binary mixtures other than the ethanol/water system was further examined [241]. Figure 6.33 shows some experimental data for pervaporation of ethanol/n-heptane mixtures with different membranes. The main features of experimental results are as follows. PA (aromatic polyamide) and CTA (cellulose triacetate-diacetate blend)-2B membranes exhibited ethanol selectivity in the entire range of the feed ethanol mole fraction, whereas n-heptane permeated preferentially through Glad (polyethylene) membrane. It has to be recalled that polyamide is one of the most hydrophilic materials among polymeric membranes studied, whereas polyethylene Glad membrane is the most hydrophobic. The results for the rest of the membranes are intermediate, and the lines representing the permeate ethanol mole fraction intersect the diagonal line. The results from the CTA-2A and CTA-2B films are different, indicating that the separation of liquid mixtures depends not only on the hydrophilicity of the membrane material, but also on the pore size of the membrane. As for the permeation rate, PA and CTA-2B membranes that are ethanol selective showed an increase in the total permeation rate with an increase in the feed mole fraction. Glad, on the other hand, showed a decrease in the total permeation rate with an increase in the feed ethanol mole fraction. For the rest of the membranes, the total permeation rate increased in the low ethanol mole fraction range, with an increase in the feed ethanol mole fraction, and remained constant in the intermediate ethanol mole fraction range, then decreased. Figure 6.34 shows the results of the calculation for different B_j/B_i ratios, using Equations 6.135 and 6.145. Note that subscript i indicates ethanol, while j indicates heptane in the figure. Thus, a small B_j/B_i ra-

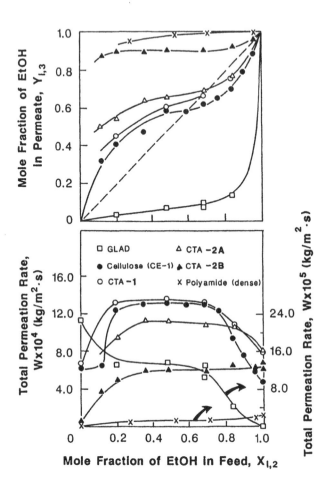

Figure 6.33. Permeate composition and permeation rate vs. feed composition for pervaporation of ethanol/n-heptane mixtures. (Upstream pressure, atmospheric; downstream pressure, PA: not available; CE-1: 1579 Pa, CTA-1: 2000 Pa, CTA-2A: 2000 Pa, CTA-2B: 1800 Pa, Glad: 733 Pa; temperature, PA: 25°C, other membranes: 23°C.) (Reproduced from [241] with permission.)

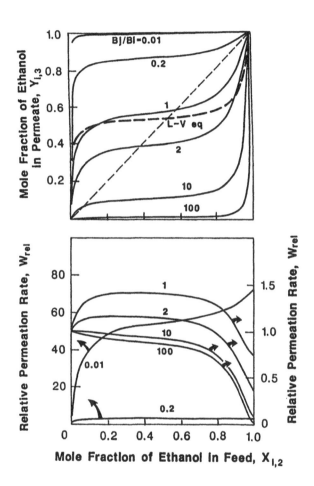

Figure 6.34. Permeate composition and relative permeation rate vs. feed composition for pervaporation of ethanol/n-heptane mixtures. (Upstream pressure, atmospheric; downstream pressure, 0 Pa; temperature, 23°C.) (Reproduced from [241] with permission.)

tio should correspond to an ethanol-selective membrane, whereas a large B_j/B_i ratio corresponds to a n-heptane-selective membrane. When the B_j/B_i ratio is 0.01, corresponding to a highly alcohol-selective membrane, the relative permeation rate increases with an increase in the feed ethanol mole fraction. When the B_j/B_i ratio is 100, corresponding to a highly n-heptane-selective membrane, the trend is reversed. When the above ratio is 1.0, the relative permeation rate increases, remains constant, and then decreases, as the feed ethanol mole fraction increases. Thus, all the main features of the experimental results could be reproduced by the theoretical calculation.

Quantitative agreement between the theoretical prediction and experimental data was also examined. Experimental data from PA, CTA-1, and Glad membranes were chosen for this purpose. From Equations 6.135 and 6.145, it was found that the parameters B_i/δ and B_j/δ can be calculated from a set of experimental data, including the permeation flux and the mole fraction of the ith component in the permeate, at one particular feed mole fraction of the ith component, and the data for other feed compositions can be calculated using the parameters so obtained. Table 9 B_i/δ and B_j/δ, and the experimental data from which the above parameters were calculated. In Figure 6.35 the agreement of calculated values with the experimental ones is examined. The agreement is excellent, testifying to the validity of the equations and the associated parameters. Figure 6.36 shows the theoretical prediction for the influence of the downstream pressure on pervaporation of ethanol/n-heptane mixtures. The figure indicates that a downstream pressure below 4000 Pa (30 mmHg) affects the permeate composition very little. But the pressure effect seems more significant on the total permeation rate. The latter effect is, however, least when the B_j/B_i ratio is unity. The predicted effect of downstream pressure was tested experimentally. In order to represent membranes of high n-heptane selectivity (B_j/B_i ratio = 100), and those which are slightly ethanol selective (B_j/B_i ratio = 1.0; the feed ethanol mole fraction $\leq$ 0.58), a Glad membrane and a CTA-2A membrane, respectively, were chosen. Experimental results at a fixed feed ethanol mole fraction of 0.482 are shown in Figure 6.37. The figure indicates that the total permeation rate of the Glad membrane decreased remarkably with an increase in the downstream pressure, while the permeation rate changed only little with respect to the CTA-2A membrane, which is in good agreement with the theoretical prediction.

It has been shown in the theory that the permeation rate of pure vapor can be expressed by Equation 6.122. The latter equation was tested experimentally by vapor permeation of pure ethanol at 79°C using a cellulose membrane. The feed ethanol vapor was in equilibrium with liquid ethanol, and therefore the feed ethanol vapor was the saturation vapor pressure of ethanol at 79°C (781 mm Hg). The downstream pressure was changed from nearly equal to 0 mm

Table 9. Parameters Pertinent to Pervaporation of Ethanol/n-Heptane System [241]

| Membrane | Experimental Data | | | Pressure | Calculated Values | |
	$X_{i,2}$	$Y_{i,3}$	W (kg/m²)	p_3 (Pa)	B_i/δ (mol/m² s Pa²)	B_j/δ
PA	0.531	0.984	1.24×10^{-5}	0	5.765×10^{-12}	1.317×10^{-13}
CTA-1	0.482	0.609	1.36×10^{-3}	2000	3.630×10^{-10}	3.050×10^{-10}
GLAD	0.482	0.069	1.38×10^{-4}	733	2.810×10^{-12}	5.086×10^{-11}

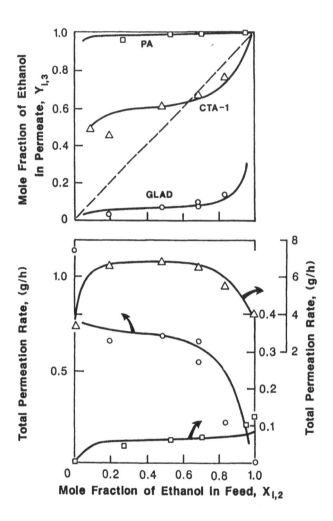

Figure 6.35. Comparison of theoretical predictions (lines) with experimental data (symbols) for pervaporation of ethanol/*n*-heptane mixtures. (Upstream pressure, atmospheric; downstream pressure, PA: 0 Pa, CTA-1: 2000 Pa, Glad: 733 Pa; temperature, PA: 25°C, CTA-1 and Glad: 23°C.) (Reproduced from [241] with permission.)

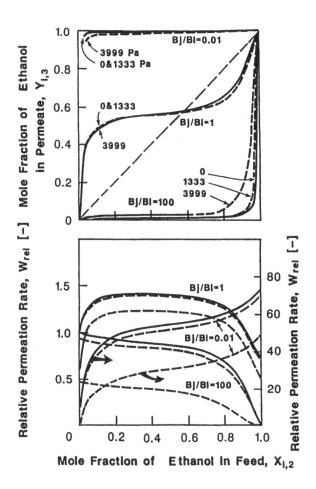

Figure 6.36. Effect of downstream pressure on pervaporation of ethanol/*n*-heptane mixtures (theoretical). (Upstream pressure, atmospheric; temperature, 23°C.) (Reproduced from [241] with permission.)

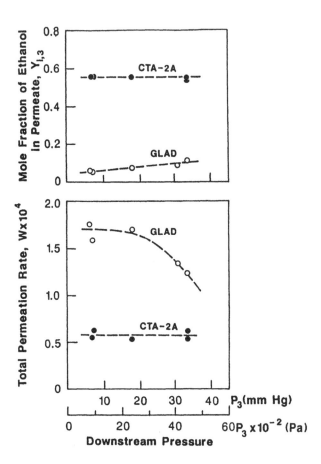

Figure 6.37.   Influence of downstream pressure on pervaporation of ethanol/*n*-heptane mixtures (experimental). (Upstream pressure, atmospheric; temperature, 23°C; ethanol mole fraction in feed, 0.482.) (Reproduced from [241] with permission.)

Hg to 760 mm Hg. As the experimental results shown in Figure 6.38 indicate, the permeation rate decreased parabolically with an increase in the downstream pressure. Thus, experimental data support Equation 6.122. As for pervaporation of a binary mixtures, Equations 6.135 and 6.145 were derived on the assumption that evaporation of the feed liquid takes place near the feed liquid-membrane interface. If this is a valid assumption, we should be able to obtain the same experimental data for both pervaporation and vapor permeation. This was tested experimentally by filling the feed chamber of the permeation cell with vapor that was in equilibrium with a feed liquid mixture of ethanol/heptane at 22°C. A CE(cellulose)-2 membrane was used. The downstream pressure was kept at 800 Pa (6 mm Hg). The results are illustrated in Figure 6.39. Comparing the vapor permeation data in Figure 6.39 with those for pervaporation by the CE-2 membrane in Figure 6.33, the shape of the lines correlating the permeate mole fraction of ethanol vs. the feed mole fraction of ethanol is exactly the same. The permeation rate in Figure 6.39, however, was about one tenth of that in Figure 6.33, although the shape of the correlation curve is the same, indicating the change in the morphology of the polymeric material, depending on whether it is in contact with liquid or vapor.

It is believed that the liquid-filled portion of the membrane is highly swollen and nonselective, while the vapor-filled portion of the membrane is dry and selective. When a nonselective layer and a selective layer are connected in series, the latter layer being behind the former layer, a possibility of concentration polarization arises at the boundary of two layers. This is called concentration polarization inside the membrane. Figure 6.40 shows a maximum in the profile of acetic acid mole fraction across the membrane observed by Tyagi et al. during the pervaporation of an acetic acid/water mixture, with an aromatic polyamide membrane [251]. The mole fraction of acetic acid is defined as the mole fraction of acetic acid in the liquid acetic acid/water mixture in the membrane, without including the membrane polymer. The figure clearly indicates the presence of concentration polarization inside the membrane.

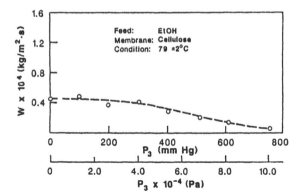

Figure 6.38. Permeation rate vs. downstream pressure for vapor permeation of pure ethanol (experimental). (Membrane, CE-1; temperature, 79°C (Reproduced from [241] with permission.)

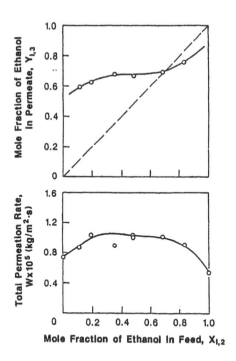

Figure 6.39. Permeate composition and permeation rate vs. feed composition for vapor permeation of ethanol/n-heptane mixtures (Experimental). (Membrane, CE-1; downstream pressure, 800 Pa; temperature, 23°C.) (Reproduced from [241] with permission.)

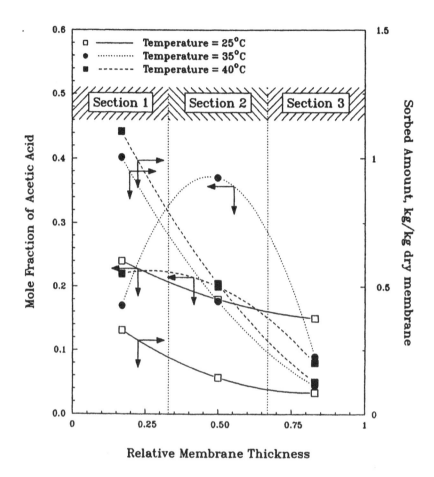

Figure 6.40. Concentration polarization occurring in the membrane. (Reproduced from [251] with permission.)

Nomenclature for Chapter 6

6.1 Reverse Osmosis

A = constant characterizing electrostatic repulsive force, m

A_t = specific surface area of the polymer particle, m²/kg

a = activity of the solute

$a_\pm$ = activity of symmetric electrolytes

B = constant characterizing van der Waals attractive force, m³

b = frictional function, —

C = constant characterizing repulsive force due to the overlap of electron cloud, m¹²

C_A = dimensionless solute concentration at the pore outlet, —

c_A = molar solute concentration, mol/m³

c_{Ai} = molar solute concentration at distance $\underline{d}$ from the interface, mol/m³

c_{Ab} = molar solute concentration of the bulk solution at infinite distance, mol/m³

D = constant characterizing steric repulsion at the interface, m

D_{water} = Stokes radius of water, m

D = dielectric constant of water, $78.54 \times 8.854 \times 10^{-12}$, F/m

D_{AB} = diffusivity of solute in water, m²/s

D' = dielectric constant of membrane material, F/m

$\underline{d}$ = distance from the interface, m

f = solute separation based on the solute concentration in the bulk feed, —

f' = true value of solute separation by the membrane pore, —

g = density of the permeate solution, kg/m^3

h_i = quantity defined by Equation 6.63, —

K = partition coefficient, —

k = mass transfer coefficient, m/s

M_B = molecular weight, kg/mol

m = solute molality, mol of solute/kg of solvent

$m_\pm$ = molality of electrolytes, mol/kg of solvent

N = Avogadro's number

PR = product permeation rate through given area of membrane surface, kg/h

PWP = pure-water permeation rate through given area of membrane surface, kg/h

p = pressure, Pa

q_w = molar flow rate of solvent water, mol/s

R = gas constant, 8.314 J/mol K

R_a = radius of the water channel in the membrane pore, m

R_b = membrane pore radius, m

$\overline{R}_b$ = average pore radius, m

r = radial distance in the cylindrical coordinate, m

T = absolute temperature, K

t_i = thickness of the pure-water layer, m

$u_B(r)$ = solution velocity profile in the cylindrical pore, m/s

$V_{R'}$ = chromatography retention volume, m^3

$[V_{R'}]_{min}$ = minimum retention volume, m^3

$[V_{R'}]_{water}$ = retention volume of heavy water, m^3

V_s = volume of the interfacial water in the stationary phase, m^3

v_s = permeation velocity, m/s

Subscripts

W	=	work required to move a charged particle from vacuum to a solvent media, J/mol
$Y(R_b)$	=	normal pore size distribution function, 1/m
Z	=	ionic valence
α	=	activity coefficient
$\alpha(\rho)$	=	dimensionless solution velocity profile in a cylindrical pore, —
β_1	=	dimensionless solution viscosity, —
β_2	=	dimensionless operating pressure, —
Γ	=	surface excess of the solute, mol/m^2
δ	=	effective pore length, m
ε	=	electric charge, 1.602×10^{-19}, C
η	=	viscosity of the solution in the pore, Pa s
λ_f	=	D/R_b, —
ρ	=	solution density, kg/m^3, or dimensionless radial distance, —
σ	=	interfacial tension, N/m, or the standard deviation of the normal pore size distribution, m
$\Phi(\rho)$	=	$\phi/\mathbf{R}T$, —
ϕ	=	potential in the interfacial force field, J/mol
χ_{AB}	=	quantity characterizing the solute-solvent friction, J s/m^2 mol
χ_{AM}	=	quantity characterizing the friction between solute and membrane pore wall, J s/m^2 mol
1	=	bulk feed solution
2	=	concentrated boundary solution
3	=	membrane permeate

6.2 Pore Model for Membrane Gas Transport

A = solute or constant interfacial area

i = ith Gaussian normal pore size distribution

T = constant temperature

A_G = gas permeability, $kmol/m^2$ s Pa

A_1 = constant for a given membrane related to the porous structure, m^2

A_2 = constant related to the surface transport, $kmol/m^3$ s Pa^2

C_R = coefficient of resistance for the transport of adsorbed molecules, kg/s m^2

$\bar{c}$ = mean speed of the gas molecules, m/s

d = collision diameter of gas, m

G_1 = constant defined by Equation 6.103

G_2 = constant defined by Equation 6.104

G_3 = constant defined by Equation 6.105

I_1 = quantity defined by Equation 6.106

I_2 = quantity defined by Equation 6.107

I_3 = quantity defined by Equation 6.108

I_4 = quantity defined by Equation 6.110

I_5 = quantity defined by Equation 6.111

Q_g = molar flow rate by the gas-phase flow mechanism through a given membrane area, kmol/s

Q_k = molar flow rate by the Knudsen mechanism through a given membrane area, kmol/s

Q_s = molar flow rate by the surface flow mechanism through a given membrane area, kmol/s

Q_{sl} = molar flow rate by the slip flow mechanism through a given membrane area, kmol/s

Q_t	=	total molar flow rate through a given membrane area, kmol/s
Q_v	=	molar flow rate by the viscous flow mechanism through a given membrane area, kmol/s
k	=	the Boltzmann constant, 1.381×10^{-23} J/K
k_H	=	adsorption constant pertaining to Henry's isotherm, kmol/kg Pa
M	=	molecular weight
$N(R_b)$	=	normal pore size distribution function, 1/m
N_t	=	total number of pores on a given membrane surface
p	=	gas pressure, Pa
$\bar{p}$	=	$(p_2 + p_3)/2$
q_k	=	molar flow rate by the Knudsen mechanism through a single pore, kmol/s
q_{sl}	=	molar flow rate by the slip flow mechanism through a single pore, kmol/s
q_v	=	molar flow rate by the viscous flow mechanism through a single pore, kmol/s
$\mathbf{R}$	=	gas constant, 8.314 J/mol K
R	=	pore radius, m
R_{max}	=	maximum pore radius, m
$\bar{R}$	=	average pore radius, m
S	=	effective membrane area, m^2
T	=	absolute temperature, K
δ	=	effective pore length, m
η	=	viscosity, Pa s
λ	=	mean free path, m

ρ_{app} = apparent density of the membrane, kg/m³

σ = standard deviation of the pore size distribution, m

τ = tortuosity factor for pores

Subscripts

2 = high-pressure side of the membrane

3 = low-pressure side of the membrane

6.3 Pore Model for Pervaporation

A = a constant related to the liquid transport, mol/m s Pa

B = a constant related to the vapor transport, mol/m s Pa²

J = molar permeation flux, mol/m² s

J_{liq} = J for liquid transport, mol/m² s

J_{vap} = J for vapor transport, mol/m² s

M = molecular weight

p_2 = upstream pressure, Pa

p_3 = downstream pressure, Pa

p_* = saturation vapor pressure, Pa

$p_{j,pure,*}$ = saturation vapor pressure of jth component in equilibrium with pure liquid of jth component, Pa

W = weight permeation flux, kg/m² s

W_{rel} = relative weight flux, —

Y_3 = mole fraction in the permeate vapor, —

Y_* = mole fraction in the saturated vapor, —

δ = the total pore length, m

δ_a = the length of the pore where liquid transport dominates, m

δ_b = the length of the pore where vapor transport dominates, m

304

Subscripts

$$i = i\text{th component of the mixture}$$
$$j = j\text{th component of the mixture}$$

7

Membrane Modules

7.1 Module Types

7.1.1 Plate and Frame Module

This type of module appeared in the earliest stage of industrial membrane applications. The structure is simple and the membrane replacement is easy. As illustrated in Figure 7.1, spacer-membrane-support plates are stacked alternately and pressed from both ends by oil pressure. The feed flows in the module, inwards and outwards, alternately, enabling the entire membrane surface to be covered by the feed stream. The membrane permeate is collected from each support plate. The spacer surface is made uneven in order to promote turbulence of the feed fluid and minimize concentration polarization. The module diameter is 20 to 30 cm. The total membrane area in one module is up to 19 m^2, according to the DDS design, depending on the height of the module. Since the permeate is collected, at each support plate, from two neighboring membranes, it is easy to find and replace severely damaged membranes. Figures 7.2 and 7.3 represent commercial plate and frame modules produced by DDS.

7.1.2 Spiral-Wound Module

The basic structure of this module is illustrated in Figure 7.4. A permeate spacer is sandwiched between two membranes, the porous support side of the membrane being faced to the permeate spacer. Three edges of the membranes are sealed with glue to form a membrane envelope, the open end being connected to a central tube with holes. The membrane leaf so produced is wound spirally around the central tube together with a feed spacer. In order to make the leaf length shorter, several membrane leaves are wound simultaneously as illustrated in Figure 7.5. Usually two to six leaves are used for a module of 4-in. diameter, and 4 to 30 leaves in a module of 8-in. diameter. A polypropylene or polyethylene net 0.2 to 2.0 mm thick is used for the feed spacer, whereas

Internal Flow

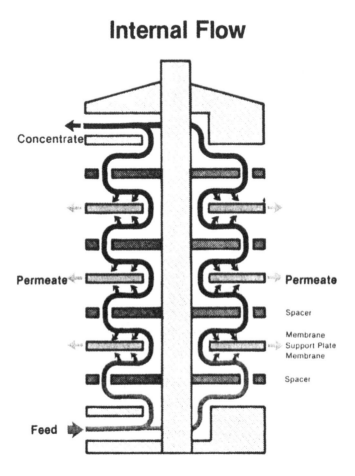

Figure 7.1. Schematic representation of DDS plate and frame module. (From DDS Technical Bulletin, with permission.)

Module 30

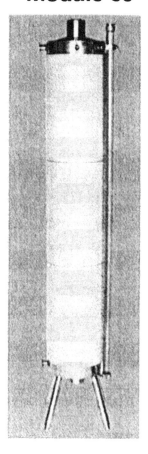

Figure 7.2. Commercial plate and frame module. (From DDS Technical Bulletin, with permission.)

Plates and Spacers

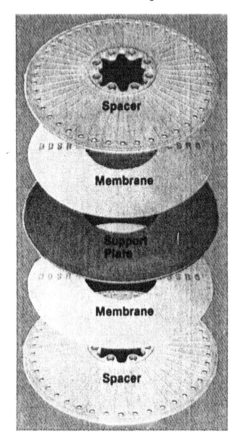

Figure 7.3. Spacer, membrane, and support plate in a commercial plate and frame module. (From DDS Technical Bulletin, with permission.)

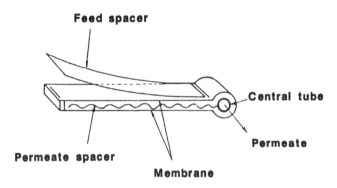

Figure 7.4. Basic structure of spiral-wound module.

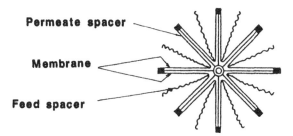

Figure 7.5. Multileaves spiral-wound module.

polyester cloth 0.2 to 1.0 mm thick, hardened with melamine or epoxy resin, is used for the permeate spacer. In the UOP module the feed flows through the feed spacer, parallel to the central tube, whereas the permeate flows through the permeate spacer, spirally, perpendicular to the feed flow direction, and is collected into the central tube. The structure of the Osmonics module is illustrated in Figures 7.6 and 7.7. In the Toray module, on the other hand, both feed and permeate flow spirally inwards (Figure 7.8). The latter structure enables the formation of a uniform flow pattern and minimizes concentration polarization. The spiral-wound module is featured by

1. High pressure durability

2. Compactness

3. Minimum membrane contamination

4. Minimum concentration polarization

5. Minimum pressure drop at the permeate channel

The module diameter is usually 4 or 8 in. and the length is 40 to 60 in.

7.1.3   Tubular Module

In this type of module a number of membranes of tubular shape are encased in a container. For example, 18 tubes are connected in series by headers at both ends of the Nitto NTR-1500-P18A module. Figures 7.9 and 7.10 show the structure of the module. Cellulose acetate membranes are formed in the internal wall of the support tube of 12-mm internal diameter. The tubular membranes so prepared are inserted into plastic tubes with many holes, which are mounted in a module container. The feed liquid flows inside the tube, and the permeate flows from the inside to the outside of the membrane tube and is collected at the permeate outlet. There are also tubular modules in which the feed is supplied to the outside of the membrane tube. The main features of the tubular module are

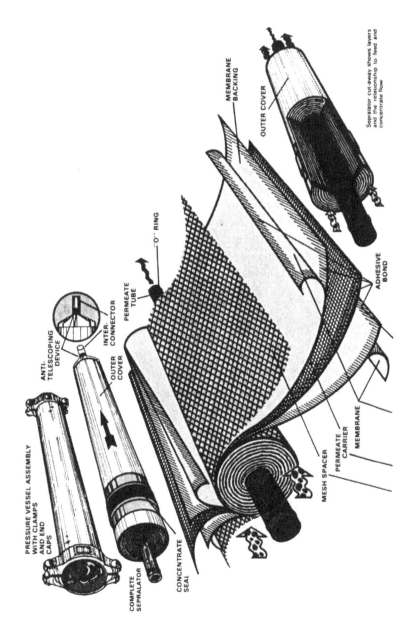

Figure 7.6. Structure of osmonics-type spiral-wound module. (From Technical Bulletin of Osmonics, Inc., with permission.)

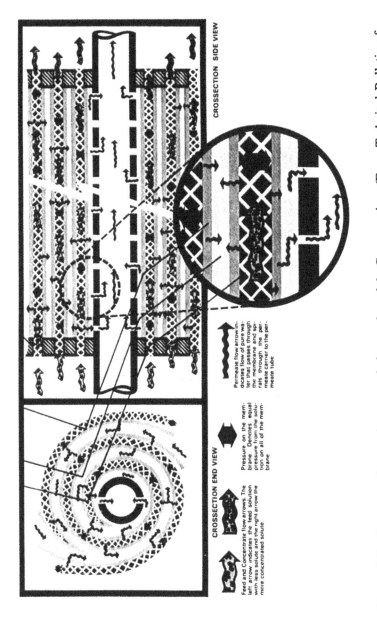

Figure 7.7. Structure of osmonics-type spiral-wound module Cross-section. (From Technical Bulletin of Osmonics, Inc., with permission.)

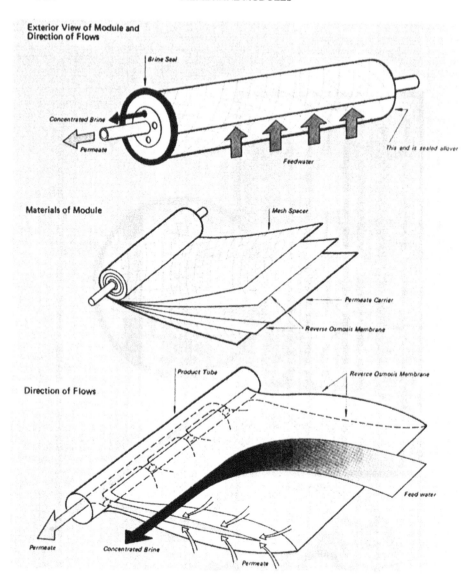

Figure 7.8. Structure of Toray-type spiral-wound module. (From Technical Bulletin of Toray Industries, Inc., with permission.)

Figure 7.9. Structure of Nitto-Denko-type tubular module. (From Technical Bulletin of Nitto Electric Industrial Co., Ltd., with permission.)

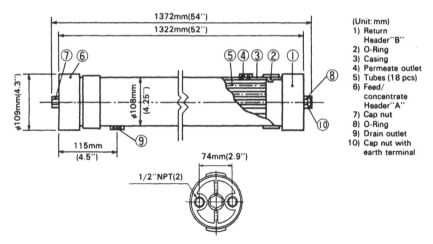

Figure 7.10. Picture of a Nitto-Denko-type tubular module. (From Technical Bulletin of Nitto Electric Industrial Co., Ltd., with permission.)

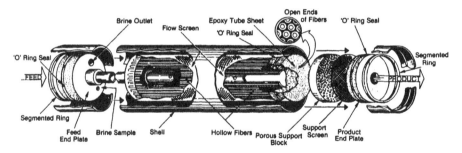

Figure 7.11. Design of Du Pont's hollow fiber module. (From Technical Bulletin of E.I. du Pont de Nemours Inc., with permission.)

1. The module can be operated with simple pretreatment of feed liquid.

2. Membrane contamination can be minimized by the high feed flow rate.

3. Contaminated membrane surfaces can easily be washed.

4. Membrane replacement is easy.

5. The membrane area/module space ratio is small.

6. Energy consumption per amount of liquid treated is high.

7.1.4   Hollow Fiber Module

Hollow fiber membranes are fibers 0.1 to 1.5 mm long with a hollow space inside (Figure 7.11). The feed is supplied to either the inside or outside of the fiber, and the permeate passes through the fiber wall to the other side of the

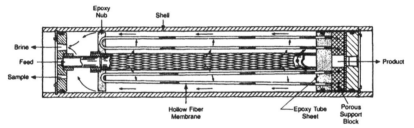

Figure 7.12. Design of Du Pont's hollow fiber module assembly. I. (From Technical Bulletin of E.I. du Pont de Nemours Inc., with permission.)

fiber. The fiber wall has a structure of the asymmetric membrane, the active skin layer being placed to the feed side. A bundle of hollow fibers are mounted in a pressure vessel, and the open end of U-shaped fibers are potted into a head plate (Figure 7.12). In a typical example of the DuPont permeator, the feed solution is distributed from the central distributor (Figure 7.13) and flows radially through the hollow fiber bundle. The permeate, on the other hand, flows inside the hollow fiber parallel to the distributor and is collected at the open ends of the fiber. The hollow fiber module is featured by a very large (membrane surface area/module space) ratio, and therefore the size of the hollow fiber module is smaller than other modules for a given performance capacity. However, the regeneration of fibers is very difficult once they are contaminated and degenerated.

7.2   Module Calculation

7.2.1   Reverse Osmosis

Hollow Fiber Module
There are several equations established for the design of hollow fiber reverse osmosis modules. Among those, equations derived by Gill et al. are the most rigorous, but complicated [252]–[254], while those developed by Ohya et al. are simple, but do not include the pressure change in hollow fibers [255]–[258]. Taniguchi [259], Darwish et al. [260], and Abdel-Jawad and Darwish [261] basically follow Ohya's approach. Rautenbach's approach includes all important aspects of hollow fiber modules, and the equations involved are relatively simple [262]. Therefore, the following development of hollow fiber design equations is based on his approach. In this approach a cylindrical hollow fiber bundle is considered as a continuous phase. A distribution tube is located at the center of the cylinder, and the feed solution is distributed evenly in the radial direction (see Figure 7.14). Looking closely into the bundle, each hollow fiber of length 2L forms a U tube, the fiber being bent at its center. Both ends of the fiber are open to the atmosphere (see Figure 7.15). The feed solution flows in the hollow

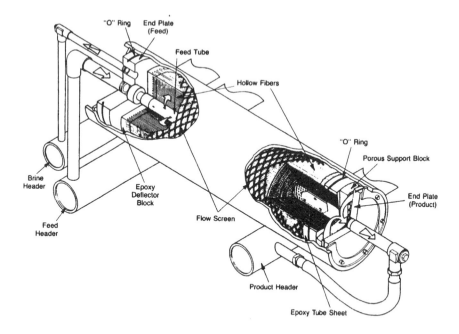

Figure 7.13. Design of Du Pont's hollow fiber module assembly. II. (From Technical Bulletin of E.I. du Pont de Nemours Inc., with permission.)

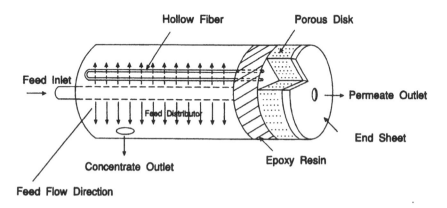

Figure 7.14. Schematic diagram of hollow fiber module.

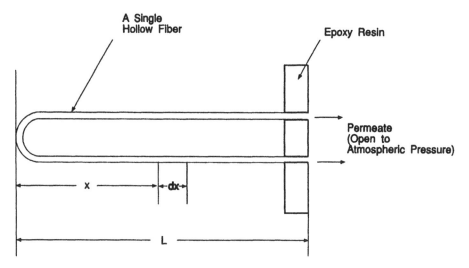

Figure 7.15. Schematic diagram of single hollow fiber.

fiber bundle on the shell side, and the permeate flows through the hollow fiber wall into the bore side. The permeate moves in the bore of the hollow fiber to the longitudinal direction and leaves the module from both ends of the hollow fiber, which are open to the atmosphere.

When model equations are developed to describe the module performance, pressure, solute concentration, and flow rate of the feed solution are considered to be the functions of the radial coordinate, r, alone. On the other hand, pressure, solute concentration, and the flow rate of the permeate solution inside ithe hollow fiber are considered to be the functions of both longitudinal and radial coordinates, (x, r).

First, we should look at a hollow fiber at the radial coordinate, r. According to the schematic diagram illustrated in Figure 7.16, the permeate mass balance for a small segment between the distance x from the end of the hollow fiber and $x + dx$, becomes as follows.

$$\text{Mass input} = \tfrac{\pi}{4}d_i^2 u_3 + \pi d_e dx v_w$$
$$\text{Mass output} = \tfrac{\pi}{4}d_i^2 \left(u_3 + \frac{\partial u_3}{\partial x}dx\right)$$

At the steady state, input and output are balanced; therefore,

$$\frac{\pi}{4}d_i^2 u_3 + \pi d_e dx v_w = \frac{\pi}{4}d_i^2 \left(u_3 + \frac{\partial u_3}{\partial x}dx\right) \tag{7.1}$$

where d_i and d_e are the inside diameter and effective diameter of a hollow fiber tube (m), u_3 is the permeate flow velocity in the bore of the hollow fiber tube

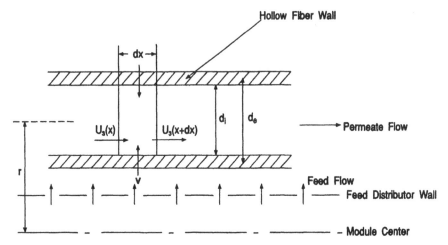

Figure 7.16. Mass balance in a hollow fiber tube.

(m/s), and v_w is the permeation velocity (m/s) of solvent (water) calculated on the basis of the effective diameter. Rearranging,

$$\frac{\pi}{4}d_i^2\frac{\partial u_3}{\partial x} = \pi d_e v_w \tag{7.2}$$

Therefore,

$$\frac{\partial u_3}{\partial x} = 4\frac{d_e}{d_i^2}v_w \tag{7.3}$$

The material balance at the same hollow fiber segment is as follows

$$\text{Input} = \frac{\pi}{4}d_i^2\left(u_3\bar{c}_{A3}\right) + \pi d_e dx J_A$$

where $\bar{c}_{A3}$ is the solute concentration on the bore side of the hollow fiber at the longitudinal distance x.

$$\text{Output} = \frac{\pi}{4}d_i^2\left[\left(u_3\bar{c}_{A3}\right) + \frac{\partial\left(u_3\bar{c}_{A3}\right)}{\partial x}dx\right]$$

At steady state, input and output should be balanced; therefore,

$$\frac{\pi}{4}d_i^2\left(u_3\bar{c}_{A3}\right) + \pi d_e dx J_A = \frac{\pi}{4}d_i^2\left[\left(u_3\bar{c}_{A3}\right) + \frac{\partial\left(u_3\bar{c}_{A3}\right)}{\partial x}dx\right] \tag{7.4}$$

Dividing both sides of the equation by dx and rearranging,

$$\frac{\pi}{4}d_i^2\frac{\partial(u_3\bar{c}_{A3})}{\partial x} = \pi d_e J_A \qquad (7.5)$$

Therefore,

$$\frac{\partial(u_3\bar{c}_{A3})}{dx} = \frac{4d_e}{d_i^2}J_A \qquad (7.6)$$

and

$$\frac{\partial u_3}{\partial x}\bar{c}_{A3} + u_3\frac{\partial\bar{c}_{A3}}{\partial x} = \frac{4d_e}{d_i^2}J_A \qquad (7.7)$$

Combining Equations 7.3 and 7.7,

$$\frac{4d_e}{d_i^2}v_w\bar{c}_{A3} + u_3\frac{\partial\bar{c}_{A3}}{\partial x} = \frac{4d_e}{d_i^2}J_A \qquad (7.8)$$

Rearranging,

$$\frac{\partial\bar{c}_{A3}}{\partial x} = \frac{4d_e}{u_3d_i^2}(J_A - v_w\bar{c}_{A3}) \qquad (7.9)$$

The pressure drop occurring while the solution passes through the above hollow fiber segment can be written as

$$\frac{\partial p_3}{\partial x} = -\frac{32}{d_i^2}\eta u_3 \qquad (7.10)$$

The molar solvent flux, J_B (mol of water per m^2 s), the molar flux of the solute, J_A (mol/m^2 s), and the concentration polarization occurring at the boundary layer on the feed side of the membrane can be written, by using Equations 5.4, 5.50, and 5.45 of Chapter 5, as

$$J_B = A\left[(p_2 - p_3) - b(c_{A2} - c_{A3})\right] \qquad (7.11)$$

$$J_A = B(c_{A2} - c_{A3}) \qquad (7.12)$$

$$\frac{c_{A2} - c_{A3}}{c_{A1} - c_{A3}} = \exp\left(\frac{v_w}{k}\right) \qquad (7.13)$$

Furthermore,

$$v_w = \frac{J_B}{c} \qquad (7.14)$$

$$c_{A3} = \frac{J_A}{v_w} \tag{7.15}$$

where c is the total molar concentration (mol/m^3) of permeate solution, which is nearly equal to the molar concentration of solvent (water). It is assumed in Equation 7.11 that the osmotic pressure is proportional to the molar concentration of solution. Some explanation is in order to explain the difference between c_{A3} and $\bar{c}_{A3}$. The quantity c_{A3} is the concentration in the solution permeating through the hollow fiber wall of the small segment the length of which is dx. This permeate solution joins the solution coming from the preceding section of the hollow fiber, i.e., from distance zero to distance x, and flows into the next segment of the hollow fiber as a solution whose concentration is $\bar{c}_{A3}$. These two concentrations are therefore different.

The (differential) Equations 7.3, 7.9, and 7.10 can be solved together with the membrane transport Equations 7.11 to 7.13, when a set of data for the feed solution, i.e., c_{A1} and p_2 ($p_2 = p_1$, since there is no pressure drop from the bulk to the boundary solution), are given under the following boundary conditions:

$$u_3(x = 0, r) = 0 \tag{7.16}$$
$$\bar{c}_{A3}(x = 0, r) = c_{A3} \tag{7.17}$$
$$p_3(x = L, r) = 101,325 \text{ Pa } (= \text{atmospheric pressure}) \tag{7.18}$$

It should be noted that Equations 7.16 and 7.17, as boundary conditions, make the right side of Equation 7.9 indeterminate at $x = 0$, since both the numerator and the denominator become zero. L'Hopital's rule should be applied in such a case. According to the rule, instead of Equation 7.9,

$$\left(\frac{\partial \bar{c}_{A3}}{\partial x}\right)_{x=0} = \frac{4 d_e}{d_i^2} \frac{\left([\partial (J_A - v_w \bar{c}_{A3})]/\partial x\right)_{x=0}}{(\partial u_3/\partial x)_{x=0}} \tag{7.19}$$

From Equation 7.3,

$$\left(\frac{\partial u_3}{\partial x}\right)_{x=0} = 4 \frac{d_e}{d_i^2} v_w(0) \tag{7.20}$$

Therefore,

$$\left(\frac{\partial \bar{c}_{A3}}{\partial x}\right)_{x=0} = \frac{1}{v_w(0)} \left[\frac{\partial (J_A - v_w \bar{c}_{A3})}{\partial x}\right]_{x=0} \tag{7.21}$$

and

$$(\partial \bar{c}_{A3})_{x=0} = \frac{1}{v_w(0)} \partial (J_A - v_w \bar{c}_{A3})_{x=0} \tag{7.22}$$

Then

$$\bar{c}_{A3}(dx) - \bar{c}_{A3}(0)$$

$$= \frac{1}{v_w(0)} \left(\left[J_A(dx) - v_w(dx)\bar{c}_{A3}(dx) \right] - \left[J_A(0) - v_w(0)\bar{c}_{A3}(0) \right] \right) \tag{7.23}$$

Because of the boundary condition (Equation 7.17), $\bar{c}_{A3} = c_{A3}$ at $x = 0$. Then from Equation 7.15,

$$J_A(0) - v_w(0)\bar{c}_{A3}(0) = 0 \text{ at } x = 0$$

Therefore, from Equation 7.23,

$$\bar{c}_{A3}(dx) = \bar{c}_{A3}(0) + \frac{1}{v_w(0)} \left[J_A(dx) - v_w(dx)\bar{c}_{A3}(dx) \right] \tag{7.24}$$

Rearranging,

$$\bar{c}_{A3}(dx) \left[1 + \frac{v_w(dx)}{v_w(0)} \right] = \bar{c}_{A3}(0) + \frac{J_A(dx)}{v_w(0)} \tag{7.25}$$

Therefore,

$$\bar{c}_{A3}(dx) = \frac{\bar{c}_{A3}(0) + \left[J_A(dx)/v_w(0) \right]}{1 + \left[v_w(dx)/v_w(0) \right]} \tag{7.26}$$

For a small increment, Δx,

$$\bar{c}_{A3}(\Delta x) \simeq \frac{\bar{c}_{A3}(0) + \left[J_A(\Delta x)/v_w(0) \right]}{1 + \left[v_w(\Delta x)/v_w(0) \right]} \tag{7.27}$$

As mentioned earlier, the numerical quantities for c_{A1} and p_1 are required to solve (differential) Equations 7.3, 7.9, and 7.10, together with membrane transport Equations 7.11 to 7.13. Those numerical values can be calculated as functions of radial distance, in the following way.

The volumetric flow rate at the outlet of a hollow fiber located at radial distance r is given by $\frac{\pi}{4}d_i^2 u_3(x = L, r)$. Similarly, the amount of the solute eluted from the outlet of the hollow fiber is given by $\frac{\pi}{4}d_i^2 u_3(x = L, r)\bar{c}_{A3}(x = L, r)$. Suppose the number of the hollow fibers in a unit area of the bundle cross-section is n_h; the total number of hollow fibers between the radial distance r and $r + dr$ is

$$n_h \left[\pi(r + dr)^2 - \pi r^2 \right] \simeq 2\pi r n_h dr \tag{7.28}$$

The volume of the permeate solution collected through all hollow fibers in this cross-sectional segment is given as $2\pi r n_h dr \times \frac{\pi}{4} d_i^2 u_3(x = L, r)$, and the amount of the solute collected is $2\pi r n_h dr \times \frac{\pi}{4} d_i^2 u_3(x = L, r)\bar{c}_{A3}(x = L, r)$. Then the following mass balance, material balance, and pressure drop equations can be established for the feed solution that is flowing on the shell side of the hollow fiber bundle. The mass balance at the cross-sectional segment from the radial distance r to $r + dr$ is

$$\text{Mass input} = 2\pi r L u_1$$

where L is the length of the hollow fiber (m), and u_1 is the velocity of feed solution flowing radially in the hollow fiber bundle (m/s).

$$\text{Mass output} = 2\pi(r + dr)L\left(u_1 + \frac{du_1}{dr}dr\right) + 2\pi r n_h dr \times \frac{\pi}{4}d_i^2 u_3(x = L, r)$$

At steady state, input and output should be balanced; then

$$2\pi r L u_1 = \left[2\pi(r + dr)L\left(u_1 + \frac{du_1}{dr}dr\right)\right] + 2\pi r n_h dr \times \frac{\pi}{4}d_i^2 u_3(x = L, r) \tag{7.29}$$

Rearranging,

$$0 \simeq 2\pi dr L u_1 + 2\pi r L\frac{du_1}{dr}dr + 2\pi r n_h dr \times \frac{\pi}{4}d_i^2 u_3(x = L, r) \tag{7.30}$$

Division of both sides of Equation 7.30 by $2\pi L dr$, and further rearrangement of the equation, yields

$$u_1 + r\frac{du_1}{dr} = -\frac{\pi r n_h d_i^2}{4L}u_3(x = L, r) \tag{7.31}$$

or

$$\frac{d(ru_1)}{dr} = -\frac{\pi r n_h d_i^2}{4L}u_3(x = L, r) \tag{7.32}$$

Similarly, from the material balance, we obtain

$$\frac{d(ru_1c_{A1})}{dr} = -\frac{\pi r n_h d_i^2}{4L}u_3(x = L, r)\bar{c}_{A3}(x = L, r) \tag{7.33}$$

where c_{A1} is the feed concentration (mol/m^3). As for the pressure drop of the feed solution, the following equation is valid [263]:

$$\frac{dp_1}{dr} = -30\eta S_v^2 \frac{(1-\varepsilon)^2}{\varepsilon^3} u_1 \tag{7.34}$$

where

$$S_v = \pi n_h d_e \tag{7.35}$$

and

$$\varepsilon = 1 - \frac{\pi n_h d_e^2}{4} \tag{7.36}$$

The (differential) Equations 7.32 to 7.34 can be solved under the boundary conditions

$$u_1 = u_1^0 \quad \text{at } r = R_i \tag{7.37}$$
$$c_{A1} = c_{A1}^0 \quad \text{at } r = R_i \tag{7.38}$$
$$p_1 = p_1^0 \quad \text{at } r = R_i \tag{7.39}$$

where R_i is the radius of the collection tube (m), and the superscript 0 indicates the properties of the feed solution when it enters the hollow fiber bundle.

Let us now solve (differential) Equations 7.32 to 7.34 established in the hollow fiber bundle (the radial coordinate, r, being the independent variable) and the differential Equations 7.3, 7.9, and 7.10 established on the bore side of the hollow fiber (the axial coordinate, x, being the independent variable), simultaneously. Although (differential) Equations 7.3, 7.9, and 7.10 are written in the form of partial differential equations, they are actually ordinary differential equations, since x is the only independent variable at a fixed r. Quantities d_e, d_i, L, and n_h are known as hollow fiber and hollow fiber bundle geometry. A and B are membrane transport parameters and are known for a given membrane. k is the mass transfer coefficient, which depends on the module geometry and the solution flow rate.

Let us assume that k is known and b, a parameter for the osmotic pressure, is also known. Let us also assume that u_1^0, c_{A1}^0 and p_1 are given. The following steps can apply for the module calculation.

· Step 1. Assume a numerical value for $p_3(0, R_i)$.

· Step 2. Using the numerical values of u_1^0, c_{A1}^0, and p_1^0, which were given, and the numerical values A, B, and k, which were also given, for the membrane and the membrane module, calculate v_w, c_{A2}, c_{A3}, and J_A and J_B from (transport) Equations 7.11 to 7.15. These values correspond to a position $(x = 0, r = R_i)$ on the cylindrical coordinate.

· Step 3. The derivatives $\frac{\partial u_3}{\partial x}$ and $\frac{\partial p_3}{\partial x}$ are calculated. From the boundary condition (Equation 7.16) $\frac{\partial p_3}{\partial x} = 0$.

· Step 4. For a preset size of the increment Δx, $u_3(\Delta x, R_i)$, and $p_3(\Delta x, R_i)$ are calculated by

$$u_3 (\Delta x, R_i) = \frac{\partial u_3}{\partial x} \Delta x + u_3 (0, R_i)$$

$$= \frac{\partial u_3}{\partial x} \Delta x \qquad (7.40)$$

$$p_3 (\Delta x, R_i) = \frac{\partial p_3}{\partial x} \Delta x + p_3 (0, R_i)$$

$$= p_3 (0, R_i) \qquad (7.41)$$

· Step 5. Using the value for $p_3(\Delta x, R_i)$ obtained above (transport) Equations 7.11 to 7.15 are solved for v_w, c_{A2}, c_{A3}, J_A, and J_B. These values correspond to a position $(\Delta x, R_i)$ on the cylindrical coordinate. Note that solutions will be equal to those corresponding to the position $(0, R_i)$, since there is no change in p_3 from $x = 0$ to $x = \Delta x$. Then

$$J_A (\Delta x, R_i) = J_A (0, R_i)$$
$$v_w (\Delta x, R_i) = v_w (0, R_i)$$

Using the above relations in Equation 7.27,

$$\bar{c}_{A3} (\Delta x, R_i) \simeq \frac{\bar{c}_{A3} (0, R_i) + \left[J_A / (0, R_i) / v_w (0, R_i) \right]}{1 + \left[v_w (0, R_i) / v_w (0, R_i) \right]} \qquad (7.42)$$

Remember,

$$J_A (0, R_i) / v_w (0, R_i) = \bar{c}_{A3} (0, R_i) \qquad (7.43)$$

Hence,

$$\bar{c}_{A3} (\Delta x, R_i) \simeq \bar{c}_{A3} (0, R_i) = c_{A3} (0, R_i) \qquad (7.44)$$

Since

$$\bar{c}_{A3} (\Delta x, R_i) = \bar{c}_{A3} (0, R_i) \qquad (7.45)$$

$$\frac{\partial \bar{c}_{A3}}{dx} = 0 \text{ at } x = 0 \tag{7.46}$$

Equation 7.46 is always applicable when L'Hopital's rule is used.

· Step 6. The partial derivatives $\frac{\partial u_3}{\partial x}$, $\frac{\partial \bar{c}_{A3}}{\partial x}$, and $\frac{\partial p_3}{\partial x}$ are calculated using Equations 7.3, 7.9, and 7.10.

· Step 7. u_3, $\bar{c}_{A3}$, and p_3 are calculated for the position $(2\Delta x, R_i)$ by

$$u_3 (2\Delta x, R_i) = \frac{\partial u_3}{\partial x}\Delta x + u_3 (\Delta x, R_i) \tag{7.47}$$

$$\bar{c}_{A3} (2\Delta x, R_i) = \frac{\partial \bar{c}_{A3}}{\partial x}\Delta x + \bar{c}_{A3} (\Delta x, R_i) \tag{7.48}$$

$$p_3 (2\Delta x, R_i) = \frac{\partial p_3}{\partial x}\Delta x + p_3 (\Delta x, R_i) \tag{7.49}$$

Steps 2 to 7 are repeated until x becomes $x = L$.

· Step 8. $p_3(x = L, R_i)$ so obtained should satisfy the boundary condition, Equation 7.18. Otherwise we go back to Step 1 and repeat the whole process until Equation 7.18 can be satisfied. When it is satisfied we know the numerical values for $u_3(L, R_i)$, $\bar{c}_{A3}(L, R_i)$, and $p_3(L, R_i)$.

· Step 9. The derivatives $d[ru_1(r)]/dr$, $d[ru_1(r)c_{A1}(r)]/dr$, and dp_1/dr are calculated from Equations 7.32, 7.33, and 7.34 for $r = R_i$. Equations 7.35 and 7.36 are used to calculate S_v and ε, which are necessary in Equation 7.34.

· Step 10. Using the derivatives calculated above, the derivatives such as $\frac{u_1}{dr}$ and $\frac{c_{A1}}{dr}$ for $r = R_i$ can be calculated by

$$\frac{du_1(r)}{dr} = \frac{1}{r}\left[\frac{d\left[ru_1(r)\right]}{dr} - u_1(r)\right] \tag{7.50}$$

$$\frac{dc_{A1}(r)}{dr} = \frac{1}{ru_1(r)}\left[\frac{d\left(ru_1(r)c_{A1}(r)\right)}{dr} - \right.$$
$$\left. \left[u_1(r)c_{A1}(r) - r\frac{du_1(r)}{dr}c_{A1}(r)\right]\right] \tag{7.51}$$

Then for a predetermined size of the increment Δr, $u_1 (R_i + \Delta r)$, $c_{A1} (R_i + \Delta r)$, and $p_1 (R_i + \Delta r)$ can be obtained by

$$u_1 (R_i + \Delta r) = u_1^0 + \frac{du_1(r)}{dr} \times \Delta r \tag{7.52}$$

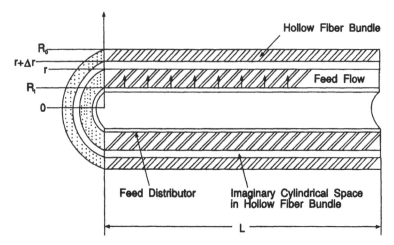

Figure 7.17.   Calculation in the hollow fiber bundle.

$$c_{A1}(R_i + \Delta r) = c_{A1}^0 + \frac{dc_{A1}(r)}{dr} \times \Delta r \qquad (7.53)$$

$$p_1(R_i + \Delta r) = p_1^0 + \frac{dp_1(r)}{dr} \times \Delta r \qquad (7.54)$$

· Step 11. Steps 1 − 8 are repeated to know the values $u_3(L, R_i + \Delta r)$, $\bar{c}_{A3}(L, R_i + \Delta r)$, and $p_3(L, R_i + \Delta r)$.

· Step 12. Step 9 and Step 10 are repeated to obtain numerical values for $u_1(R_i + 2\Delta r)$, $c_{A1}(R_i + 2\Delta r)$, and $p_1(R_i + 2\Delta r)$ until r becomes $r = R_o$. Thus, the profiles of $u_3, \bar{c}_{A3}, p_3, u_1, c_{A1}$, and p_1, in the module, are obtained (see Figure 7.17).

Example 1

The following numerical values are given for a reverse osmosis module that treats aqueous sodium chloride solution. Generate the profiles of $u_3, \bar{c}_{A3}, p_3, u_1$, c_{A1}, and p_1 in the hollow fiber module. Calculate the sodium chloride concentration in the permeate from the module. Calculate the volumetric flow rate of the permeate.

$$
\begin{aligned}
u_1^0 &= 0.35 \text{ m/s} \\
c_{A1}^0 &= 0.1 \text{ kmol/m}^3 \\
p_1^0 &= 2758 \text{ kPa (gauge)} \\
n_h &= 5 \times 10^7 \text{ 1/m}^2 \\
k &= 1.24 \times 10^{-6} \text{ m/s} \\
A &= 0.2 \times 10^{-7} \text{ kmol/m}^2\text{s kPa} \\
B &= (D_{Am}K_A/\delta) = 0.075 \times 10^{-7} \text{ m/s}
\end{aligned}
$$

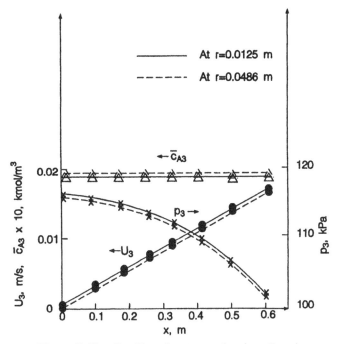

Figure 7.18. Profiles of u_3, $\bar{c}_3$, and p_3 in x direction.

$$
\begin{aligned}
c &= c_1 = c_2 = c_3 = 55.3\ \text{kmol/m}^3 \\
R_i &= 0.0125\ \text{m} \\
R_o &= 0.0486\ \text{m} \\
L &= 0.6\ \text{m} \\
d_e &= 0.000105\ \text{m} \\
d_i &= 0.0001\ \text{m} \\
\eta_{\text{water}} &= 0.8941 \times 10^{-6}\ \text{kPa s} \\
\text{Osmotic pressure coefficient} &= 4.6374 \times 10^3\ \text{kPa m}^3/\text{kmol}
\end{aligned}
$$

Using the computer program given in the Appendix, the profiles are calculated. The results are shown in Figures 7.18 and 7.19. The sodium chloride concentration in the permeate of the module is 0.00189 kmol/m³. From Figure 7.18, u_3 is 0.017 m/s at the fiber end irrespective of the radial distance r. Then the volumetric flow rate of the permeate is

$$
\begin{aligned}
\frac{\pi^2}{4} \left(R_0^2 - R_i^2\right) \left(d_i^2\right) n_h u_3 \\
&= \frac{3.1416^2}{4} \left(0.0486^2 - 0.0125^2\right) \left(0.0001^2\right) \left(5 \times 10^7\right) (0.017) \\
&= 4.63 \times 10^{-5} \text{m}^3/\text{s}
\end{aligned}
$$

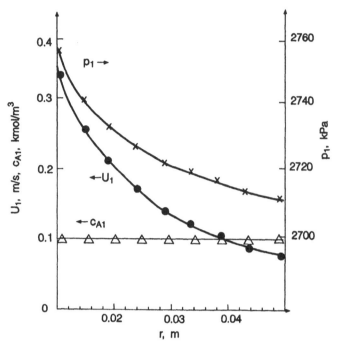

Figure 7.19. Profiles of u_1, c_{A1}, and p_1 in r direction.

Spiral-Wound Module

The analysis of the spiral-wound module was done by Ohya and Taniguchi [257]. No pressure drop on either side of the membrane and the concurrent flow of the feed and permeate were assumed. Furthermore, Ohya and Taniguchi made the following four assumptions:

1. Immediate and complete mixing of the permeate with the main flow in the permeate channel is assumed.

2. Plug flows in both feed and permeate channels are assumed.

3. Transport of the solute by diffusion in both feed and permeate channels are neglected both in parallel and perpendicular directions to that of the main flow.

4. No pressure drop occurs to the direction perpendicular to that of the main flow.

Rautenbach et al. adopted these assumptions in developing their equations for module design.

Figure 7.20 illustrates the rectangular coordinate established on a membrane of a spiral-wound module. The feed solution flows along the x-axis. It enters the rectangular membrane sheet at $x = 0$ and leaves at $x = L$. The permeate solution flows along the y-axis. The flow starts from one end of the membrane

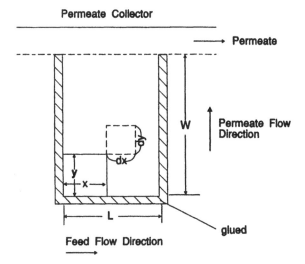

Figure 7.20. Rectangular coordinate established on a membrane of spiral-wound module.

envelope at $y = 0$ and ends at $y = W$, where the permeate is collected into a collection tube. The channel thicknesses for the feed and the permeate are h_f and h_p, respectively. The following mass balance, material balance, and pressure drop equations can be established at a small rectangular segment of an area $dxdy$ at a position (x, y). In the feed channel,

$$\text{Mass input} = \varepsilon_f h_f dy u_1(x)$$

where ε_f is the void ratio of the feed channel and equal to (total channel volume – spacer volume)/total channel volume.

$$\text{Mass output} = \varepsilon_f h_f dy u_1(x + dx) + \varepsilon_f dx dy v_w$$

It is assumed that only a fraction of the rectangular area $(dxdy)$ of the membrane is used for permeation. Looking into Figure 7.21, the area effective for the membrane permeation is equal to area $dxdy$ – area covered by the spacer (which is indicated by the shadowed area in Figure 7.21). Assuming that the height of the spacer is the same as that of the feed channel, the effective area of the membrane to the total area becomes equal to ε_f. The mass balance should be established at a steady state. Then

$$\varepsilon_f h_f dy u_1(x) = \varepsilon_f h_f dy u_1(x + dx) + \varepsilon_f dx dy v_w \qquad (7.55)$$

Since

$$u_1(x + dx) \simeq u_1(x) + \frac{du_1(x)}{dx} dx \qquad (7.56)$$

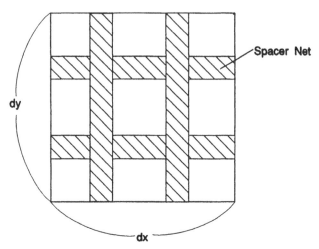

Figure 7.21. Effective membrane area.

substituting Equation 7.56 for $u_1(x + dx)$ in Equation 7.55 and rearranging,

$$\frac{du_1(x)}{dx} = -\frac{v_w}{h_f} \qquad (7.57)$$

In the permeate channel,

$$
\begin{aligned}
\text{Mass input} \quad &= \quad \varepsilon_p h_p dx u_3(y) + \varepsilon_f dx dy v_w \\
\text{Mass output} \quad &= \quad \varepsilon_p h_p dx u_3(y + dy)
\end{aligned}
$$

The mass balance should be established at a steady state. Then

$$\varepsilon_p h_p dx u_3(y) + \varepsilon_f dx dy v_w = \varepsilon_p h_p dx u_3(y + dy) \qquad (7.58)$$

Since

$$u_3(y + dy) \simeq u_3(y) + \frac{du_3(y)}{dy} dy \qquad (7.59)$$

substituting Equation 7.59 for $u_3(y + dy)$ in Equation 7.58 and rearranging,

$$\frac{du_3(y)}{dy} = \frac{v_w}{h_p} \frac{\varepsilon_f}{\varepsilon_p} \qquad (7.60)$$

The following equations for the material balance in the feed channel can be derived in exactly the same way.

$$\frac{d(u_1(x)c_{A1}(x))}{dx} = -\frac{J_A}{h_f} \quad (7.61)$$

Since

$$\frac{d[u_1(x)c_{A1}(x)]}{dx} = \frac{du_1(x)}{dx}c_{A1}(x) + u_1(x)\frac{dc_{A1}(x)}{dx} \quad (7.62)$$

$$= -\frac{v_w}{h_f}c_{A1}(x) + u_1(x)\frac{dc_{A1}(x)}{dx} \quad (7.63)$$

substituting Equation 7.61 for the left-hand side of Equation 7.63 and rearranging,

$$\frac{dc_{A1}(x)}{dx} = -\frac{J_A - v_w c_{A1}(x)}{u_1(x)h_f} \quad (7.64)$$

Similarly, for the permeate channel the material balance equation can be established as

$$\frac{d\bar{c}_{A3}(y)}{dy} = \frac{J_A - v_w \bar{c}_{A3}(y)}{u_3(y)h_p}\frac{\varepsilon_f}{\varepsilon_p} \quad (7.65)$$

At $y = 0$, L'Hopital's rule can apply. Then

$$\frac{d\bar{c}_{A3}(y)}{dy} = \frac{d(J_A - v_w \bar{c}_{A3})/dy}{du_3(y)/dy}\frac{\varepsilon_f}{h_p \varepsilon_p} \quad (7.66)$$

and

$$\frac{d\bar{c}_{A3}(y)}{dy} = 0 \quad (7.67)$$

As for the pressure drop in the feed channel,

$$\frac{dp_1(x)}{dx} = \frac{\rho}{2}u_1(x)^2\frac{\lambda_f}{2h_f} \quad (7.68)$$

where

$$\lambda_f = 2000 Re^{1.00} \quad (7.69)$$

$$Re = \frac{\rho u_1(x)d_h}{\eta} \quad (7.70)$$

And in the product channel,

$$\frac{dp_3(y)}{dy} = \frac{\rho}{2} u_3(y)^2 \frac{\lambda_p}{2h_p} \tag{7.71}$$

where

$$\lambda_p = 1075 Re^{0.78} \tag{7.72}$$

$$Re = \frac{\rho u_3(y) d_p}{\eta} \tag{7.73}$$

The same membrane transport equations, from Equations 7.11 to 7.15, that were used for the analysis of hollow fiber module, can be used in the spiral-wound module; i.e.,

$$J_B = A\left[(p_2 - p_3) - b(c_{A2} - c_{A3})\right] \tag{7.74}$$

$$J_A = B(c_{A2} - c_{A3}) \tag{7.75}$$

$$\frac{c_{A2} - c_{A3}}{c_{A1} - c_{A3}} = \exp\left(v_w/k\right) \tag{7.76}$$

$$v_w = J_B/c \tag{7.77}$$

$$c_{A3} = J_A/v_w \tag{7.78}$$

Furthermore, the mass transfer coefficient used in the concentration polarization equation (Equation 7.76) can be obtained by the following correlation:

$$Sh = 1.065 \left(\frac{h_f}{L_{sp}} \frac{\eta_{sp}}{2 - \eta_{sp}}\right)^{1/2} Re^{1/2} Sc^{1/3} \tag{7.79}$$

where

$$Sh = \frac{k d_h}{D_{AB}} \tag{7.80}$$

$$Re = \frac{\rho u_1(x) d_h}{\eta} \tag{7.81}$$

$$Sc = \frac{\mu}{\rho D_{AB}} \tag{7.82}$$

where η_{sp} and L_{sp} are the mixing efficiency of spacer (0.5) and the spacer mesh width (m), respectively. The (differential) Equations 7.57, 7.60, 7.64, 7.65, 7.68, and 7.71 can be solved together with (transport) Equations 7.74 to 7.78, under the following boundary conditions:

$$u_1(x = 0, y) = u_1^0 \tag{7.83}$$

$$c_{A1}(x = 0, y) = c_{A1}^0 \tag{7.84}$$

$$p_1(x = 0, y) = p_1^0 \tag{7.85}$$

$$u_3(x, y = 0) = 0 \tag{7.86}$$

$$\bar{c}_{A3}(x, y = 0) = c_{A3} \tag{7.87}$$

$$p_3(x, y = W) = 10,1325 \text{ Pa (= atmospheric pressure)} \tag{7.88}$$

The steps involved are as follows:

· Step 1. The boundary conditions at the entrance of the feed channel, u_1^0, c_{A1}^0, and p_1^0, are given.

· Step 2. The pressure on the permeate side at the sealed corner of the membrane envelope, $p_3(0, 0)$, is assumed.

· Step 3. The transport equations are solved, and numerical values v_w, c_{A2}, c_{A3}, J_B, and J_A are obtained. The mass transfer coefficient, k, necessary for the above calculation is obtained from Equation 7.79 together with Equations 7.80, 7.81, and 7.82.

· Step 4. The derivatives $\frac{du_3(y)}{dy}$ and $\frac{dp_3(y)}{dy}$ are obtained from Equations 7.60 and 7.71 for a position on the rectangular coordinate (0, 0).

· Step 5. $u_3(\Delta y)$ and $p_3(\Delta y)$ for a preset increment Δy can be calculated by

$$u_3(\Delta y) = 0 + \frac{du_3(y)}{dy} \times \Delta y \tag{7.89}$$

$$p_3(\Delta y) = p_3(0, 0) + \frac{dp_3(y)}{dy} \times \Delta y \tag{7.90}$$

$\bar{c}_{A3}(\Delta y)$ is obtained applying L'Hopital's rule.

· Step 6. The derivatives $\frac{du_3(y)}{dy}$, $\frac{d\bar{c}_{A3}(y)}{dy}$, and $\frac{dp_3}{dy}$ are obtained for a position $(0, \Delta y)$ on the rectangular coordinate, using $u_3(\Delta y)$, $\bar{c}_{A3}(\Delta y)$, and $p_3(\Delta y)$ values calculated at Step 5. Equation 7.65 can be used instead of L'Hopital's rule, this time. $u_3(2\Delta y)$ and $p_3(2\Delta y)$ are obtained using the respective derivatives together with equations similar to Equations 7.89 and 7.90. For the calculation of $\bar{c}_{A3}(2\Delta y)$, use

$$\bar{c}_{A3}(2\Delta y) = \bar{c}_{A3}(\Delta y) + \frac{d\bar{c}_{A3}}{dy} \times \Delta y \tag{7.91}$$

These calculations are repeated until $u_3(0, W)$, $\bar{c}_{A3}(0, W)$, and $p_3(0, W)$ are obtained.

· Step 7. $p_3(0, W)$ is examined to see if it satisfies the boundary condition Equation 7.88. If Equation 7.88 is satisfied, we can go to Step 8; otherwise we have to go back to Step 2 and reassume a numerical value for $p_3(0, 0)$.

· Step 8. All numerical values for u_1, c_{A1}, p_1, u_3, $\bar{c}_{A3}$, and p_3 are now firmly determined at the point $(0, 0)$. Using these numerical values, the derivatives $\frac{du_1(x)}{dx}$, $\frac{dc_{A1}(x)}{dx}$, and $\frac{dp_1(x)}{dx}$ can be calculated at a point $(0, 0)$ on the rectangular coordinate, by (differential) Equations 7.57, 7.64, and 7.68 together with (the membrane transport) Equations 7.74 to 7.78. Then for the preset increment Δx, $u_1(\Delta x, 0)$, $c_{A1}(\Delta x, 0)$, and $p_1(\Delta x, 0)$ can be calculated by

$$u_1(\Delta x, 0) = u_1^0 + \frac{du_1(x)}{dx} \times \Delta x \qquad (7.92)$$

$$c_{A1}(\Delta x, 0) = c_{A1}^0 + \frac{dc_{A,1}(x)}{dx} \times \Delta x \qquad (7.93)$$

$$p_1(\Delta x, 0) = p_1^0 + \frac{dp_1(x)}{dx} \times \Delta x \qquad (7.94)$$

· Step 9. All numerical values for u_1, c_{A1}, p_1, u_3, $\bar{c}_{A3}$, and p_3 are also known at the point $(0, \Delta y)$, from Steps 2 to 7. Therefore, similar calculations as in Step 8 can be executed, and the numerical values for $u_1(\Delta x, \Delta y)$, $c_{A1}(\Delta x, \Delta y)$, and $p_1(\Delta x, \Delta y)$ can be calculated, etc., until $u_1(\Delta x, W)$, $c_{A1}(\Delta x, W)$, and $p_1(\Delta x, W)$ are obtained.

· Step 10. Steps 2 to 7 are repeated to solve for the numerical values of $u_3(\Delta x, 0), \bar{c}_3(\Delta x, 0), p_3(\Delta x, 0), u_3(\Delta x, \Delta y), \bar{c}_{A3}(\Delta x, \Delta y), p_3(\Delta x, \Delta y), \ldots,$ $u_3(\Delta x, W), \bar{c}_{A3}(\Delta x, W)$, and $p_3(\Delta x, W)$.

· Step 11. We repeat Steps 8 and 9 to obtain the numerical values for u_1, c_{A1}, and p_1 at the points $(2\Delta x, 0)$, $(2\Delta x, \Delta y)$, etc., until $(2\Delta, W)$ is reached. Then we repeat Steps 2 to 7 for the calculation of u_3, $\bar{c}_{A3}$, and p_3 at points $(2\Delta x, 0)$, $(2\Delta x, \Delta y) \ldots (2\Delta x, W)$. This procedure is repeated until the calculation is completed by achieving (L, W).

7.2.2 Membrane Gas Separation

Hollow Fiber
A mathematical formulation for the gas separation by hollow fibers operating in a cocurrent mode was developed by Pan [264], [265]. The assumptions made in his model were

1. The membrane permeability is independent of the pressure and concentration.

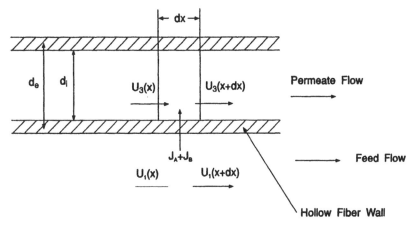

Figure 7.22. Hollow fiber working in cocurrent mode for gas separation.

2. The pressure drop in the feed gas is negligible.

3. The pressure drop of the permeate flow in the fiber is governed by the Hagen–Poiseuille equation.

Under the above assumptions, the following mass and material balance equations and the pressure drop equations can be derived. The method of deriving such equations is analogous to that employed for the formulation of reverse osmosis modules. Figure 7.22 illustrates schematically a hollow fiber

$$\frac{du_3(x)}{dx} = \frac{4d_e}{d_i^2}(J_A + J_B) \qquad (7.95)$$

working in a cocurrent mode. Mass balance for the permeate is where $u_3(x)$ is the molar velocity (mol/m^2 s) of permeate in the hollow fiber tube. Other symbols have been defined in the foregoing sections. Material balance for the permeate is

$$\frac{d\bar{X}_{A3}(x)}{dx} = 0 \text{ when } x = 0 \text{ from L'Hopital's rule} \qquad (7.96)$$

$$\text{otherwise} = \frac{4d_e}{u_3 d_i^2}\left(J_A\bar{X}_{B3}(x) - J_B\bar{X}_{A3}(x)\right) \qquad (7.97)$$

Pressure drop of the permeate is

$$\frac{dp(x)^2}{dx} = \frac{64\mu RT}{d_i^2}u_3^2 \qquad (7.98)$$

Mass balance for the feed is

$$\mathbf{u}_1(x) + \mathbf{u}_3(x) = \mathbf{u}_1^0 \tag{7.99}$$

Material balance for the feed is

$$\mathbf{u}_1(x)X_{A1}(x) + \mathbf{u}_3(x)\overline{X}_{A3}(x) = X_{A1}^0 \mathbf{u}_1^0 \tag{7.100}$$

Pressure for the feed is

$$p_1(x) = p_1^0 \tag{7.101}$$

The following equations apply for the mole fractions:

$$X_{A1} + X_{B1} = 1 \tag{7.102}$$
$$\overline{X}_{A3} + \overline{X}_{B3} = 1 \tag{7.103}$$
$$X_{A3} + X_{B3} = 1 \tag{7.104}$$

The transport equations for the gas permeation are

$$J_A = \frac{P_A}{\delta}(p_1 X_{A1} - p_3 X_{A3}) \tag{7.105}$$

$$J_B = \frac{P_B}{\delta}(p_1 X_{B1} - p_3 X_{B3}) \tag{7.106}$$

$$X_{A3} = \frac{J_A}{J_A + J_B} \tag{7.107}$$

The (mass balance) Equations 7.95 and 7.99, (material balance) Equations 7.97 and 7.100, and the equations for pressure drop, Equations 7.98 and 7.101, can be solved together with (membrane transport) Equations 7.105, 7.106, and 7.107, under the boundary conditions

$$\mathbf{u}_1(0) = \mathbf{u}_1^0 \tag{7.108}$$
$$X_{A1}(0) = X_{A1}^0 \tag{7.109}$$
$$p_1(0) = p_1^0 \tag{7.110}$$
$$\mathbf{u}_3(0) = 0 \tag{7.111}$$
$$\overline{X}_{A3}(0) = X_{A3}(0) \tag{7.112}$$
$$p_3(L) = p_3^L \tag{7.113}$$

Spiral-Wound Module

The approach to the analysis of the spiral-wound module for the gas separation is the same as that of reverse osmosis. The following mass balance, material balance, and the pressure drop equations can be established. Mass balance is

$$\frac{\partial u_1(x, y)}{\partial x} = -\frac{2(J_A + J_B)}{h_f} \tag{7.114}$$

$$\frac{\partial u_3(x, y)}{\partial y} = \frac{2(J_A + J_B)}{h_p} \tag{7.115}$$

Material balance is

$$\frac{\partial X_{A1}(x, y)}{\partial x} = -\frac{1}{u_1(x, y)h_f} \left[J_A X_{B1}(x, y) - J_B X_{A1}(x, y) \right] \tag{7.116}$$

$$\frac{\partial \overline{X}_{A3}(x, y)}{\partial y} = 0 \text{ when } y = 0 \text{ from L'Hopital's rule} \tag{7.117}$$

$$\text{otherwise} = \frac{1}{u_3(x, y)h_p} \left[J_A \overline{X}_{B3}(x, y) - J_B \overline{X}_{A3}(x, y) \right] \tag{7.118}$$

Pressure drop is

$$\frac{\partial^2 p_1(x, y)}{\partial x^2} = \frac{24\mu RT}{h_f^2} u_1(x, y) \tag{7.119}$$

$$\frac{\partial^2 p_3(x, y)}{\partial y^2} = \frac{24\mu RT}{h_p^2} u_3(x, y) \tag{7.120}$$

The membrane transport equations are

$$J_A = \frac{P_A}{\delta} \left[p_1(x, y)X_{A1}(x, y) - p_3(x, y)X_{A3}(x, y) \right] \tag{7.121}$$

$$J_B = \frac{P_B}{\delta} \left[p_1(x, y)X_{B1}(x, y) - p_3(x, y)X_{B3}(x, y) \right] \tag{7.122}$$

$$X_{A3} = \frac{J_A}{J_A + J_B} \tag{7.123}$$

The equations for mass balance, Equations 7.114 and 7.115, those for material balance, Equations 7.116 and 7.117 or 7.118, and those for pressure drop, Equations 7.119 and 7.120, can be solved together with (membrane transport) Equations 7.121 to 7.123, under the boundary conditions

$$u_1(0, y) = u_1^0 \tag{7.124}$$

$$X_{A1}(0, y) = X_{A1}^0 \tag{7.125}$$

$$p_1(0, y) = p_1^0 \tag{7.126}$$

$$\mathbf{u}_3(x, 0) = 0 \tag{7.127}$$

$$\overline{X}_{A3}(x, 0) = X_{A3}(x, 0) \tag{7.128}$$

$$p_3(x, L) = p_3^L \tag{7.129}$$

Nomenclature for Chapter 7

A = pure-water permeation constant, mol/m^2 s Pa

B = solute transport pareameter, m/s

b = osmotic pressure coefficient, Pa m^3/mol

c = total molar concentration, nearly equal to concentration of solvent (water), mol/m^3

c_{A1} = solute concentration in the bulk feed solution, mol/m^3

c_{A2} = solute concentration in the concentrated boundary solution, mol/m^3

c_{A3} = solute concentration in the solution permeating through the hollow fiber wall, mol/m^3

D_{AB} = diffusion coefficient of solute A in solvent B, m^2/s

$\bar{c}_{A3}$ = solute concentration on the bore side of the hollow fiber at the longitudinal distance x or in the permeate channel at the distance y in case of spiral-wound module, mol/m^3

d_e = effective diameter of hollow fiber tube, m

d_h = $2h_f$

d_i = inside diameter of hollow fiber tube, m

d_p = $2h_p$

h_f = thickness of feed flow channel, m

h_p = thickness of permeate flow channel, m

J_A = flux of solute, or component A through the membrane, mol/m^2 s

J_B = flux of solvent, or component B through the membrane, mol/m^2 s

k = mass transfer coefficient, m/s

L = length of the hollow fiber, or length of the feed channel in the spiral-wound module, m

L_{sp} = width of the feed spacer mesh, m

n_h = total number of hollow fibers in a unit area of the bundle cross-section, $1/\text{m}^2$

P_A = permeability coefficient of the component A, mol m/m^2 s Pa

P_B = permeability coefficient of the component B, mol m/m^2 s Pa

p_1 = pressure on the feed side, Pa

p_2 = p_1

p_3 = pressure on the permeate side, Pa

R =   gas constant, 8.314 J/mol K

R_i =   radius of the collection tube, m

R_o =   radius of the hollow fiber bundle, m

Re =   Reynolds number, —

r =   distance in the radial direction in the hollow fiber bundle, m

S_v =   quantity defined by Equation 7.35

Sc =   Schmidt number, —

Sh =   Sherwood number, —

T =   absolute temperature, K

u_1 =   velocity of feed solution flowing radially in the hollow fiber bundle or flowing in the feed channel in case of spiral-wound module, m/s

u_3 =   permeate flow velosity in the hollow fiber bore, or in the permeate spacer in case of spiral-wound module, m

$\mathbf{u}_1$ =   molar velocity of feed gas mixture per unit area of the hollow fiber cross-section, mol/m^2 s

$\mathbf{u}_3$ =   molar velocity of permeate per unit area of the hollow fiber cross-section, mol/m^2 s

v_w =   permeation velocity of solvent (water), m/s

X_{A1} =   mole fraction of the component A in the feed gas mixture, —

X_{B1} =   mole fraction of the component B in the feed gas mixture, —

X_{A3} =   mole fraction of the component A in the gas mixture permeating through the membrane, —

X_{B3} =   mole fraction of the component B in the gas mixture permeating through the membrane, —

$\overline{X}_{A3}$ =   mole fraction of the component A in the gas mixture flowing in the permeate channel, —

$\overline{X}_{B3}$ =   mole fraction of the component B in the gas mixture flowing in the permeate channel, —

x =   distance in the direction of permeate flow in case of hollow fiber, or distance in the direction of feed flow in case of spiral-wound module, m

y =   distance in the direction of permeate flow in case of spiral-wound module, m

W =   length of the permeate channel in the spiral-wound module, m

δ =   membrane thickness, m

ε = void ratio, —

ε_f = void ratio of the feed flow channel, —

ε_p = void ratio of the permeate flow channel, —

η = viscosity, Pa s

η_{sp} = mixing efficiency of spacer, $\simeq 0.5$

λ_f = quantity defined by Equation 7.69

λ_p = quantity defined by Equation 7.72

ρ = solution density, kg/m^3

Superscripts

0 = properties of the feed solution (or gas mixture) at the module inlet

L = quantities at the hollow fiber outlet

8

Concentration Profile in a Laminar Flow Channel

8.1 Calculation of the Concentration Profile

The concept of concentration polarization was given earlier, and a simple model based on the film theory was developed to describe the phenomenon mathematically. In this chapter an attempt is made to develop a differential equation of the solute material balance, the solution of which will rigorously establish a concentration profile of the solute in the vicinity of the solution-membrane boundary. It is hoped that this approach will furnish a deeper understanding of the concentration polarization phenomenon.

The concentration profile in a laminar flow channel between two reverse osmosis membranes has been established by numerical calculation and reported [266]–[276]. The following derivation is based on the approach of Kimura and Sourirajan [51].

Suppose a solution flows through a channel between two flat membranes placed in parallel. Assume that the membrane width is much larger than the distance between the two membranes. The solution permeates both the upper and the lower membranes (Figure 8.1). The solution flow is therefore considered to be restricted to x-y directions on Figure 8.1 and to be symmetric on both sides of the center line. Assume also that the solution flow is laminar, and the effect of the convective flow can be neglected. Furthermore, the diffusion constant of the solute, D_{AB} (m^2/s), is assumed to be independent of the solute concentration, and the solute concentration at the channel inlet is assumed to be c_A^0 (mol/m^3) over the entire cross-section of the channel inlet. The mass balance of the solute can then be established as follows. Considering a square-shaped area of $dxdy$, as shown in Figure 8.2, which has a $1 - m$ depth in the z direction,

$$\text{Solute influx} = \Delta y\, (uc_A)_{x=x} \text{ at } x = x$$

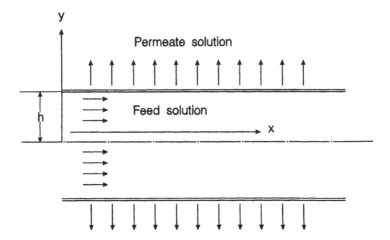

Figure 8.1.   Feed solution channel between two flat membranes placed in a parallel position.

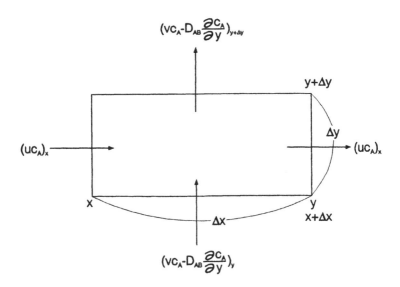

Figure 8.2.   Mass balance of solute.

where u is the solution flow velocity (m/s) in the x direction, and c_A is the solute concentration (mol/m^3).

$$\Delta x \left(v c_A - D_{AB} \frac{\partial c_A}{\partial y} \right)_{y=y} \quad \text{at } y = y$$

where v is the solution velocity (m/s) in the y direction, and $v c_A$ and $-D_{AB} \frac{\partial c_A}{\partial y}$ are convective and diffusive components of the solute flow (mol/m^2 s), respectively.

$$\text{Solute outflux} = \Delta y \, (u c_A)_{x+\Delta x} \text{ at } x = x + \Delta x$$

Considering an increase in the quantity, $u c_A$, with an increment in distance, Δx, the solute outflux can be given as

$$\Delta y \left[(u c_A)_{x=x} + \frac{\partial (u c_A)}{\partial x}\bigg|_{x=x} \cdot \Delta x \right]$$

Similarly, at $y = y + \Delta y$,

$$\Delta x \left[\left(v c_A - D_{AB} \frac{\partial c_A}{\partial y} \right)_{y=y} + \frac{\partial \left(v c_A - D_{AB} \frac{\partial c_A}{\partial y} \right)}{\partial y}\bigg|_{y=y} \cdot \Delta y \right]$$

Since influx and outflux should be balanced at a steady state,

$$\Delta y \, (u c_A)_{x=x} + \Delta x \left(v c_A - D_{AB} \frac{\partial c_A}{\partial y} \right)_{y=y} =$$
$$\Delta y \left[(u c_A)_{x=x} + \frac{\partial (u c_A)}{\partial x}\bigg|_{x=x} \cdot \Delta x \right]$$
$$+ \Delta x \left[\left(v c_A - D_{AB} \frac{\partial c_A}{\partial y} \right)_{y=y} + \frac{\partial \left(v c_A - D_{AB} \frac{\partial c_A}{\partial y} \right)}{\partial y}\bigg|_{y=y} \cdot \Delta y \right] \quad (8.1)$$

Rearranging,

$$\frac{\partial (u c_A)}{\partial x}\bigg|_{x=x} \cdot \Delta x \Delta y + \frac{\partial \left(v c_A - D_{AB} \frac{\partial c_A}{\partial y} \right)}{\partial y}\bigg|_{y=y} \cdot \Delta x \Delta y = 0 \quad (8.2)$$

Division of Equation 8.2 by $\Delta x \Delta y$ yields

$$\frac{\partial (u c_A)}{\partial x} + \frac{\partial \left(v c_A - D_{AB} \frac{\partial c_A}{\partial y} \right)}{\partial y} = 0 \tag{8.3}$$

The above differential equation describes the concentration profile of the solute in the channel. The boundary conditions are

$$c_A = c_A^0 \quad \text{when } x = 0 \tag{8.4}$$

$$\frac{\partial c_A}{\partial y} = 0 \quad \text{when } y = 0 \tag{8.5}$$

$$v c_A - D_{AB} \frac{\partial c_A}{\partial y} \bigg|_{y=h} = J_A \quad \text{when } y = h \tag{8.6}$$

Combining Equations 50, 55 and 59 of Chapter 5 with the boundary condition Equation 8.6, under the assumption of $J_B \gg J_A$ and assuming that the total molar concentration, c, is practically constant, Kimura and Sourirajan derived

$$D_{AB} \frac{\partial c_A}{\partial y} = v c_A \left[\frac{v}{v + (D_{Am} K_A / \delta)} \right] \tag{8.7}$$

In order to solve (differential) Equation 8.3 under boundary condition Equations 8.4, 8.5, and 8.6, the velocity profiles u and v, as a function of the rectangular coordinate (x, y), have to be known. Berman derived the following equations to describe the velocity profile in a channel between two flat porous plates placed in parallel [277].

$$u = \left(\frac{3}{2} \right) \bar{u} \left(1 - \frac{y^2}{h^2} \right) \left[1 - \frac{N_F}{420} \left(2 - 7 \frac{y^2}{h^2} - 7 \frac{y^4}{h^4} \right) \right] \tag{8.8}$$

$$v = v' \left[\left(\frac{y}{2h} \right) \left(3 - \frac{y^2}{h^2} \right) - \frac{N_F y}{280 h} \left(2 - \frac{3 y^2}{h^2} + \frac{y^6}{h^6} \right) \right] \tag{8.9}$$

where $\bar{u}$ is the average velocity (m/s) in the x direction at a distance x from the channel inlet, and v' is the velocity (m/s) in the y direction at the channel wall. This velocity is equal to the permeation velocity of the solution and nearly equal to the permeation velocity of solvent water, v_w. The dimensionless parameter N_F is defined as $N_F = h v' / v$ and is a kind of Reynolds number. In general, the terms involving the parameter N_F are very small and can be ignored. Then

Equations 8.8 and 8.9 can be approximated by

$$u = \left(\frac{3}{2}\right)\bar{u}\left(1 - \frac{y^2}{h^2}\right) \tag{8.10}$$

$$v = v_w\left(\frac{y}{2h}\right)\left(3 - \frac{y^2}{h^2}\right) \tag{8.11}$$

Equations 8.3, 8.4, 8.5, and 8.7 can be written in the following dimensionless form:

$$\frac{\partial}{\partial X}(UC) + \frac{\partial}{\partial Y}\left[VC\alpha_0 - \left(\frac{\partial C}{\partial Y}\right)\right] = 0 \tag{8.12}$$

with boundary conditions

$$C = 1 \text{ at } X = 0 \tag{8.13}$$

$$\frac{\partial C}{\partial Y} = 0 \text{ at } Y = 0 \tag{8.14}$$

$$\alpha_0 \frac{\partial C}{\partial Y} = V_w C_2\left[\frac{V_w}{\kappa + V_w}\right] \text{ at } Y = 1 \tag{8.15}$$

where

$$\alpha_0 = D_{AB}/(v_w^0 h) \tag{8.16}$$

$$U = u/\bar{u}^0 \tag{8.17}$$

$$V = v/v_w^0 \tag{8.18}$$

$$V_w = v_w/v_w^0 \tag{8.19}$$

$$X = \frac{v_w^0}{\bar{u}^0}\frac{x}{h} \tag{8.20}$$

$$Y = y/h \tag{8.21}$$

$$C = c_A/c_{A1}^0 \tag{8.22}$$

$$C_2 = c_{A2}/c_{A1}^0 \tag{8.23}$$

$$\kappa = \frac{(D_{Am}K_A/\delta)}{v_w^0} \tag{8.24}$$

The meaning of each symbol should be self-explanatory from the preceding section and the description of the system. The superscript 0 indicates the quantities at the channel inlet, and v_w^0 is defined as an imaginary permeation velocity of solvent (water) at the channel inlet, under the condition $C_3 = c_{A3}/c_{A1}^0 = 0$ and $C_2 = 1$ (meaning that the solute does not permeate through the membrane at all, and there is no concentration polarization). The following two dimensionless quantities are also defined:

$$\gamma = \frac{\pi_{A1}^0}{p_2 - p_3} \tag{8.25}$$

$$\theta = \frac{(D_{Am}K_A/\delta)}{v_w^*} \tag{8.26}$$

where v_w^* is the pure-water permeation rate (m/s), and π_{A1}^0 is the osmotic pressure corresponding to c_{A1}^0.

Among dimensionless quantities γ, θ, and κ, there is the following relationship

$$\kappa = \frac{\theta}{1 - \gamma} \tag{8.27}$$

The above equation can be derived by the combination of Equations 8.43, 8.44, and 8.49, which will appear later in this chapter.

Equations 8.10 and 8.11 can also be written in dimensionless form, as

$$U = \frac{3}{2}(1 - \Delta)(1 - Y^2) \tag{8.28}$$

$$V = V_w \left(\frac{Y}{2}\right)(3 - Y^2) \tag{8.29}$$

Note that the relation

$$\bar{u} = \bar{u}^0(1 - \Delta) \tag{8.30}$$

was used in the derivation of the above equation, where Δ is the fraction recovery into the permeate of the solution entering the channel inlet.

A dimensionless (differential) Equation 8.12 and the Equations 8.28 and 8.29 should be solved simultaneously under the dimensionless boundary conditions, i.e., Equations 8.13, 8.14, and 8.15. In order to solve these equations simultaneously, quantities V_w and Δ in Equations 8.28 and 8.29 should be known.

An assumption was made that the total molar concentration c is constant, and $c_A = cX_A$. An assumption was also introduced that the osmotic pressure of the solution is proportional to the mole fraction of the solute:

$$\pi = BX_A \tag{8.31}$$

Using the above assumption, Equation 47 of Chapter 5 can be written as

$$J_B = A\left[(p_2 - p_3) - B(X_{A2} - X_{A3})\right] \tag{8.32}$$

$$= A(p_2 - p_3)\left[1 - \frac{BX_{A1}^0}{p_2 - p_3}\left(X_{A2}/X_{A1}^0 - X_{A3}/X_{A1}^0\right)\right] \tag{8.33}$$

Since

$$BX_{A1}^0 = \pi_{A1}^0 \tag{8.34}$$

$$BX_{A1}^0/(p_2 - p_3) = \gamma \tag{8.35}$$

Introducing the following dimensionless quantities,

$$C_2 = X_{A2}/X_{A1}^0 \tag{8.36}$$

$$C_3 = X_{A3}/X_{A1}^0 \tag{8.37}$$

Equation 8.33 can be written as

$$J_B = A(p_2 - p_3)[1 - \gamma(C_2 - C_3)] \tag{8.38}$$

Under the assumption of Equation 58, Equation 54 (both from Chapter 5) can be written as

$$J_A = c\left(\frac{D_{Am}K_A}{\delta}\right)(X_{A2} - X_{A3}) \tag{8.39}$$

Furthermore, since $N_B \gg N_A$ in most solutions,

$$X_{A3} \simeq J_A/J_B \tag{8.40}$$

Then combining Equations 8.38, 8.39 and 8.40 and rearranging,

$$X_{A3} = \frac{c(D_{Am}K_A/\delta)(X_{A2} - X_{A3})}{A(p_2 - p_3)[1 - \gamma(C_2 - C_3)]} \tag{8.41}$$

Therefore,

$$C_3 = \frac{X_{A3}}{X_{A1}^0} = \frac{c(D_{Am}K_A/\delta)(X_{A2}/X_{A1}^0 - X_{A3}/X_{A1}^0)}{A(p_2 - p_3)[1 - \gamma(C_2 - C_3)]} \tag{8.42}$$

Furthermore, since

$$A(p_2 - p_3)/c = v_w^* \tag{8.43}$$

where v_w^* is the pure-water permeation velocity,

$$\frac{D_{Am}K_A}{\delta}\frac{A(p_2 - p_3)}{c} = \frac{D_{Am}K_A}{\delta}v_w^* = \theta \tag{8.44}$$

Therefore, Equation 8.42 can be written as

$$C_3 = \frac{\theta(C_2 - C_3)}{1 - \gamma(C_2 - C_3)} \tag{8.45}$$

Solving the above equation in terms of C_3,

$$C_3 = \frac{\sqrt{(1 + \theta - \gamma C_2)^2 + 4\gamma\theta C_2} - (1 + \theta - \gamma C_2)}{2\gamma} \tag{8.46}$$

The permeation rate of the solvent water, v_w, on the other hand, is

$$v_w = J_B/c \tag{8.47}$$

Equation 8.38 can then be written as

$$v_w = \frac{A(p_2 - p_3)}{c}[1 - \gamma(C_2 - C_3)] \tag{8.48}$$

Recall that $v_w = v_w^0$ when $x = 0$ and under the conditions $C_2 = 1$ and $C_3 = 0$; therefore,

$$v_w^0 = \frac{A(p_2 - p_3)}{c}(1 - \gamma) \tag{8.49}$$

Then

$$V_w = \frac{v_w}{v_w^0} = \frac{[1 - \gamma(C_2 - C_3)]}{1 - \gamma} \tag{8.50}$$

As for the fraction recovery Δ,

$$\Delta = \int_0^X V_w dX \tag{8.51}$$

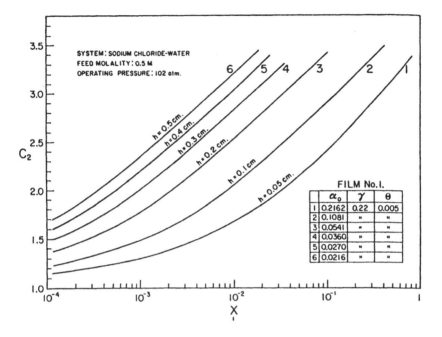

SYSTEM: SODIUM CHLORIDE-WATER
FEED MOLALITY: 0.5 M
OPERATING PRESSURE: 102 atm.

FILM No. I.		
α_0	γ	θ
1	0.2162	0.22	0.005
2	0.1081	"	"
3	0.0541	"	"
4	0.0360	"	"
5	0.0270	"	"
6	0.0216	"	"

Figure 8.3. Effect of channel thickness, h, on C_2 vs. X. (Feed, aqueous sodium chloride solution; concentration, 0.5 molal; operating pressure, 10,335 kPa.) (Reproduced from [105] with permission.)

and

$$\Delta = \frac{\bar{u}^0 - \bar{u}}{\bar{u}^0} = 1 - \frac{\bar{u}}{\bar{u}^0} = 1 - \bar{U} \qquad (8.52)$$

are valid.

The dimensionless (partial differential) Equation 8.12 and Equations 8.28, 8.29, 8.46, 8.50, and 8.52 can be solved simultaneously, under the boundary conditions Equations 8.13, 8.14, and 8.15. Kimura and Sourirajan made numerical calculations. The dimensionless quantities C_2, V_w, and C_3 or $\bar{C}_3$ are given as functions of X, for different combinations of α_0, γ, and θ in Figures 8.3, 8.4, and 8.5. The figures clearly show how the concentration polarization, the permeation rate, and the permeate composition change as the feed solution advances in the flow channel.

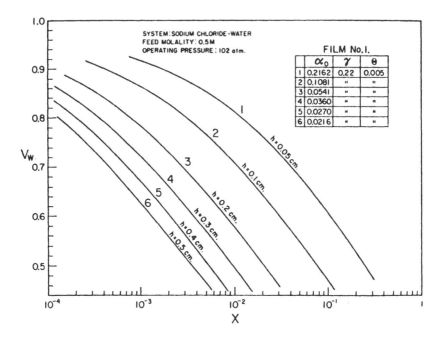

Figure 8.4.   Effect of channel thickness, h, on V_w vs. X. (Feed, aqueous sodium chloride solution; concentration, 0.5 molal; operating pressure, 10,335 kPa.) (Reproduced from [105] with permission)

8.2   Calculation of the Mass Transfer Coefficient

The mass transfer coefficient was calculated for the system where laminar flow develops between two parallel flat membranes, also by Kimura and Sourirajan [51].

A dimensionless parameter including the mass transfer coefficient is defined as

$$\lambda = \frac{k}{\left(D_{Am}K_A/\delta\right)} \tag{8.53}$$

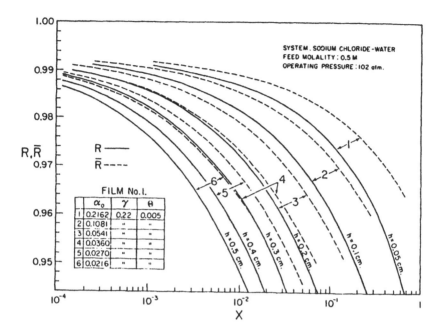

Figure 8.5.   Effect of channel thickness, h, on C_3 or $\overline{C}_3$ vs. X. (Feed aqueous sodium chloride solution; concentration, 0.5 molal; operating pressure, 10,335 kPa; $R = 1 - C_3$ and $\overline{R} = 1 - \overline{C}_3$.) (Reproduced from [105] with permission.)

Since in the solution of A in B, $X_A \ll 1$ is valid in most cases, Equation 60 of Chapter 5 can be approximated by

$$J_B = kc \ln \left(\frac{X_{A2} - X_{A3}}{X_{A1} - X_{A3}} \right) \tag{8.54}$$

Combining Equations 8.40, 8.39, and 8.54,

$$X_{A3} = \frac{(D_{Am} K_A / \delta)}{k} \frac{(X_{A2} - X_{A3})}{\ln\left[(X_{A2} - X_{A3}) / (X_{A1} - X_{A3}) \right]} \tag{8.55}$$

and

$$C_3 = \frac{X_{A3}}{X_{A1}^0} = \frac{1}{\lambda} \frac{(C_2 - C_3)}{\ln\left[(C_2 - C_3) / (C_1 - C_3) \right]} \tag{8.56}$$

Furthermore,

$$C_3 = \frac{C_2}{1 + \lambda \ln \left[(C_2 - C_3) / (C_1 - C_3) \right]} \qquad (8.57)$$

Considering the mass balance of the solute,

$$C_1 = \frac{1 - \int_0^X C_3 V_w dX}{1 - \Delta} \qquad (8.58)$$

The above equation can be justified, since $C_1(1 - \Delta)$ represents the amount of solute that remains in the feed solution stream after the solution has passed a dimensionless distance X, whereas $\int_0^X C_3 V_w dX$ represents the amount of the solute permeating through the membrane from dimensionless distance zero to X. Defining the Sherwood number, N_{Sh}, as

$$N_{Sh} = \frac{4kh}{D_{AB}} \qquad (8.59)$$

$$N_{Sh} = 4 \cdot \frac{k}{(D_{Am} K_A / \delta)} \cdot \frac{v_w^0 h}{D_{AB}} \cdot \frac{(D_{Am} K_A / \delta)}{v_w^0} \qquad (8.60)$$

Using Equations 8.53, 8.16, 8.24, and 8.56,

$$N_{Sh} = \frac{4\lambda\kappa}{\alpha_0} \qquad (8.61)$$

$$= \frac{4\kappa (C_2 - C_3)}{\alpha_0 C_3 \ln \left[(C_2 - C_3) / (C_1 - C_3) \right]} \qquad (8.62)$$

It was shown earlier that the quantities C_2, C_3, V_w, and Δ can be obtained as functions of X when operational parameters γ and θ are given. Since C_1 can be calculated using C_3, V_w and Δ in Equation 8.58, N_{Sh} as well as k can be calculated as a function of distance, from Equation 8.62, for the solution flow through a channel between two flat membranes.

Kimura and Sourirajan showed that Equation 8.62 can be simplified to

$$\frac{N_{Sh}}{4} = \frac{1}{\alpha_0 \ln (C_2/C_1)} \qquad (8.63)$$

when $\gamma = \theta = 0$ [106]. Dresner developed the following equation to calculate the ratio C_2/C_1 when $\gamma = \theta = 0$:

$$\frac{C_2}{C_1} = 1 + 1.536(\xi_F)^{1/3} \text{ when } \xi_F \leq 0.02 \tag{8.64}$$

$$= 1 + \xi_F + 5\left(1 - \exp(-\sqrt{\xi_F/3})\right) \text{ when } \xi_F \geq 0.02 \tag{8.65}$$

where

$$\xi_F = \frac{X}{3\alpha_0^2} \tag{8.66}$$

Dresner also approximated C_2/C_1 by

$$\frac{C_2}{C_1} = 1 + \frac{1}{3\alpha_0^2} \tag{8.67}$$

when ξ_F is very large. Combining Equation 8.63 with Equations 8.64, 8.65, and 8.66, we obtain

$$\frac{\alpha_0 N_{Sh}}{4} = \frac{1}{\ln\left[1 + 1.536(\xi_F)^{1/3}\right]} \text{ when } \xi_F \leq 0.02 \tag{8.68}$$

$$= \frac{1}{\ln\left[1 + \xi_F + 5\left(1 - \exp(-\sqrt{\xi_F/3})\right)\right]} \text{ when } \xi_F \geq 0.02 \tag{8.69}$$

and when ξ_F is very large,

$$\frac{\alpha_0 N_{Sh}}{4} = \frac{1}{\ln\left[1 + (1/3\alpha_0^2)\right]} \tag{8.70}$$

Figure 8.6 illustrates the results of the numerical calculation in which the differential equations were rigorously solved. They are shown by different symbols corresponding to different α_0 values. The numerical calculation was per-

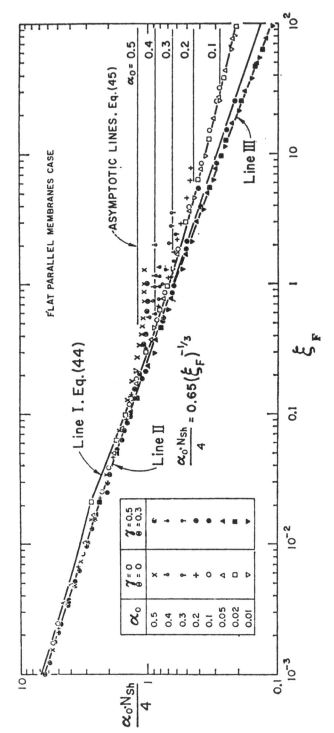

Figure 8.6. Relation between N_{Sh} and ξ_F for the laminar flow between two flat membranes placed in a parallel position. (Line III corresponds to $\gamma = 0.5$, $\theta = 0.3$.) (Reproduced from [107] with permission.)

formed for two sets of parameters including ($\gamma = 0$, $\theta = 0$) and ($\gamma = 0.5$ and $\theta = 0.3$). Line I on the figure shows the results of the approximated calculation by Equations 8.68 and 8.69 when $\gamma = \theta = 0$, and the asymptotic lines show the results from Equation 8.70. The numerical solutions seem to be well approximated by Line I and the asymptotic lines. Furthermore, all the calculated values can be well represented by Line II, which is the result of the calculation by the following equation:

$$\frac{\alpha_0 N_{Sh}}{4} = 0.65(\xi_F)^{-1/3} \tag{8.71}$$

The agreement between Line II and the numerical calculation is especially good when $\xi_F < 1$. Even when $\xi_F > 1$, the results of numerical solutions seem to scatter around Line II.

Equation 8.71 can also be derived from the Levesque equation [107],

$$\frac{kx}{D_{AB}} = \frac{x}{0.893} \left(\frac{b_F}{9D_{AB}x}\right)^{1/3} \tag{8.72}$$

and its modified version,

$$\frac{N_{Sh}}{4} = \frac{h}{0.893} \left(\frac{b_F}{9D_{AB}x}\right)^{1/3} \tag{8.73}$$

When a turbulent flow is developed between two parallel plates, b_F is given by

$$b_F = \frac{\left(\overline{u}^0\right)^2 (0.079) (N_{Re})^{-1/4}}{2\nu} \tag{8.74}$$

Combining Equations 8.73 and 8.74 and rearranging,

$$\frac{\alpha_0 N_{Sh}}{4} = 0.08 (N_{Re})^{1/4} (\xi_F)^{-1/3} \tag{8.75}$$

The above equation enables us to calculate the mass transfer coefficient k as a function of the Reynolds number and the distance from the channel inlet when a turbulent flow is developed near the channel entrance.

Nomenclature for Chapter 8

A = pure-water permeability constant, mol/m^2 s Pa

B = proportionality constant for the osmotic pressure, Pa

b_F = velocity gradient of the flow to the x direction at the membrane surface, 1/s

C = dimensionless concentration, —

c = total molar concentration of the solution, mol/m^3

c_A = molar concentration of the solute, mol/m^3

D_{AB} = solute diffusivity, m^2/s

$D_{Am}K_A/\delta$ = solute transport parameter, m/s

h = half of the solution channel height, m

J_A = molar flux of the solute, mol/m^2 s

J_B = molar flux of solvent, mol/m^2 s

k = mass transfer coefficient, m/s

N_F = hv'/ν

N_{Re} = Reynolds number, —

N_{Sh} = Sherwood number, —

p = pressure, Pa

U = dimensionless velocity to x direction, —

$\overline{U}$ = defined as $\bar{u}/\bar{u}^0$

u = velocity in x direction, m/s

$\bar{u}$ = average velocity in x direction at a distance x from the channel entrance, m/s

V = dimensionless velocity in y direction, —

V_w = defined as v_w/v_w^0

v = velocity in y direction, m/s

v' = velocity in y direction at the channel wall, m/s

v_w = permeation velocity of solvent water, m/s (in the presence of the solute in the feed solution)

v_w^* = pure water permeation velocity, m/s (in the absence of the solute in the feed solution)

X = mole fraction, —

X = dimensionless distance in x direction from the channel entrance, —

x = distance in x direction from the channel entrance, m

Y = dimensionless distance in y direction from the center of the channel, —

y = distance in y direction from the center of the channel, m

z = distance in z direction, m

α_0 = dimensionless quantity defined by Equation 8.16, —

γ = dimensionless quantity defined by Equation 8.25, —

Δ = fraction product recovery, —

θ = dimensionless quantity defined by Equation 8.26, —

κ = dimensionless quantity defined by Equation 8.24, —

λ = dimensionless quantity defined by Equation 8.53, —

ν = kinematic viscosity, m^2/s

ξ_F = dimensionless quantity defined by Equation 8.66, —

π = osmotic pressure, Pa

Superscripts

0 = at the channel entrance

Subscripts

1 = bulk feed solution

2 = solution-membrane boundary

3 = permeate solution

9

Membrane Reactors

9.1 Outline of Membrane Reactors

The membrane reactor is gaining increasing attention in the chemical and biotechnology industries. In particular, the membrane reactor was extensively studied for application in fermentation processes, as a membrane bioreactor [278]. Various membrane separation processes can be used in a membrane bioreactor, although most of the work was concerned with ultrafiltration and microfiltration, in combination with the fermenter. The fermentation broth from a fermenter is allowed to flow continuously through a cross-flow membrane filtration module where the microorganisms are rejected by the membrane and recycled to the fermenter. As a result, high cell densities can be achieved. On the other hand, the product, which often affects the bioreaction adversely as a reaction inhibitor, is removed from the fermentation broth. This makes the continuous operation possible at high productivities. Two to five g/l h alcohol can be produced by the conventional batch reactor, while 100 g/l h can be achieved by the continuous system in which membrane filtration is incorporated. However, since both substrates and products pass through the membrane, their separation is not possible by this system.

Another membrane process that can be used in combination with a fermenter is pervaporation. The fermentation broth is supplied as feed to a pervaporation module, and a vacuum is applied to the permeate side of the membrane. When an ethanol-selective membrane is used, the bioreaction product ethanol passes through the membrane as permeate vapor, which is liquefied later by compression. Microorganisms, substrate, and nutrients are retained on the feed side. The product ethanol can therefore be purified, and since the membrane is ethanol selective, it can even be concentrated in the permeate. Corresponding to the feed ethanol concentration of 5.5 wt%, 30.6% of ethanol in the permeate can be achieved [279]. Thus, ethanol can be both purified and concentrated. Concentrated ethanol can further be dehydrated by pervaporation, using a water-selective membrane.

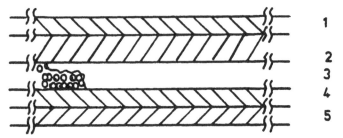

Figure 9.1. Schematic representation of sandwiched-membrane bioreactor. (Reproduced from [281] with permission.)

An example described in detail in this chapter shows that a reverse osmosis membrane can also be used for the separation of the product from the mixture of microorganisms, substrates, and nutrients. Often, microorganisms and enzymes that act as biocatalysts are immobilized to membranes, mostly to hollow fiber membranes.

Recently, the membrane reactor was extended from bioreaction to ordinary chemical reactions. For example, when pervaporation processes are coupled with an esterification reaction, the shift in the reaction equilibrium can occur by the removal of byproduct water from the reaction vessel, resulting in higher conversion rates and productivities. Furthermore, the recent progress in preparation techniques and the performance of inorganic membranes enabled the use of the membrane reactor for the reaction that requires a high temperature. For example, inorganic membranes that are permeable to hydrogen have been studied (by several groups) for the removal of hydrogen from the reaction vessel in which hydrogen is produced as a byproduct. It is expected that the coupling of reaction and separation by a membrane that constitutes a part of the reaction vessel will find wider applications in the chemical and biotechnology industries, in the future. The following example describes a membrane fermenter in which a reverse osmosis and a ultrafiltration membrane are used.

9.2   Analysis of a Membrane Bioreactor

A bioreactor in which living yeast cells are sandwiched between a cellulose acetate ultrafiltration membrane and a cellulose acetate reverse osmosis membrane (Figure 9.1) was proposed [280], [281]. A solution containing glucose substrate and nutrients is in contact with the ultrafiltration (UF) membrane, and pressure is applied on the substrate solution. The substrate permeates the UF membrane freely, together with solvent (water), and arrives at the cell layer where the bioreaction starts to occur, leading to the ethanol product. When the solution is forced by pressure out of the cell layer from the side that is in contact with the RO membrane, permselection between glucose and product ethanol occurs. While ethanol permeates almost freely through the RO membrane, the latter is practically impermeable to the glucose substrate [198]. Therefore, one is able to obtain product ethanol solution without much contamination from glucose and nutrients. The advantages of such a bioreactor, over the conventional membrane bioreactors, are

1. High cell concentration within a limited reactor space

2. Forced convective flow of substrate to the cell layer, which is much faster than diffusive transport

3. Removal of product ethanol and CO_2 from the cell layer, and prevention of product inhibition

A similar device for membrane immobilization has been proposed by Michels [282], as a membrane-moderated immobilized cell bioreactor. A continuous-type bioreactor system is schematically shown in Figure 9.2. The feed solution is pumped into the reactor at a speed sufficient to produce fluid turbulence in the proximity of the membrane boundary layer, so that the development of the concentrated boundary layer due to concentration polarization was prevented. After passing a bioreactor, the feed solution was recycled to the feed solution vessel.

The transport analysis is based on the approach in which the transport of the substrate glucose through a series of barrier layers is considered [283]–[285]. The consumption of glucose due to bioreaction in the cell layer should be superimposed to the membrane transport. Also, the transport of the bioreaction product, ethanol, through the barrier layer should be considered. A quasistationary approach is adopted. The above approach assumes that the time required for substrate glucose and ethanol to reach a local equilibrium is much smaller than the time for change in the cell number. In order to facilitate the understanding of the equations developed below, the symbols shown in Figure 9.3 will be used. The letters a, b, c, and d indicate the four barrier layers. The meaning of the barrier layers (b, UF membrane; c, cell layer; and d, reverse osmosis membrane) is self-explanatory. The barrier layer a is the boundary layer developed on the feed side of the ultrafiltration membrane, due to the concentration polarization effect. The numbers 1 and 5 indicate feed and permeate, respectively. The numbers 2, 3, and 4 indicate barrier boundaries. Capital letters A and B are also used as subscripts to indicate solute (either glucose or ethanol) and solvent (water), respectively. All other symbols are defined in the end of this chapter. The molar flux of water is given by following three equations:

$$J_B = A_b (p_2 - p_3) \tag{9.1}$$

$$J_B = A_c (p_3 - p_4) \tag{9.2}$$

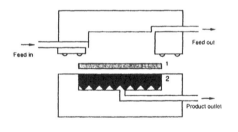

t; Sandwiched Cell Layer 2; Porous Stainless Steel

Membrane Bioreactor Cell

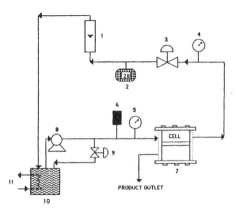

Figure 9.2. Illustration of a continuous system. [(1) Flowmeter, (2) ther-
mometer, (3) back-pressure regulator, (4 and 5) pressure gauge, (6) accumulator,
(7) membrane bioreactor, (8) high-pressure pump, (9) valve, (10) feed tank, (11)
heat exchanger.] (Reproduced from [285] with permission.)

$$J_B = A_d (p_4 - p_5 - \pi_4 + \pi_5) \qquad (9.3)$$

where J, A, and p are molar flux (mol/m^2 s), pure-water permeability constant
(mol/m^2 s Pa), and pressure (Pa). Note that Equation 9.3 is for a reverse osmosis
membrane, and therefore the effect of the osmotic pressure was included. As
in Figure 9.3a, the molar flux of substrate glucose through the concentrated
boundary layer (barrier layer a) is denoted as J_A^g, and the same glucose should
flow through the ultrafiltration membrane (barrier layer b) and the cell layer
(barrier layer c). However, glucose is produced at the rate of r_g (mol/m^3 s)
(or consumed at the rate of r_g when r_g is negative) in the cell layer. When the
effective volume of the cell layer and the effective membrane area are V (m^3)
and S (m^2), respectively, an additional glucose flux of $r_g V/S$ is generated in

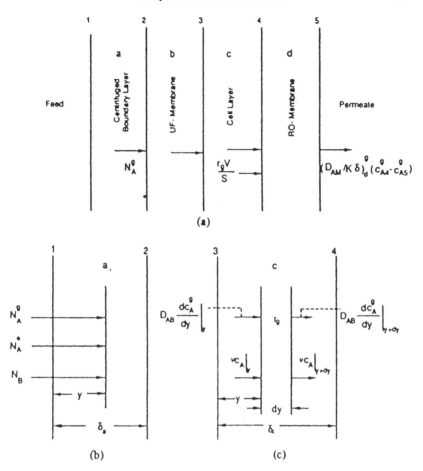

Figure 9.3. Schematic illustration of barrier layers involved in UF/cell layer/RO system. (Reproduced from [285] with permission.)

the cell layer. Therefore, at the cell layer-reverse osmosis membrane boundary (boundary 4), the glucose flux should be equal to $J_A^g + r_g V/S$. The glucose flux in the reverse osmosis membrane (barrier layer d), on the other hand, can be given by a Fick's-type equation, such as $(D_{Am}K_A/\delta)_d^g(c_{A4}^g - c_{A5}^g)$. Equating the flux on the cell layer side of the boundary 4 to that on the reverse osmosis membrane side of the boundary 4, the following equation can be obtained:

$$J_A^g + r_g V/S = (D_{Am}K_A/\delta)_d^g \left(c_{A4}^g - c_{A5}^g \right) \qquad (9.4)$$

As Equation 9.4 indicates, the molar flux of glucose through the reverse osmosis membrane is equal to $J_A^g + r_g V/S$. The total molar flux is equal to $J_A^g + r_g V/S + J_A^e + r_e V/S + J_B$, where J_A^e and r_e have the same meaning as J_A^g and r_g, but are applicable to ethanol solute, and J_B is the molar flux of the solvent water. Usually, the molar flux of solvent is far greater than those of solutes; therefore, the total molar flux is approximated by J_B. Then the glucose mole fraction in the permeate from the reverse osmosis membrane becomes

$$X_{A5}^g \simeq \left(J_A^g + r_g V/S\right)/J_B \tag{9.5}$$

Furthermore,

$$v = \text{the total molar flux }/c \simeq J_B/c \tag{9.6}$$

where v is the solution permeation velocity (m/s), and c is the total molar concentration (mol/m^3).

Now let us consider the transport at the concentrated boundary layer (barrier layer a). In the concentrated boundary layer, the molar flux of glucose at a distance y from the boundary 1 (see Figure 9.3b) can be written as

$$J_A^g = -\left(D_A^g\right)_a c\frac{dX_A^g}{dy} + X_A^g \left(J_A^g + J_A^e + J_B\right) \tag{9.7}$$

where $\left(D_A^g\right)_a$ means the diffusivity (m^2/s) of solute glucose in the boundary layer (barrier layer a). Solving the above equation with the boundary conditions

$$X_A^g = X_{A1}^g \text{ when } y = 0 \tag{9.8}$$

$$X_A^g = X_{A2}^g \text{ when } y = \delta_a \tag{9.9}$$

where δ_a is the thickness of the barrier layer a (m), we obtain

$$\ln \frac{X_{A2}^g - \left[J_A^g/\left(J_A^g + J_A^e + J_B\right)\right]}{X_{A1}^g - \left[J_A^g/\left(J_A^g + J_A^e + J_B\right)\right]} = \frac{J_A^g + J_A^e + J_B}{\left[\left(D_A^g\right)_a/\delta_a\right]c} \tag{9.10}$$

Since $J_B \gg J_A^g + J_A^e$, Equation 9.10 becomes

$$\ln \frac{X_{A2}^g - \left(J_A^g/J_B\right)}{X_{A1}^g - \left(J_A^g/J_B\right)} \simeq \frac{J_B}{\left[\left(D_A^g\right)_a/\delta_a\right]c} \tag{9.11}$$

Combining Equations 9.5, 9.6, and 9.11,

$$\ln \frac{X_{A2}^g - X_{A5}^g + \left(r_g V/S\right)/J_B}{X_{A1}^g - X_{A5}^g + \left(r_g V/S\right)/J_B} = \frac{v}{k_a^g} \tag{9.12}$$

where k_a^g is the mass transfer coefficient (m/s) of glucose in the barrier layer a, and

$$k_a^g = \left(D_A^g\right)_a / \delta_a \tag{9.13}$$

Both numerator and denominator of Equation 9.12 are multiplied by c and we obtain

$$\ln \frac{c_{A2}^g - c_{A5}^g + \left(r_g V/S\right)/v}{c_{A1}^g - c_{A5}^g + \left(r_g V/S\right)/v} = \frac{v}{k_a^g} \tag{9.14}$$

or

$$\frac{c_{A2}^g - c_{A5}^g + \left(r_g V/S\right)/v}{c_{A1}^g - c_{A5}^g + \left(r_g V/S\right)/v} = \exp\left(\frac{v}{k_a^g}\right) \tag{9.15}$$

Equation 9.15 describes the transport of substrate glucose through the concentrated boundary layer. An equation similar to Equation 9.15 should be applicable to the ultrafiltration membrane (barrier layer b), and

$$\frac{c_{A3}^g - c_{A5}^g + \left(r_g V/S\right)/v}{c_{A2}^g - c_{A5}^g + \left(r_g V/S\right)/v} = \exp\left(\frac{v}{k_b^g}\right) \tag{9.16}$$

where k_b^g is the mass transfer coefficient (m/s) of solute glucose in the barrier layer b. Finally, the transport through the cell layer (barrier layer c) is described by a differential equation:

$$\left(D_A^g\right)_c c \frac{d^2 X_A^g}{dy^2} - \frac{d\left(J_A^g + J_A^e + J_B\right) X_A^g}{dy} + r_g = 0 \tag{9.17}$$

where $\left(D_A^g\right)_c$ is the diffusivity of glucose (m^2/s) in the barrier layer c. By using the relation $J_B \gg J_A^g + J_A^e$ and Equation 9.6, the above equation becomes

$$\left(D_A^g\right)_c \frac{d^2 c_A^g}{dy^2} - v \frac{dc_A^g}{dy} + r_g = 0 \tag{9.18}$$

The meaning of the above equation is shown schematically in Figure 9.3c. The solution for Equation 9.18 is given as

$$c_A^g = C_1^g \exp\left(\frac{v}{\left(D_A^g\right)_c} y\right) + \left(\frac{r_g}{v} y\right) + C_2^g \tag{9.19}$$

where C_1^g and C_2^g are integral constants. C_1^g and C_2^g can be determined by the following boundary conditions:

$$c_A^g = c_{A3}^g \text{ at } y = 0 \tag{9.20}$$

and

$$
\begin{aligned}
J_A^g &= -\left(D_A^g\right)_c \frac{dc_A^g}{dy} + v c_A^g \\
&= \left(D_{Am} K_A / \delta\right)_d^g \left(c_{A4}^g - c_{A5}^g\right) - r_g V / S \text{ at } y = 0
\end{aligned} \tag{9.21}
$$

where J_A^g indicates the molar flux of glucose substrate (mol/m² s) at $y = 0$ and is given by Equation 9.4. Combining Equations 9.19 and 9.21,

$$
\begin{aligned}
-\left(D_A^g\right)_c &\left(C_1^g v / \left(D_A^g\right)_c \exp\left[\frac{v}{\left(D_A^g\right)_c}(0) \right] + r_g / v \right) \\
+ v C_1^g &\exp\left[v / \left(D_A^g\right)_c (0) \right] + v (r_g / v)(0) + v C_2^g \\
&= \left(D_{Am} K_A / \delta\right)_d^g \left(c_{A4}^g - c_{A5}^g\right) - r_g V / S
\end{aligned} \tag{9.22}
$$

Therefore,

$$C_2^g = \frac{\left(D_{Am} K_A / \delta\right)_d^g}{v} \left(c_{A4}^g - c_{A5}^g\right) - \frac{r_g V}{vS} + \frac{\left(D_A^g\right)_c r_g}{v^2} \tag{9.23}$$

Furthermore,

$$V / S = \delta_c \tag{9.24}$$

where δ_c is the thickness (m) of the barrier layer c. Defining the mass transfer coefficient of glucose in barrier layer c by

$$\frac{\left(D_A^g\right)_c}{\delta_c} = k_c^g \tag{9.25}$$

Equation 9.23 becomes

$$C_2^g = \frac{\left(D_{Am} K_A / \delta\right)_d^g}{v} \left(c_{A4}^g - c_{A5}^g\right) - \frac{r_g V}{vS} + \frac{k_c^g r_g V}{v^2 S} \tag{9.26}$$

Combining Equations 9.19 and 9.20,

$$C_1^g + C_2^g = c_{A3}^g \tag{9.27}$$

Equations 9.26 and 9.27 allow us to calculate integration constants C_1^g and C_2^g. Furthermore,

$$c_A^g = c_{A4}^g \ at \ y = \delta_c \tag{9.28}$$

Therefore,

$$c_{A4}^g = C_1^g \exp\left(\frac{v}{(D_A^g)_c}\delta_c\right) + \frac{r_g}{v}\delta_c + C_2^g \tag{9.29}$$

By using the relations given by Equations 9.24 and 9.25,

$$c_{A4}^g = C_1^g \exp(v/k_c^g) + r_g V/vS + C_2^g \tag{9.30}$$

Equations similar to Equations 9.4, 9.15, 9.16, 9.30, 9.27, 9.26, and 9.5 can be developed for product ethanol, with superscript e, indicating that they are the properties of ethanol:

$$J_A^e + r_e V/S = \left(\frac{D_{Am}K_A}{\delta}\right)_d^e (c_{A4}^e - c_{A5}^e) \tag{9.31}$$

$$\frac{c_{A2}^e - c_{A5}^e + (r_e V/S)/v}{c_{A1}^e - c_{A5}^e + (r_e V/S)/v} = \exp\left(\frac{v}{k_a^e}\right) \tag{9.32}$$

$$\frac{c_{A3}^e - c_{A5}^e + (r_e V/S)/v}{c_{A1}^e - c_{A5}^e + (r_e V/S)/v} = \exp\left(\frac{v}{k_b^e}\right) \tag{9.33}$$

$$c_{A4}^e = C_1^e \exp(v/k_c^e) + r_e V/vS + C_2^e \tag{9.34}$$

where

$$C_1^e + C_2^e = c_{A3}^e \tag{9.35}$$

and

$$C_2^e = \frac{(D_{Am}K_A/\delta)_d^e}{v}(c_{A4}^e - c_{A5}^e) - \frac{r_e V}{vS} + \frac{k_c^e r_e V}{v^2 S} \tag{9.36}$$

Furthermore,

$$X_{A5}^e \simeq \frac{J_A^e + r_e V/S}{J_B} \tag{9.37}$$

As for the rate of biocatalytic reactions for glucose consumption and ethanol production, the following equations are applicable:

$$r_g = -\frac{1}{Y_{X/S}}\frac{dX}{dt} - mX \tag{9.38}$$

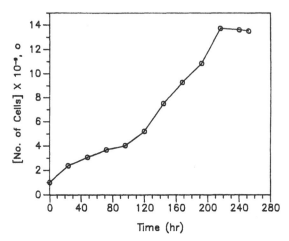

Figure 9.4. Increase in the total cell number in the cell layer. [Operating pressure, 2758 kPa (gauge) (= 400 psig); initial cell number, 1.0×10^9; feed glucose concentration, 20%.] (Reproduced from [285] with permission.)

$$r_e = -\frac{1}{Y_{P/S}} r_g \qquad (9.39)$$

Equations 9.1–9.4, 9.15, 9.16, 9.30, 9.5, 9.31–9.34, and 9.37–9.39 have so far been developed for the transport of water, substrate glucose, and product ethanol and for the rate of biocatalytic reactions. There are altogether 15 equations. Examining the quantities involved in those equations, we find that p_3, p_4, J_B, J_A^g, c_{A2}^g, c_{A3}^g, c_{A4}^g, c_{A5}^g, r_g, J_A^e, c_{A2}^e, c_{A3}^e, c_{A4}^e, c_{A5}^e, and r_e are unknowns. Since there are 15 equations and 15 unknowns, these equations can be solved for the unknowns.

The numerical calculation based on the above equations can be executed by following Steps 1 to 6.

· Step 1. By using the experimental data on cell number growth, such as shown in Figure 9.4, dX/dt is obtained at time $t = 0$.

· Step 2. From the above growth rate, all 15 unknowns are calculated from 15 equations of transport and biocatalytic reaction rate. The initial feed concentration of the substrate glucose, $[c_{A1}^g]^0$, and the product ethanol, $[c_{A1}^e]^0$, are used for the calculation. The numerical answers so obtained correspond to time $t = 0$, and they are shown in the bracket $[--]^0$.

· Step 3. During the following time interval, Δt, an amount of glucose, $[J_A^g]^0 S \Delta t$, is lost from the feed. Since the initial amount of substrate glucose is $V_1^0 [c_{A1}^g]^0$, the amount left in the feed is $V_1^0 [c_{A1}^g]^0 - [J_A^g]^0 S \Delta t$. Similarly, the volume of the solution ($\simeq$ the volume of solvent) left in the feed is

$V_1^0 - M_B J_B S \Delta t / \rho_B$. However, the volume of the feed solution is so large that it is hardly changed by the loss of solvent due to membrane permeation. Then the solution volume can be assumed to be constant at a value of V_1^0. The feed glucose concentration is therefore $(V_1^0 [c_{A1}^g]^0 - [J_A^g]^0 S \Delta t)/V_1^0$ at $t = \Delta t$. Similarly, the ethanol concentration in the feed solution is $-([J_A^e]^0 S \Delta t)/V_1^0$.

· Step 4. By using the cell number growth data, dX/dt at time Δt is obtained and designated as $[dX/dt]^1$.

· Step 5. All 15 unknowns are calculated by using the feed glucose and ethanol concentrations obtained in Step 3, and $[dX/dt]^1$ obtained in Step 4. All numerical answers so obtained correspond to time $t = \Delta t$ and are shown in the bracket $[--]^1$.

· Step 6. The above steps are repeated, and we obtain $[--]^2$, $[--]^3$, ..., $[--]^n$, corresponding to $t = 2\Delta t, t = 3\Delta t, \ldots, t = n\Delta t$. Thus, the solvent flux ($\simeq$ solution flux), J_B, the permeate concentration of glucose, c_{A5}^g, and the permeate concentration of ethanol, c_{A5}^e, at time $i\Delta t$, can be obtained as $[J_B]^i$, $[c_{A5}^g]^i$, and $[c_{A5}^e]^i$. The total permeate volume from time 0 to time $i\Delta t$ is $(M_B S \Delta t / \rho_B) \sum [J_B]^i$.

It is obvious from the above explanation that the experimental cell growth data are necessary in the above approach to calculate membrane reactor performance. The objective of this work is to show that the cell growth data can be connected to the reactor performance data by a model in which the membrane transport and the bioreaction rate equations are coupled.

Some experimental results for the performance of a continuous biocatalytic reactor are shown in Figures 9.5 and 9.6. Figure 9.4 shows that the total number of cells in the cell layer has increased from 1.0×10^9 to 1.4×10^{10}, during the reaction period. This figure is used to evaluate dX/dt for the theoretical calculation of the concentration of substrate glucose and product ethanol in the permeate, and the product permeation rate. Figure 9.5 shows the experimental glucose and ethanol concentration in the permeate. The main features of the figure are that glucose concentration decreases, whereas the ethanol concentration increases, with reaction time. Figure 9.6 shows the total amount of the permeate collected and the permeation rate vs. reaction time.

MEMBRANE REACTORS

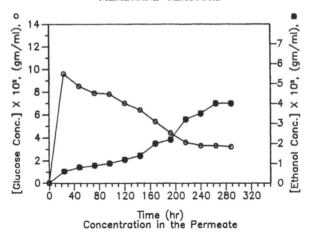

Figure 9.5. Glucose and ethanol concentration in the permeate. [Operating pressure, 2758 kPa (gauge) (= 400 psig); initial cell number, 1.0×10^9; feed glucose concentration, 20%.] (Reproduced from [285] with permission.)

Numerical calculations were performed by using the transport and bioreaction rate equations and the associated numerical parameters. The following numerical parameters are required for the calculations:

1. A_b, A_c, and A_d; $\left(D_{Am}K_A/\delta\right)_d^g$, $\left(D_{Am}K_A/\delta\right)_d^e$

2. Osmotic pressures π_4 and π_5, which correspond to the solution compositions on both sides of the reverse osmosis membrane (at the boundaries 4 and 5)

3. Cell volume V and effective membrane area S

4. Mass transfer coefficients through barrier layers a, b, and c, for glucose and ethanol, k_a^g, k_b^g, k_c^g, k_a^e, k_b^e, and k_c^e.

It should be especially noted that the membrane transport parameters A, $(D_{Am}K_A/\delta)$, and k, and the cell layer geometry parameters V and S, define the membrane reactor system, and the reactor performance can be calculated for a given cell growth rate under a set of operating conditions. All necessary parameters are listed in Table 1 [285]. They were obtained either from the literature [283], [284] or from experiments. As for the osmotic pressure, it must be a complicated function of the solution composition, since the solution involves a mixture of solutes, including substrate glucose, several nutrients, and product

Table 1. Parameters Used for the Simulation [285]

Parameters	Values
A_b	3.239×10^{-6} mol/m^2 s Pa
A_c	2.770×10^{-8} mol/m^2 s Pa
A_d	1.750×10^{-7} mol/m^2 s Pa
p_2	2859.3 kPa
p_5	101.352 kPa
B^g	1.429×10^5 kPa
$(D_{Am}K_A/\delta)_d^g$	1.280×10^{-8} m/s
$(D_{Am}K_A/\delta)_d^e$	3.564×10^{-5} m/s
k_a^g	1.33×10^{-5} m/s
k_b^g	1.16×10^{-6} m/s
k_c^g	3.04×10^{-4} m/s
k_a^e	2.064×10^{-5} m/s
k_b^e	1.801×10^{-6} m/s
k_c^e	4.719×10^{-4} m/s
S	6×10^{-4} m^2
V	2×10^{-6} m^3
$Y_{P/S}$	0.375
$Y_{X/S}$	0.06
m	2.778×10^{-4} s^{-1}

ethanol. An assumption is made, however, that the osmotic pressure is contributed from substrate glucose alone. The reason for the above assumption is as follows. The concentrations of nutrients involved in this study are so small that their contributions to the osmotic pressure can be ignored. As for the contribution of product ethanol, the concentration is high, and its contribution to the osmotic pressure cannot be ignored. However, the rejection of ethanol by the cellulose acetate reverse osmosis membrane is so low that the ethanol concentrations on both sides of the reverse osmosis membrane can be considered to be practically equal. Then the ethanol contribution to π_4 (the osmotic pressure at the cell layer side of the membrane) and to π_5 (the osmotic pressure at the permeate side of the membrane) can be considered equal and does not affect $\pi_4 - \pi_5$. Therefore, the osmotic pressure can be considered to be a function of the glucose concentration, c_A^g, alone. The osmotic pressure of the glucose solution is given in Table 2 [286]. Furthermore, it can be approximated as a

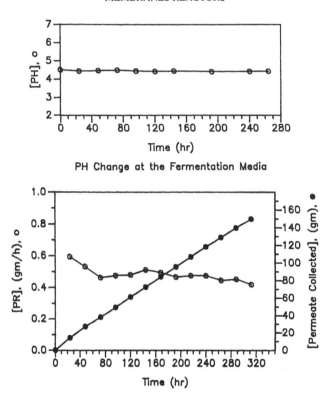

Figure 9.6. Amount of permeate collected and permeation rate. [Operating pressure, 2758 kPa (gauge) (= 400 psig); initial cell number, 1.0×10^9; feed glucose concentration, 20%.] (Reproduced from [285] with permission.)

linear function of a glucose mole fraction, such as $\pi^g = B^g \times X_A^g$. The numerical value of B^g is also listed in Table 1. Since all numerical parameters necessary for the solution of 15 transport and biocatalytic reaction rate equations were available, the numerical solution was attempted by using a library program (IMSL) for the simultaneous solution of nonlinear equations, following Steps 1 to 6 described earlier. The results are compared in Figure 9.7 with the experimental values. Glucose and ethanol concentration data at different barrier boundaries vs. reaction time are also shown in Figure 9.8. The agreement is reasonably good, testifying to the validity of the model equations used and the associated parameters. A set of transport and biocatalytic equations also allow us to back-calculate the cell growth rate (and therefore the total cell number when an initial cell number is given) when we know the glucose and ethanol concentration in the permeate. Calculations were carried out for the case of an operating pressure of 2068 kPa (gauge) (300 psig), an initial feed glucose concentration of 12%, and an initial cell number of 1×10^9 cells, with respect

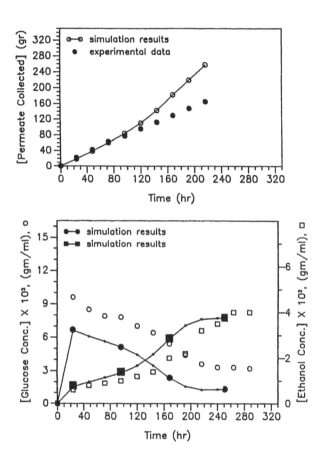

Figure 9.7. Comparison of calculated (lines) and experimental (symbols) values. (Reproduced from [285] with permission.)

Table 2. Osmotic Pressure and
Molar Density Data for D-Glucose [286]

Molality	Mole fraction $(\times 10^3)$	Osmotic pressure (kPa)	Molar density $(\times 10^{-4} \text{mol/m}^3)$
0	0	0	5.535
0.1	1.798	259	5.494
0.2	3.590	517	5.446
0.3	5.375	776	5.394
0.4	7.154	1034	5.342
0.5	8.927	1293	5.291
0.6	10.693	1517	5.245
0.7	12.453	1744	5.195
0.8	14.207		5.152
0.9	15.955		5.105

to reverse osmosis and ultrafiltration membranes used for the experiments. The calculated results of the total cell number in a cell layer are shown in Figure 9.9, with an experimental final cell number. The agreement in the final cell number is very good. Glucose and ethanol concentration data at different barrier boundaries are shown in Figure 9.10.

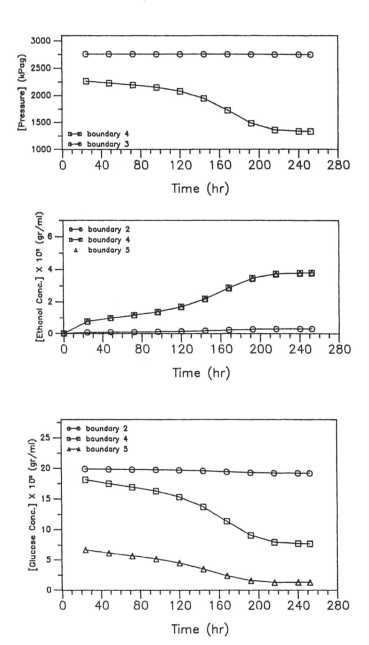

Figure 9.8. Some results of simulation (change in pressure and glucose and ethanol concentrations at different barrier boundaries). (Reproduced from [285] with permission.)

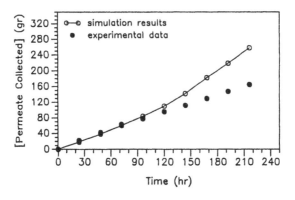

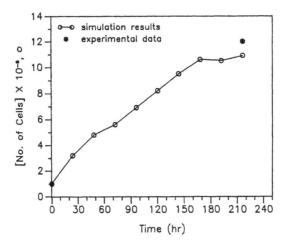

Figure 9.9. Calculated values for the total cell number in a cell layer. [Operating pressure, 2068 kPa (gauge) (= 300 psig); initial cell number, 1.0×10^9; feed glucose concentration, 12%.] (Reproduced from [285] with permission.)

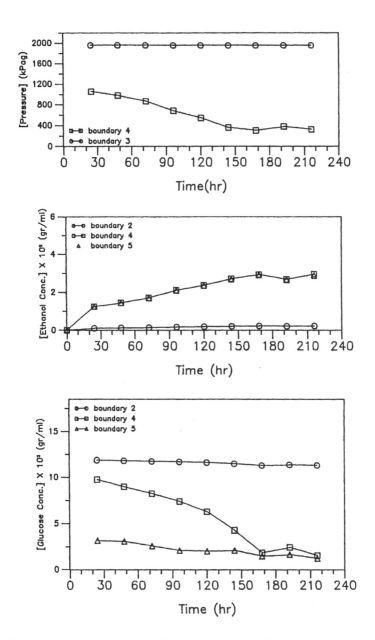

Figure 9.10.   Some results of simulation (change in pressure and glucose and ethanol concentration at different barrier boundaries). (Reproduced from [285] with permission.)

Nomenclature to Chapter 9

A = pure-water permeability constant, mol/m^2 s Pa

B = proportionality constant between osmotic pressure and mole fraction, Pa

c = total molar concentration of solution including solutes and solvent, mol/m^3

c_A = molar concentration of solute A, mol/m^3

D_A = diffusivity of solute in the barrier layer, m^2/s

D_{Am} = diffusivity of solute in the reverse osmosis membrane, m^2/s

$D_{Am}K_A/\delta$ = solute transport parameter, m/s

J_A = solute flux, mol/m^2 s

J_B = flux of solvent water, mol/m^2 s

K_A = partition coefficient of solute between solution and reverse osmosis membrane phases, —

k = mass transfer coefficient, m/s

M_B = molecular weight of solvent, —

m = cell maintenance coefficient, 1/s

p = pressure, Pa

r_e = ethanol production rate, mol/m^3 s

r_g = glucose consumption rate, mol/m^3 s

S = effective area of cell layer, m^2

t = time, s

V = effective volume of cell layer, m^3

V_1^0 = feed solution volume, m^3

v = permeation velocity through barrier layers, m/s

X = total cell number

X_A = mole fraction of solute, —

$Y_{P/S}$ = product yield coefficient

$Y_{X/S}$ = cell mass yield coefficient

y = distance to the flow direction, m

δ in $(D_{Am}K_A/\delta)$ = effective thickness of reverse osmosis membrane, m

δ = thickness of barrier layer, m

ρ_B = density of solvent water, kg/m^3

π = osmotic pressure, Pa

Superscripts

g = properties of glucose

e = properties of ethanol

Subscripts

$1-5$ = as illustrated in Figure 9.3

$a-d$ = as illustrated in Figure 9.3

A = solute (glucose and ethanol)

B = solvent (water)

10

Applications

10.1 Reverse Osmosis

10.1.1 Seawater Desalination by Reverse Osmosis

The announcement of the asymmetric cellulose acetate membrane by Loeb and Sourirajan in 1960 made the membrane desalination process industrially practical and opened up the avenue for membrane applications to various separation processes. The membrane desalination process has grown steadily and become well accepted in the marketplace. It was estimated that at the beginning of the 1980s the market share for reverse osmosis was about 23%, with multistage flush distillation (MSF) having a two-thirds share. By the end of the 1980s the situation had completely changed. Reverse osmosis is currently the major contributor to the seawater desalination, with 85% market share, and MSF was reduced to 3%. Economics, energy consumption, and the advancement of the technology are the causes of this drastic reversal [287]. As for the energy required to produce 3.79 m^3 (1000 gal) of freshwater, comparison was made for different seawater desalination processes, by Channabasappa in 1976 [288]. It was concluded that the reverse osmosis process requires much less energy than the MSF process. While 1.06×10^9 J (1,000,000 BTU) is necessary for the MSF process, reverse osmosis with a turbine for energy recovery needs only 2.71×10^8 J (257, 000 BTU), roughly one fourth that of the MSF process. It is estimated today that about 12 million m^3/day of reverse osmosis water is manufactured for desalting plants. About one tenth of this capacity is to desalt seawater and highly brackish water [287]. According to another estimate by Kremen et al., the total operating capacity of seawater reverse osmosis plants worldwide was approximately 0.114 million m^3/day (30 million gal/day) in 1983. At that time the cost of water produced by the RO process was estimated to be (in 1980 dollars) $0.79 to 1.58 per cubic meter ($3 to $6 per 1000 gal) [287]. In the third quarter of 1990, the total installed and contracted capacity was estimated to be 0.379 million m^3/day (100 million gal/day). Interestingly

383

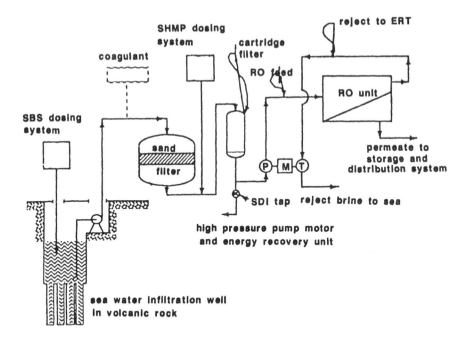

Figure 10.1. Schematic flow diagram of ERCROS seawater RO plant. (Reproduced from [289] with permission.)

enough, the cost of water produced by the RO process was estimated to be (in 1990 dollars) $0.926 per cubic meter ($3.51 per 1000 gal). These results reflect a number of factors that favorably affected the RO desalination process, especially the improvement of the technology and increasing experience in the construction and operation of the plant.

The membranes currently used for seawater desalination are aromatic polyamide membrane (hollow fiber, Du Pont), cellulose acetate membranes (hollow fiber, Toyobo), thin-film composite membranes (spiral-wound, UOP), thin-film composite membranes (spiral-wound, Dow), and thin-film composite membranes (spiral-wound, Toray).

Despite many technological advancements, the basic plant design for the seawater desalination process has not changed. An outline of the process is made in the following pages, according to the description of the ERCROS seawater RO plant where the FilmTec FT 30 membrane was used [289]. The flow diagram of the process is schematically presented in Figure 10.1. The feed water is taken in from a seawater infiltration well, located in porous volcanic rock. The well is located about 100 m from the shoreline. The seawater level is about 24 m below the ground level. Two centrifugal pumps set on a rock shelf just above the seawater, pump the feed seawater to the ground. Sodium bisulfite (SBS) is dosed into the well for disinfection. The raw seawater is pumped from the well to three sand filters in parallel. Then sodium hexametaphosphate (SHMP) is injected prior to the 5μ cartridge filter that follows the sand filter, to prevent precipitation of magnesium sulfate. The rate of the SHMP dose is to maintain 5 to 10 ppm in the feed seawater. The silt density index (SDI), which is a turbidity indicator by colloidal particles, is then monitored. SDI values are reported to be consistently in the range of 0.5 to 1.0, which means the feed seawater is very clean. This is because raw seawater is taken in from a well. In order to pressurize the feed seawater, two high-pressure pump-motor-turbine sets are installed. One is operating, and the other is standby. The pump-motor-turbine set enables the recovery of energy inherent to the high-pressure brine rejected at the RO unit. Before being released, the high-pressure brine drives the turbine (reverse-running pump) mounted on a common shaft with the electric motor and the high-pressure pump. The high-pressure feed seawater is piped to the RO vessel rack. According to the description by Dow, the rack has three vertical rows of pressure vessels, where each row is six vessels high. Twelve pressure vessels in two vertical rows are connected in parallel to form the first pass or array. Brine from these 12 pressure vessels is then collected into an intermediate manifold to the second pass or array of the remaining six pressure vessels in parallel. The brine rejected from the second pass is collected in a second manifold and piped to the energy recovery turbine from which it discharges into a drain trench in the building floor and ultimately to an outfall in the sea. Each of the 18 pressure vessels contains six FilmTech seawater spiral-wound elements of 8-in. diameter and 40 in. long, producing 1000 m^3/day (265,000 gal/day). As the operation details listed in Table 1 indicate, the feed seawater TDS (total dissolved solid) is about 38,000 mg/l. Since 45% is recovered as freshwater of 325 mg/l, the TDS in the rejected brine amounts to 69,000 mg/l, the osmotic pressure of which is approximately 775 psi. This means the effective driving force at the end of the reverse osmosis unit is only 225 psi, since the operating pressure is 1000 psig.

10.1.2 Removal of Small Organic Molecules from Water

It was in the earlier part of the 1980s when membranes exhibiting extremely high separations for organic solutes from water were developed. It is expected

Table 1. Lanzarote SWRO Plant
Design and Operating Parameters [289]

Capacity	=	1000 m³/day = 184 GPM
Feed pressure	=	1000 psig
Seawater TDS	=	38,000 mg/l
Permeate TDS	=	500 mg/l, design
Permeate TDS	=	325 mg/l, actual
Recovery	=	45%
Energy consumption (excluding auxiliaries)	=	5.8–6.0 kWh/m³

Pretreatment

Sand pressure filters		
Effective size of sand	=	0.55 mm
Uniformity coefficient	=	1.3
Sand bed depth	=	1 m
Filtration rate	=	5 gal/min ft²
Cartridge filter mesh size	=	5 μm
Chemical dosing		
SBS dosing rate	=	30–40 ppm
SHMP dosing rate	=	5–10 ppm
Coagulant (if required)	=	1–2 ppm

that the importance of these membranes will grow as the general public becomes more aware of the hazardous effect of small organic molecules (sometimes carcinogenic) in our drinking water. Among others, PA-300 (UOP) and PEC-1000 (Toray) are the most effective. Both are thin-film composite membranes. The thin film in the PA-300 membrane is a polyamide film formed by the interfacial reaction of isophthaloyl chloride with polyetheramine on a porous polysulfone membrane [290]. The PEC-1000 membrane is a thin film (30 nm) of cross-linked polyethers obtained by an in-situ polymerization of tri(hydroxyethyl) isocyanate (THEIC) and furfuryl alcohol (FA), in the presence of an acid catalyst on the surface of a porous polysulfone support membrane [291]. The performance data of both PA-300 and PEC-1000 are listed in Table 2 and Table 3.

10.1.3   Nanofiltration Membranes

Reverse osmosis membranes, which exhibit sodium chloride separations slightly lower than those for desalination purposes, but much higher fluxes, were developed in the middle of the 1980s for various water treatment purposes. These membranes are called nanofiltration membranes, since the pore size of these membranes are believed to be approximately 1 nanometer (nm). Although the chemical structures of the selective layer are not necessarily disclosed, most of these membranes seem to be of composite nature with negatively charged selective skin layers. Typical examples of nanofiltration membranes are summarized as follows.

Millipore High-Flux CP$^{(R)}$ Modules [292]. These membranes were produced by depositing a thin film of a highly negatively charged polymer on a polysulfone porous substrate. Reflecting the negative charge of the selective layer, monovalent sodium cation is better separated than divalent calcium cation. The salt separation decreases with an increase in the salt concentration. At a sodium chloride separation of 95%, High Flux Millipore CP membrane modules exhibit the average flux of 7.57×10^{-4} m^3/day m^3 (module) Pa [39 gpd/ft^3 (module) psi], and the maximum value of 9.31×10^{-4} m^3/day m^3 (module) Pa [48 gpd/ft^3 (module) psi], compared with the standard cellulose acetate module output of 10 to 19 gpd/ft^3 (module) psi [1.94 to 3.68 $\times 10^{-4}$ m^3/day m^3 (module) Pa]. Chlorine and pH tolerance of the membrane is very good.

UOP TFCL-LP Module [293]. Both 4-in. and 8-in. modules are available. The salt rejection of 96% could be achieved even at 414 kPa (gauge) (60 psig). Typical experimental results of long-term tests, over 80 days operating period, are 0.895 to 0.936 m^3/m^2 day (22 to 23 gal/ft^2 day) and salt rejection of 95% at 1379 kPa (gauge) (200 psig). A test for the chlorine resistance showed that salt separation started to decrease at the chlorine exposure of 1000 ppm-h.

Film Tec NF-40 Membrane [294], [295]. This membrane is prepared by interfacial reaction of 1,3,5-benzenetricarboxilic acid chloride with piperadine on a base polysulfone material. At 1379 kPa (gauge) (200 psig), sodium chloride separation is 69%, and the pure-water permeation rate is 1.26 m^3/m^2 day (31 gal/ft^2 day). While the separation goes down to 51% for MgCl$_2$, that of Na$_2$SO$_4$ is 99%, which indicates that the membrane is negatively charged. As for organic compounds, separations of ethylene glycol, glycerol, and glucose are 27%, 43%,

Table 2. Reverse Osmosis Separation Characteristics of
PA-300 Membranes at 1000 psig and 25° C[290]

Solute	Solute conc. (ppm)	pH	Solute rejection (%)
Sodium nitrate	10,000	6.0	99.0
Ammonium nitrate	9,600	5.7	98.1
Boric acid	280	4.8	65–70
Urea	1,250	4.9	80–85
Phenol	100	4.9	93
Phenol	100	12.0	>99
Ethyl alcohol	700	4.7	90
Glycerine	1,400	5.6	99.7
DL-aspartic acid	1,500	3.2	98.3
Ethyl acetate	366	6.0	95.3
Methyl ethyl ketone	465	5.2	94
Acetic acid	190	3.8	65–70
Acetonitrile	425	6.3	>25
Acetaldehyde	660	5.8	70–75
Dimethyl phthalate	37	6.2	>95
2,4-Dichlorophenoxy acetic acid	130	3.3	>98.5
Citric acid	10,000	2.6	99.9
Alcozyme (soap)	2,000	9.3	99.3
o-Phenyl phenol	110	6.5	>99
Tetrachloroethylene	104	5.9	>93
Sodium silicate	42	8.6	>96
Sodium chromate	1,200	7.8	>99
Chromic acid	870	3.9	90–95
Cupric chloride	1,000	5.0	99.2
Zinc chloride	1,000	5.2	99.3
Trichlorobenzene	100	6.2	>99
Butyl benzoate	200	5.8	99.3

Table 3. Reverse Osmosis Separation Characteristics of
PEC-1000 Membranes at 800 psig and 25°C [291]

Solute	Solute conc. in feed (wt%)	pH	Solute rejection (%)	Water flux (m^3/m^2 day)
Methyl alcohol	5	6.9	41	0.38
Ethyl alcohol	5	5.0	92	0.30
Isopropyl alcohol	5	6.5	99.5	0.32
n-Butyl alcohol	5	5.6	99	0.30
Benzyl alcohol	4	7.0	82	0.10
Ethylene glycol	5	6.8	94	0.36
Propylene glycol	5	5.8	99.7	0.40
Glycerine	5	5.8	99.9	0.43
Phenol	1	5.2	99.0	0.24
Formic acid	5	1.9	34	0.52
Acetic acid	5	2.6	91	0.37
Propionic acid	5	2.7	97	0.39
Oxalic acid	0.5	1.8	99.1	0.69
Formaldehyde	5	3.6	69	0.30
Acetaldehyde	3	3.6	89	0.46
Acetone	4	6.7	97	0.29
Methyl ethyl ketone	4	6.6	98	0.21
Ethyl acetate	4	6.8	99.2	0.18
n-Butyl acetate	1	6.9	99.6	0.38
Tetrahydrofuran	5	6.7	99.8	0.28
1,4-Dioxane	5	6.6	99.9	0.40
Ethyldiamine	5	12.1	99.5	0.08
Aniline	1	8.3	95	0.11
Urea	1	6.9	85	0.56
N, N-Dimethylformamide	5	6.5	98	0.32
N, N-Dimethylacetamide	5	6.5	99.6	0.30
ε-Caprolactam	5	5.8	99.9	0.30
Dimethylsulfoxide	5	6.4	99.6	0.34
Lactose	5	6.1	>99.9	0.50

and 100%, respectively. Since it contains an aromatic polyamide base structure, the membrane is not chlorine resistant.

Toray UTC-20 and UTC-40 Membranes [296]. UTC-20 membranes are for low-pressure and low NaCl separation (50 to 80%), while UTC-40 membranes are for low-pressure and high NaCl separation RO operation (>90%). These membranes are also featured by having very high flux. UTC-40 membranes exhibit a flux of about $3.0 \text{ m}^3/\text{m}^2$ day at 90% NaCl separation when operated at 1470 kPa (gauge) (15 kg/cm^2 gauge) and with 1500 ppm NaCl solution. Commercial cellulose acetate membranes, on the other hand, exhibit about 0.7 m^3/m^2 day at 2940 kPa (gauge) (30 kg/cm^2 gauge) when 90% of salt rejection is obtained for 1500-ppm feed NaCl solution. UTC-20 membranes separate sucrose and raffinose almost completely, although sodium chloride separation of these membranes is only 50 to 80%. The concentration dependence of the sodium chloride separation is very strong. For example, the UTC-20 HF membrane shows NaCl separation of 50% when the feed concentration is 1500 ppm, but the separation increases to 90% when the concentration is 100 ppm, suggesting the ionic nature of the membrane. Chlorine resistance is better than for polyamide membranes, but not as high as for cellulose acetate membranes.

Nitto-Denko NTR-739 HF and NTR-729 HF Membranes [297]. These membranes are also of composite structure. The porous sublayer prepared from polysulfone material is covered with a skin layer of polyvinyl alcohol material with a carboxyl group. The membrane data are again compared with cellulose acetate membranes. The operating conditions of test runs are feed sodium chloride concentration, 1500 ppm; operating pressure, 1000 kPa (gauge) (143 psig); temperature, 25°C; and pH of the feed solution, 6 to 7. While cellulose acetate membranes showed $0.2 \text{ m}^3/\text{m}^2$ day and 91% salt rejection, NTR-739 HF membranes showed 95.5% salt rejection and $1 \text{ m}^3/\text{m}^2$ day. NTR-729 membranes showed 91% salt rejection and $1.6 \text{ m}^3/\text{m}^2$ day. The solute separation of Na_2SO_4 is much higher than that of NaCl, while that of $MgCl_2$ is lower than sodium chloride, reflecting again the anionic nature of the material of the selective skin layer. The experimental data reported by Nitto-Denko also indicate that the separation of organic solutes is generally higher than that for cellulose acetate membranes, when comparison is made for the same level of sodium chloride separations. The resistance for feed tap water containing 100 ppm of chlorine was the lowest at pH 7, where the salt rejection started to decrease after 5 to 10 h of operation.

10.1.4 Ultrapure Water Production

The demand for the production of a large amount of ultrapure water increased enormously in the past decade in semiconductor industries. It is due to the rapid

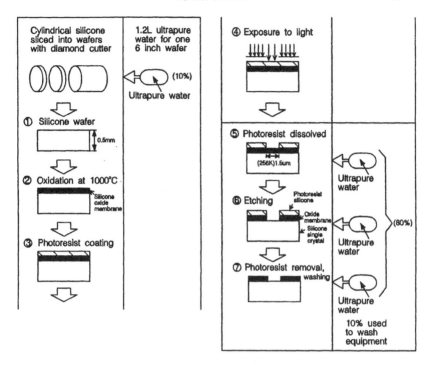

Figure 10.2.   Method of manufacturing the very large integrated circuit.

progress in the manufacturing method of the very large-scale integrated circuit (VLSI). The method of manufacturing VLSI is outlined in Figure 10.2 [298]. The process is divided into the following steps.

· Step 1. A wafer 0.5 mm thick is made from a silicon single crystal.

· Step 2. The surface of the silicon wafer is oxidized, and a silicon oxide (silica) layer is formed.

· Step 3. Organic photoresist is coated on the top of the silica layer.

· Step 4. A mask is placed on the top of the photoresist, and the surface is exposed to a light beam.

· Step 5. The part of the photoresist that was unmasked to the light beam is dissolved.

· Step 6. The silica layer under the unmasked part of the photoresist is also etched with hydrofluoric acid.

· Step 7. Arsenic or aluminum metal is doped to complete the integrated circuit.

As shown in the figure, it is necessary to apply ultrapure water in Steps 2, 5, 6, and 7 of the above process. The amount of ultrapure water required to produce one 6-in. wafer is 1.2 tons. If a small particle stays on the surface of the wafer, the shadow of the particle is formed when the wafer is exposed to the light beam, and the circuit becomes imperfect. If a bacterial particle, on the other hand, stays on the surface, it turns into C, P, and K elements while the wafer is heat-treated and causes electric shortage. The element P, when it stays on the silicon surface, forms a n-type semiconductor and lowers the quality of the integrated circuit. Therefore, water of extremely high quality is required in the washing process of the semiconductor industry. The combinations of technologies to remove dispersed colloidal particles, microorganisms, electrolyte solutes, organic solutes, and dissolved gases, outlined in Table 4, are currently used for the production of ultrapure water.

The entire system for the production of ultrapure water is divided into pretreatment, primary pure water, and subsystem. In the pretreatment step, as many dispersed particles as possible are removed by coagulation precipitation and coagulation filtration, before water is supplied to the primary pure-water step. When the silica content is high in water, it is removed by an additional ion exchange column; otherwise, water is subjected to reverse osmosis treatment to remove dissolved electrolyte and organic solutes. Cellulose acetate or aromatic polyamide membranes are normally used for this purpose. Some residual ions that are left in the water, even after the reverse osmosis treatment, are further removed by a mixed-bed polisher that consists of the mixture of several ion exchange resins. The small particles released from the ion exchange resin are removed by microfiltration. The ultrapure water so produced is called primary pure water and is stored in a primary pure-water storage tank. Urtrapure water, however, tends to dissolve many solutes and cannot be stored in the storage tank for a long time without being contaminated. Therefore, a subsystem is attached before ultrapure water is supplied to the use point. Water is disinfected by ultraviolet irradiation. Ultraviolet also oxidizes residual organic solutes into carbon dioxide and water. Ions that are dissolved into water from the connecting pipes, etc., are removed by a combination of an ultrafiltration module and a cartridge polisher that consists of the mixture of cationic and anionic ion exchange resins.

10.1.5 Reverse Osmosis Separation of Nonaqueous Solutions

Reverse osmosis separation of nonaqueous solution is not new in the literature [200]–[202], [299]–[303], [304]. However, its industrial applications have not

Table 4. Ultrapure Water Quality for Manufacture of Super-LSI [298]

Items	256 kbit Req.	256 kbit Exa.	1 Mbit Req.	4 Mbit Ass.	16 Mbit Ass.
Specific resistance, MΩ cm (25 °C)	≥17–18	≥18	≥17.5–18.0	≥18.0	≥18.0
0.2 μm	≤10–50				
0.1 μm		≤50	≤2–5		
Particles, 0.1 μm	≤50–150		≤10–20		
number/ml 0.05 μm				≤10–20	
0.03 μm					≤10
Living organisms, number/100ml	≤2–20	≤5	≤1–5	≤1	≤0.5
TOC, (μg C/l)	≤50–200	≤50	≤30–50	≤20–25	≤10
Dissolved oxygen, (μg O/l)	≤100	≤100	≤50–100	≤50–100	≤50
Silica, (μg SiO$_2$)	≤10	≤10	≤5–10	≤5	≤5
Ions, μg/l					
Na	≤1	≤0.5	≤0.1–0.5		
K		≤0.5			
Zn		≤1			
Fe	≤1	≤0.5	≤0.1–0.5		
Cu	≤1	≤1	≤0.1–0.5		
Al	≤1	≤1	≤0.1–0.5		
Cl	≤1	≤2	≤0.1–0.5		
Temperature, °C			24 ± 1°C		
Endotoxin, μg/l			≤0.1		

sufficiently been explored, primarily due to the lack of membranes suitable for the treatment of nonaqueous solvent. The development of inorganic reverse osmosis membranes would have a major impact on the field. There have been, however, several attempts at industrial applications using organic polymeric membranes.

Lubrication oil contains waxes that solidify in cold weather and cause difficulties when starting automobile engines. Therefore, dewaxing of the lubrication oil is necessary. Conventionally, it was performed by mixing solvent with lubricant oil and cooling the mixture down to −20°C, at which temperature crystallization of wax occurs. However, the solvent and lubricant oil have to be separated for solvent recovery by flashing and stripping, and the cost of the latter process is high due to its high energy consumption. Replacement by a cheaper separation process is desirable, and reverse osmosis might be an attractive alternative. Bitter et al. used silicone rubber membranes for this purpose and obtained promising results [304]. When a fluorinated siloxane polymer is used for membrane material, a flux of 1.736×10^{-5} m^3/m^2 s (1.5 m^3/m^2 day) at 4 MPa (40 bar) and a selectivity of more than 50 was obtained for a solvent mixture of methyl ethyl ketone, toluene, and a residual lubeoil fraction (volume ratio 1:1:1). The membrane was swollen 100 vol% in the solvent/lubeoil mixture. This is in contrast to the result obtained for a membrane produced from polydimethylsiloxane polymer. The membrane flux was 4.630×10^{-5} m^3/m^2 s (4 m^3/m^2 day) at 4 MPa (40 bar), and the selectivity was less than 7. The swelling was more than 300 vol%. The low swelling of the fluorinated siloxane polymer is due to the higher degree of cross-linking and the lower swelling by toluene. The fluorinated siloxane polymer was swollen 30 vol% in toluene, whereas the polydimethylsiloxane polymer was swollen 800 vol%.

A variety of polysulfone membranes were evaluated for the processing of crude oil feed stocks. The product of this process is a permeate that is the equivalent of a gas oil and a concentrate enriched in asphaltenes [305]. Membranes were laboratory produced from polysulfone and sulfonated polysulfone materials. The molecular weight cutoffs ranged from 6000 to 50,000 when they were used in an aqueous environment. A high temperature operation, from 50 to 80°C, was required because of the high viscosity of the feed crude oil. Although all membranes used in this study were ultrafiltration membranes, the performance data appear more like that of reverse osmosis membranes. A typical example is given in Figure 10.3, as molecular weight distributions of the feed crude oil and the permeate. The average molecular weight of the permeate sample was below 1000 Da, which is far lower than the nominal molecular weight cutoff of the aqueous solutions. Moreover, the molecular weight distribution of the permeate sample was strikingly similar, regardless of the molecular weight cutoff of the ultrafiltration membranes. The similarlity of the permeate composition is also indicated by Table 5, where the percentage of removal of nickel and vanadium is listed for various membranes. It is obvious that the concentrations of these metals in the permeate are almost the same for every membrane. From these observations it is concluded that the membrane separation resulted from the formation of a dynamic gel-polysulfone composite membrane. The

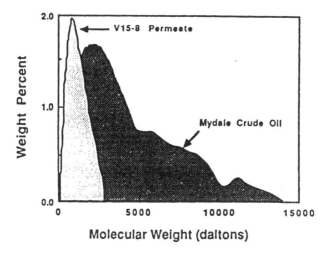

Figure 10.3. Molecular distributions of feed crude oil and permeate. (Reproduced from [305] with permission.)

Table 5. Nickel and Vanadium Removal for
Mydale Crude Permeates[a] [305]

Sample	Ni conc. (ppm)	Ni removal (%)	V conc. (ppm)	V removal (%)
Feed	29		53	
V 15-8	3	90	3	94
V16-16	2	93	3	94
U16-16	2	93	3	94
V16-16	2	93	3	94
R16-16	2	93	3	94
FS60	3	90	4	92
GR51	2	93	3	94
GR61	2	93	3	94
GF81	2	93	3	94
GS61	2	93	4	92

[a] At 55°C, 1.5MPa, $Q = 5.7$ l/min, product recovery, 40%.

major difference between various membranes was the permeation rate. A linear relation between the operating pressure and the flux was found in the case of the GR81 membrane, the pore size of which was the smallest among the membranes tested. The circulation flow rate of the feed hardly affected the flux. The flux vs. operating pressure curves for the V15-0 membrane, the pore size of which was one of the largest, showed a typical concentration polarization effect; i.e., the flux leveled off as the operating pressure was increased. The flux value of the plateau also increased with an increase in the circulation rate of the feed crude oil. The above experimental results indicate that the inherent resistance of membranes controls the membrane permeation rate at the lower operating pressures (1.0 MPa). The layer that is formed on the surface of the membrane, as a result of the concentration polarization, on the other hand, controls the permeation rate when the pore size of the membrane is very large and the operating pressure is high.

10.2   Ultrafiltration and Microfiltration

10.2.1   Making Clear Fruit Juice Using Ultrafiltration Membrane

Demand is increasing for the production of clear fruit juice as a high-quality beverage. The conventional method to degrade pectin by enzymes is effective only to a limited degree and is often insufficient to secure clarity under severe conditions. A hollow fiber ultrafiltration membrane was found to be effective for the above purpose. In particular, according to researchers at Daicel Chemical Ind., hollow fiber ultrafiltration membranes of molecular weight cutoff 30,000 (MOLSEP FUS 0381, material polyethersulfone) were found to be the best at removing pectin in mandarin orange juice, while losing a minimum amount of flavor and color [306], [307].

Figure 10.4 shows gel permeation chromatograms of mandarin orange juice, with no treatment, after enzyme treatment, and after ultrafiltration. Obviously, some juice components that were not removed completely by enzyme treatment could be removed by ultrafiltration without causing production of unnecessary new components. The results of the analysis of untreated juice and the UF-treated juice are given in Table 6. Obviously, pectin was removed completely, and some color components were also lost, while all other components were retained almost completely in the ultrafiltrate. Some operational data are given in Figures 10.5 and 10.6. Figure 10.7 shows the relation between the permeation rate and the pressure at the outlet of the hollow fiber. Note that the hollow fibers of the lowest inner diameter (0.5 mm) has shown the smallest flux. All

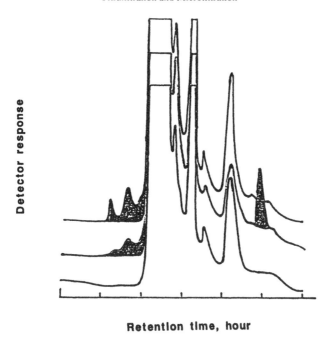

Figure 10.4. Gel permeation chromatogram of mandarin orange juice. [(Top) without treatment; (middle) with enzyme treatment; (bottom) with ultrafiltration treatment.] (Reproduced from [306] with permission.)

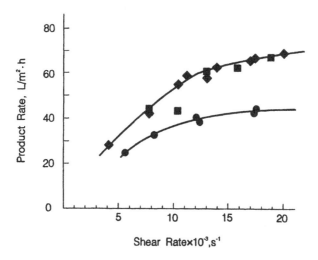

Figure 10.5. Effect of shear rate on product rate of fibers with various inner diameters. [Inner diameters, 0.5 mm (•), 0.8 mm (◇), 1.0 mm (□); outlet pressure, 1.2 kg/cm^2.] (Reproduced from [306] with permission.)

Table 6. Compositions of Mandarin Orange Juice
Before and After Ultrafiltration Treatment[a]

Items	Before UF treatment	After UF treatment
Sugar content (Bx)	10.5	10.0
Acid content (%)	0.80	0.77
Pulp (%)	7.2	—
Vitamin C (mg %)	32.1	27.4
Nitrogen in amino form (mg %)	30.8	29.0
pH	3.47	3.47
Ash content (%)	0.29	0.25
Color		
L	49.4	13.0
a	4.7	~1.5
b	29.8	3.5
Pectin (mg %)		
Soluble in water	22.0	N.D.
Soluble in salt solution	13.0	N.D.
Soluble in alkaline solution	1.2	N.D.

[a] From Daicel Chemical Industries, Ltd. Catalogue.

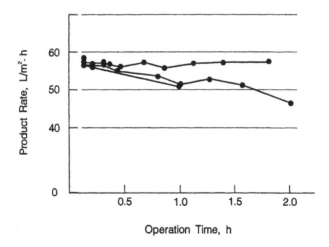

Figure 10.6. Effect of the removal of pulp particles in fruit juice by pre-treatment on the flux decline. [Without treatment, pulp content 5% (A); with filtration by 0.12 mm mesh size, pulp content 2% (B); with centrifugation at 10^4 G, pulp content 0% (C).] (Reproduced from [306] with permission.)

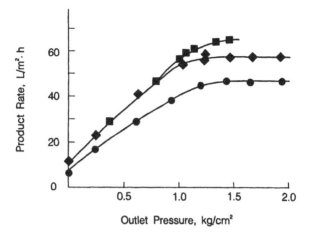

Figure 10.7. Effect of operating pressure on product rate of fibers with various diameters. [Inner diameters, 0.5 mm (•), 0.8 mm (◇), 1.0 mm (□); shear rate, $1.2 \times 10^4 \text{ s}^{-1}$; juice, mandarin orange; Bx, 11; pulp content, 2.2%; temperature, 20°C.] (Reproduced from [306] with permission.)

hollow fibers showed a typical pattern of ultrafiltration flux vs. operating pressure. Since it was found that the entrance opening of some hollow fibers was partially or completely blocked, the flux recovery was attempted by reversing the permeation direction in the membrane and flushing the hollow fiber periodically. This was accomplished by releasing the feed pressure and closing the valve of the permeate outlet. Then, the membrane flow was directed from the permeate side to the feed side, due to osmosis. Figure 10.8 shows that the flushing of hollow fibers was indeed effective to recover the permeate flux. The polyethersulfone hollow fiber membranes can also withstand high temperatures up to 95°C, and therefore hot water sterilization is applicable, rendering these membranes particularly useful for applications in food processing.

10.2.2 Cleaning of Tap Water

Tap water cleaners consisting of microfiltration hollow fiber membranes, calcium carbonate, and activated carbon bed are currently in large demand by customers in Japan [308]. Tap water in large cities has been used and cleaned many times before it reaches the kitchen. It often tastes miserable and smells of chlorine. There is a strong desire in each household for water that tastes "delicious," to be delivered from the tap. The criteria for "delicious water" are

1. It does not smell strange.

2. It does not taste strange.

3. It is transparent and without any color.

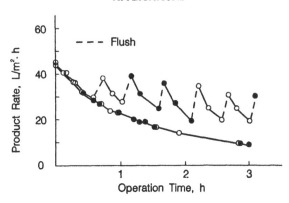

Figure 10.8. Effect of periodical flush on flux recovery. [Hollow fiber diameter, 0.8 mm; juice, mandarin orange; Bx, 11; pulp content, 2%; temperature, 20°C.] (Reproduced from [306] with permission.)

4. It does not contain any hazardous components.

5. It contains an appropriate amount of minerals.

6. It contains an appropriate amount of CO_2.

7. It is slightly acidic.

   Small household units to clean tap water were manufactured and sold. The structure of the tap water cleaner is shown in Figure 3 of Chapter 1. Tap water goes through an unwoven cloth and enters an activated carbon bed where compounds causing a chlorine-like smell, trihalomethane, and other organic compounds are removed. Then water goes through microfiltration hollow fiber membranes whose pore size ranges from 0.01 to 0.1 μm. The hollow fiber removes microorganisms, mold, and rust. Sometimes water is provided with carbon dioxide and minerals from calcium Ccarbonate.

10.3   Membrane Gas Separation

Industrial gas separation membranes are broadly classified as cellulosic and noncellulosic. While cellulosic membranes have either spiral-wound or hollow fiber configurations, most of the noncellulosic membranes have hollow fiber configurations.

   The first attempt of membrane gas separation using a dry asymmetric reverse osmosis membrane was by Sourirajan [309]. Agrawal and Sourirajan further investigated the possibility of industrial applications [310], [311]. The first ac-

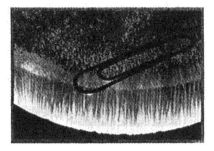

Figure 10.9. PRISM hollow fiber. (From Technical Bulletin of Permea, with permission.)

complishment in the development of industrial gas separation membranes from cellulosic material was done by Schell (US Patent 4,134,742) [312]. The patent describes the method of drying wet reverse osmosis cellulose acetate membranes. In order to preserve the structure of the wet reverse osmosis membrane, the water in the reverse osmosis membrane is exchanged with a volatile organic solvent or a solvent mixture, before being dried.

The next accomplishment was in the development of composite hollow fiber membranes, using noncellulosic materials. This was achieved by Monsanto on the basis of the resistance model of Henis and Tripodi (US Patent 4,230,463) [171]. This accomplishment led to the commercialization of the first large industrial-scale Prism separator and further widened the scope of membrane applications for gas separation processes. Similar to the asymmetric reverse osmosis membranes, the Prism hollow fiber is based on the asymmetric hollow fiber of polysulfone material (Fig 10.9). The thin outer layer (0.1 to 1% of the entire thickness) is responsible for the selectivity of the membrane. However, it is practically impossible to produce a skin layer without any pinholes that render the membrane useless because of its low selectivity. Therefore, these pinholes are filled with a coating of materials such as silicone rubber of high permeability, but low selectivity. The Prism separator was introduced into the market in 1979. Design of the Prism hollow fiber module is given in Figure 10.10. Later, in 1986, Prism Alpha separators were developed by Permea. By Prism Alpha separators, permeability higher than Prism separators was achieved without lowering the selectivity, which is the result of the control of the morphology of the porous polysulfone substrate membrane.

While most of noncellulosic membranes rely on the selectivity of glassy polymers such as polysulfone, the Ube Gas Separation System is based on polyimide produced by condensation polymerization of biphenyltetracarboxylic dianhydride and aromatic diamines. The chemical formula of the polymeric ma-

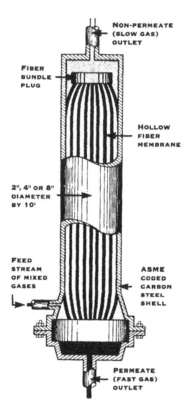

Figure 10.10. PRISM separator. (From Technical Bulletin of Permea, with permission.)

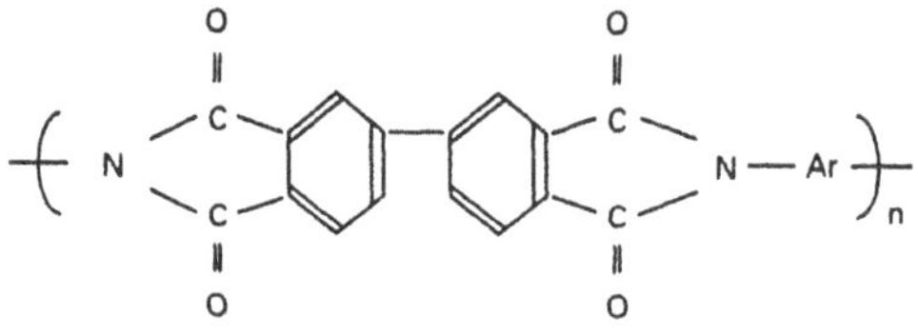

Ar shows divalent aromatic radical

Figure 10.11. Chemical formula of Ube's polyimide polymer. (From Technical Bulletin of Ube Industries, Ltd., with permission.)

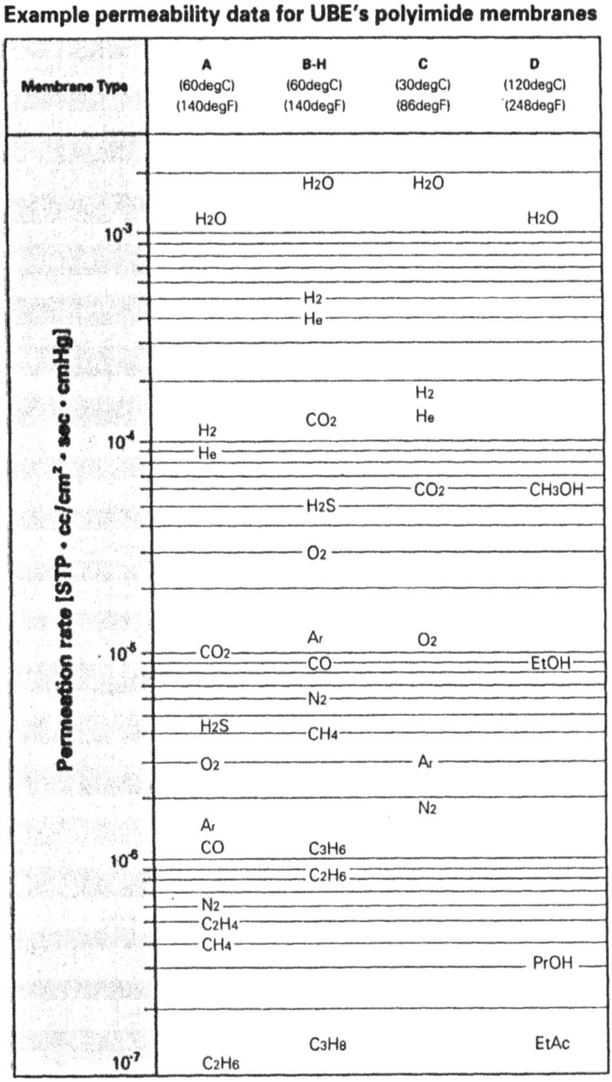

Example permeability data for UBE's polyimide membranes

Membrane Type	A (60degC) (140degF)	B-H (60degC) (140degF)	C (30degC) (86degF)	D (120degC) (248degF)
		H_2O	H_2O	
10^{-3}	H_2O			H_2O
		H_2		
		He		
				H_2
		CO_2	He	
10^{-4}	H_2			
	He			
			CO_2	CH_3OH
		H_2S		
		O_2		
		A_r	O_2	
10^{-5}	CO_2	CO		$EtOH$
		N_2		
	H_2S	CH_4		
	O_2		A_r	
			N_2	
	A_r			
10^{-6}	CO	C_3H_6		
		C_2H_6		
	N_2			
	C_2H_4			
	CH_4			
				$PrOH$
		C_3H_8		$EtAc$
10^{-7}	C_2H_6			

Permeation rate [STP · cc/cm² · sec · cmHg]

Figure 10.12. Permeability data for Ube's polyimide membranes. (From Technical Bulletin of Ube Industries Ltd., with permission.)

terial is given in Figure 10.11. The membranes are of hollow fiber type, and the gas permeates from the shell to the bore side. There are modules of different sizes from 4 Standard l/min to 100 Standard l/min. Permeabilities of the typical Ube polyimide membranes are given in Figure 10.12.

10.3.1 Hydrogen Recovery

Hydrogen recovery by membrane technology is important during the synthesis of substitute natural gas from naphtha [313]. The substitute natural gas is called rich gas and consists of H_2, CH_4, and CO_2. Hydrogen is used at various stages in this process and, for reasons of economy, recovery, and recycling of unreacted hydrogen, is desirable. Most importantly, hydrogen is used in the desulfurization of naphtha feedstock, to protect steam-forming catalysts. The hydrogen required for the desulfurization is supplied by recycling a part of a rich gas, or when a higher hydrogen concentration is necessary, membrane separation technology is used to increase hydrogen concentration. Membrane separation process is also used to adjust the hydrogen-to-carbon-monoxide ratio to 1.3 to 1, which is required for oxo synthesis [314]. A description of this process will be fully developed in a later section, in relation to membrane separation/PSA hybrid process.

Another example of hydrogen recovery is from the purge gas from an ammonium synthesis reactor, containing methane, argon, and helium gases that are inert in ammonium synthesis. Unless recovered, hydrogen is purged with inert gas or used only for its fuel value [314]. Hydrogen recovery is also important in other petroleum and petrochemical industries. Typical data on the recovery of hydrogen from refinery off-gas is given in Figure 10.13.

10.3.2 Carbon Dioxide Removal

The production of gas from oil and coal feed stock produces large quantities of carbon dioxide, which arises from the need to oxidize some of the carbon in the feed to generate the energy required to the process. CO_2 also occurs in natural gas, the concentration of which has to be reduced. Since the separated CO_2 is normally vented, a separation system must minimize the amount of CH_4 lost along with CO_2. Hence, separation of CO_2 and CH_4 gases is necessary in the process. There is also a promising future for the use of membrane separation technology in the offshore removal of CO_2 from natural gas, since in the presence of water, CO_2 can be very corrosive and attack pipe lines. The removal of CO_2 also reduces the quantity of gas to be transported on shore, particularly when the CO_2 content is high in the oil fields. The compact size of the membrane module, as compared with the absorption process, also favors membrane technology [313]. The separation of CO_2 from the associated hydrocarbon gases from oil fields that are exploited in enhanced oil recovery (EOR) is another important area of CO_2/hydrocarbons separation, particularly CO_2/methane separation. Usually, CO_2 is injected into oil reservoirs to enhance the recovery of oil. Gas released from the reservoirs is heavily diluted and con-

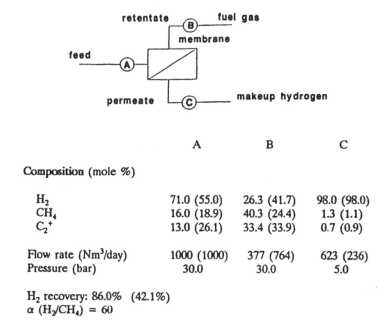

	A	B	C
Composition (mole %)			
H_2	71.0 (55.0)	26.3 (41.7)	98.0 (98.0)
CH_4	16.0 (18.9)	40.3 (24.4)	1.3 (1.1)
C_2^+	13.0 (26.1)	33.4 (33.9)	0.7 (0.9)
Flow rate (Nm^3/day)	1000 (1000)	377 (764)	623 (236)
Pressure (bar)	30.0	30.0	5.0

H_2 recovery: 86.0% (42.1%)
α (H_2/CH_4) = 60

* figures in brackets illustrate the effect of lower feed purity

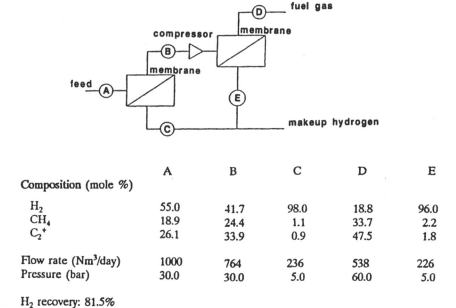

	A	B	C	D	E
Composition (mole %)					
H_2	55.0	41.7	98.0	18.8	96.0
CH_4	18.9	24.4	1.1	33.7	2.2
C_2^+	26.1	33.9	0.9	47.5	1.8
Flow rate (Nm^3/day)	1000	764	236	538	226
Pressure (bar)	30.0	30.0	5.0	60.0	5.0

H_2 recovery: 81.5%

Figure 10.13. Typical membrane performance data for the recovery of hydrogen from refinery off-gases. (Reproduced from [317] with permission.)

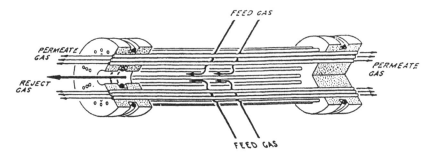

Figure 10.14. Dow Chemical Co. CYNARA membrane module. (Reproduced from [317] with permission.)

tains about 80% CO_2 at the wellhead. The separation of CO_2 is desirable for the same reasons as those given for the offshore natural gas recovery. The advantage of installing a membrane separation process, over the absorption process, has already been proven in this area. Purification of hydrocarbons from carbon dioxide is also required for such applications as upgrading biogas produced by the fermentation of biomass.

Landfills generate methane that is contaminated with carbon dioxide (approximately 50%). Landfill gas is at atmospheric pressure, is available only at low flow rates, and is expensive to upgrade. However, the gas is essentially free, and collection satisfies certain environmental interests. Also, this gas is usually more valuable than wellhead gas, since it is often located in industrialized areas. Several landfills currently use membranes to provide a high-quality gas. In this application, membranes compete directly with scrubbing, amine treatment, and PSA, and indirectly with the untreated use of the contaminated gas with low fuel value [315].

Recently, attention has been focused on the global warming effect (greenhouse effect). Among several greenhouse gases (such as CO_2, N_2O, CH_4 and CFCs), carbon dioxide is said to be responsible for half of the greenhouse effect. The removal and concentration of CO_2 from the CO_2 gas emission sources, such as power stations, steelworks, and chemical plants, and the conversion of CO_2 to useful end products are the major topics of the RITE project, started recently by the Japanese government [316]. Membrane separation processes are expected to play a key role in the removal of CO_2 from the exit gas of fuel combustion. Some examples of industrial CO_2 recovery are as follows [317]:

CYNARA System. This system uses cellulose triacetate hollow fiber modules manufactured by Dow. The usual operating pressure is from 2.07 to 4.14 MPa (gauge) (300 to 600 psig), though pressures up to 6.90 MPa (gauge) (1000 psig) can be used if necessary. Figure 10.14 illustrates the membrane module. In the module the feed gas flows radially inwards through the hollow fiber bundle,

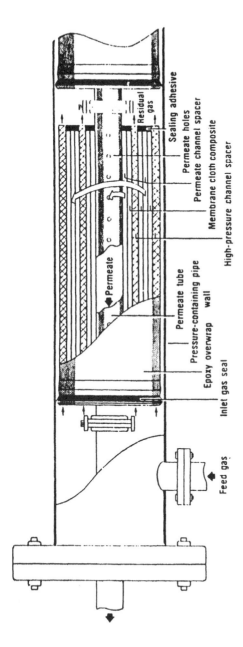

Figure 10.15. The SEPEREX spiral-wound membrane module. (Reproduced from [317] with permission.)

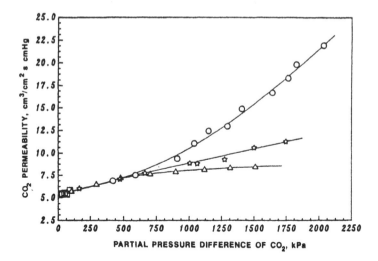

Figure 10.16. Effects of partial pressure difference of CO_2 across the membrane and feed composition on CO_2 permeability. [Downstream pressure, 101.3 kPa (1 atmospheric pressure); symbols, experimental values; lines, calculated results.] (Reproduced from [318] with permission.)

while the permeate goes through the fiber wall into the bore side of the fiber. The carbon-dioxide-enriched permeate flows within the fiber tube and reaches both ends of the fiber, to be collected. The reject gas enriched in hydrocarbon reaches the central tube and exits the module from one end of the tube.

SEPEREX System. This system also uses cellulose acetate membranes, but in spiral-wound form. CO_2 and H_2S permeate through cellulose acetate membranes 20 to 60 times faster than methane, enabling removal of highly acidic gases. Separation factors of up to 25 for CO_2/methane mixtures have been achieved for CO_2 concentrations in the range of 1.7 to 45% at the operating pressure of 6.90 MPa (gauge) (1000 psig). A Seperex spiral-wound element is illustrated in Figure 10.15.

GASEP Systems. This system was developed by Envirogenics and also uses cellulose acetate spiral-wound modules. These systems have been used for the sweetening of natural gas by removing CO_2 and H_2S, and the separation of CO_2/methane mixtures in landfills, and the separation of hydrogen in coal gasification processes.

GRACE Systems. Cellulose acetate membranes are also used in spiral-wound configuration. Their data, reproduced in Figures 10.16, 10.17, and 10.18, indicate that the permeability, defined as mol/m^2 s kPa, increases with an increase in transmembrane pressure, for both methane and CO_2. However, the

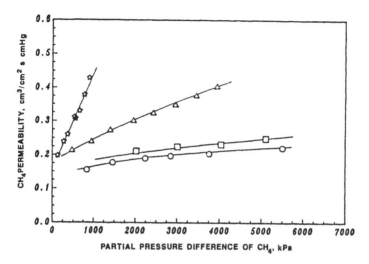

Figure 10.17. Effect of partial pressure difference of CH$_4$ across the membrane and feed composition on CH$_4$ permeability. [Downstream pressure, 101.3 kPa (1 atmospheric pressure); symbols, experimental values; lines, calculated results.] (Reproduced from [318] with permission.)

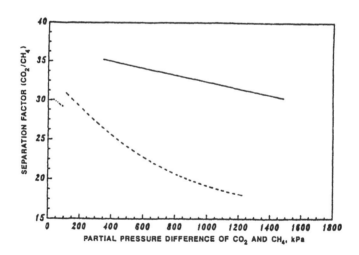

Figure 10.18. Effect of membrane plasticization on the separation factor. The separation factor is the ratio permeabilities of CO$_2$ to CH$_4$ at the same partial pressure difference across the membrane. (Both symbols and lines, experimental results; circ les and the broken line, pure gas permeation rate ratio.) (Reproduced from [318] with permission.)

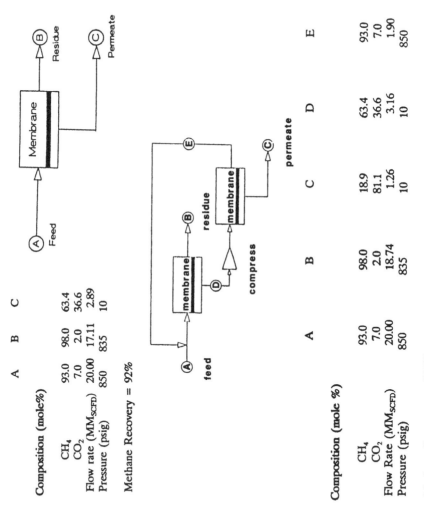

Composition (mole%)

	A	B	C
CH$_4$	93.0	98.0	63.4
CO$_2$	7.0	2.0	36.6
Flow rate (MM$_{SCFD}$)	20.00	17.11	2.89
Pressure (psig)	850	835	10

Methane Recovery = 92%

Composition (mole %)

	A	B	C	D	E
CH$_4$	93.0	98.0	18.9	63.4	93.0
CO$_2$	7.0	2.0	81.1	36.6	7.0
Flow Rate (MM$_{SCFD}$)	20.00	18.74	1.26	3.16	1.90
Pressure (psig)	850	835	10	10	850

Methane Recovery = 98.7%

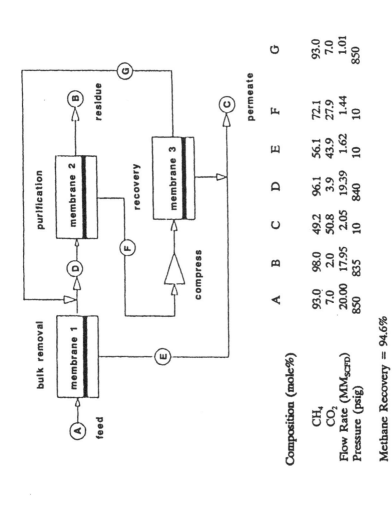

Composition (mole%)	A	B	C	D	E	F	G
CH_4	93.0	98.0	49.2	96.1	56.1	72.1	93.0
CO_2	7.0	2.0	50.8	3.9	43.9	27.9	7.0
Flow Rate (MM_{SCFD})	20.00	17.95	2.05	19.39	1.62	1.44	1.01
Pressure (psig)	850	835	10	840	10	10	850

Methane Recovery = 94.6%

Figure 10.19 Single-stage to multistage gas separation membrane processes for treating natural gas. (Reproduced from [317] with permission.)

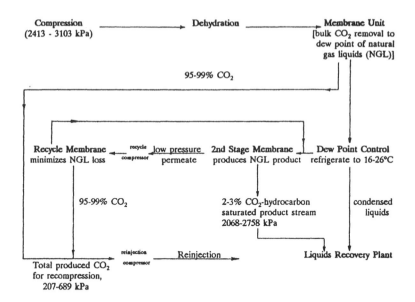

Figure 10.20. Grace carbon dioxide separation process. (Reproduced from [317] with permission.)

separation factor (defined as the ratio of the permeabilities of CO_2 to methane at the same partial pressure difference across the membrane) decreases as the CO_2 content in the feed gas mixture increases. This indicates the severe plasticization effect of cellulose acetate membranes in the presence of high-concentration CO_2 [318]. Some typical process flow sheet and experimental data of the process operation are also given in Figures 10.19 and 10.20.

10.3.3 Separation of Oxygen and Nitrogen

One of the applications of oxygen/nitrogen separation is to produce oxygen-enriched gas that is used in the medical field as a heart-lung oxygenator. In this case relatively small-size gas separation modules are needed. Oxygen-enriched gas is also used in a furnace to improve combustion. Injection of oxygen-enriched air, containing 25 to 35% oxygen, into a furnace leads to a higher flame temperature. This is particularly relevant to high-temperature furnaces used in the firing of ceramics. In tests of ceramics roasting using 28 and 26% oxygen, energy saving of 26.1 and 24.4%, respectively, can be expected.

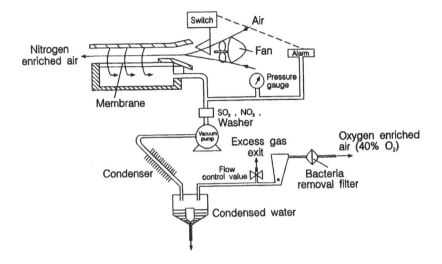

Figure 10.21. Schematic flowsheet of oxygen enrichment system for medical purposes.

Usually, membranes are preferentially permeable to oxygen gas, and nitrogen gas is left behind. Nitrogen gas also finds applications in various sectors of the chemical process industries. For example, nitrogen is used as an inert purge gas in many processes throughout the chemical industry and is used also in the synthetic natural gas industry [313]. Conventionally, nitrogen separation has been by a cryogenic separation process, but it can also be performed by membrane technology. Nitrogen can also be used for the purpose of enhanced oil recovery, instead of CO_2. Furthermore, a nitrogen blanket is an effective method of preserving fruits and vegetables, in which nitrogen-enriched air can be used. Several examples of industrial oxygen/nitrogen separation are as follows.

The Oxygen Enrichment Co. uses the technology developed by General Electric [319]. The membrane is based on a dimethylsiloxane/bisphenol A carbonate copolymer. Extremely thin membranes (10 nm) are prepared by the Langmuir-Blodgett method, and they are used to generate 30 to 40% oxygen for medical use. The flow sheet and the operational data of the separation unit are given in Figure 10.21 and Table 7, respectively [13].

Osaka Gas and the Matsushita group developed another oxygen enrichment system based on silicone material. The preparation of the thin film is also by Langmuir-Brodgett method. Their system consists of a set of equipment including a combustion blower, a combustion furnace, a suction blower, and an oxygen-enrichment chamber [13].

Table 7. Specifications of OECO Oxygen Enrichment Equipment [13]

Model	OE-2	OE-3	OE-3A
Size	27.5"H × 31.5"W × 16"D	30"H × 14"W × 16"D	
Weight	150 lbs	95 lbs	
Electric source	390 W, 115 V, 60 Hz	375 W, 115 V, 60 Hz	200 W
Operation start time	15 s	15 s	
Noise	48–52 dbA	below 50 dbA	45 dbA
Air flow rate	3–8 l/min	3–8 l/min	2–6 l/min
Composition of enriched air			
O_2	40%	40%	
CO_2	3.9 × conc. in room	3.9 × conc. in room	
NO_2, SO_2, CO	less than conc. in room	less than conc. in room	
Particles	not found	not found	

SPIRAGAS of UOP (spiral-wound module with silicone coated on a poly-sulfone porous membrane) delivers 30% oxygen air. The system operates under a feed pressure of 0.10 to 0.14 MPa (15 to 20 psia) and the operating vacuum of 0.02 to 0.028 MPa (3 to 4 psia) at 70° F. The product recovery (stage cut) for a unit is 5 to 20%, and the product flow is 10.62 standard m^3/h (9000 scfd).

The Generon system was developed by Dow Chemical Co. in 1985. From compressed air, the Generon membrane system produces an effluent product stream containing up to 99% N_2, and a membrane-permeated product stream containing up to 35% oxygen [320]. Under the design conditions of 95% nitrogen (25°C) and 725 kPa feed pressure, Generon delivers up to 9000 m^3/h of enriched nitrogen. The membrane material used is polymethylpentene (described by Dow as a proprietary polyolefine), and the membranes are in a hollow fiber configuration. Argon, oxygen, and water preferentially permeate through the membrane. The pressure drop across the module is 138 kPa; the nitrogen product leaves the system at about 690 kPa, and the oxygen-rich stream leaves the module at less than 105 kPa. The two major contaminants in air are CO_2 and water vapor, and they are concentrated in the permeate stream. The permeate product gas dew point is controllable between 203 and 233K, and the CO_2 concentration is usually less than 20 ppm [317]. A Generon system (HP 4200) is capable of producing 99.9% pure N_2, with a recovery of 22% (nitrogen out/air in).

The Prism Alpha System is a modified version of the Prism system from Permea [321]. The product flow ranges from 5.66 standard m^3/h (200 scfh) to 708 standard m^3/h (25,000 scfh). The system can operate at the feed pressure of 0.55 to 1.03 MPa (gauge)(80 to 150 psig). Special systems can be designed for pressures up to 4.14 MPa (gauge)(600 psig) or more. Compressed air passes through two coalescing filters to remove particulates and oil vapor. Then the air passes through a trim heater to provide constant temperature feed air, and then enters the separators. As it passes through the bore of the hollow fibers, oxygen and water vapor permeate rapidly to the shell side of the separators and are vented as an oxygen-enriched stream. The nitrogen product, typically from 95 to 99%, is drawn off at the opposite end of the separators, close to feed pressure.

The AVIR^{TM} system is available from A/G Technology Corporation [322]. The main features of this system are that it is compact and portable. It produces 14.16 standard m^3/h (500 scfh) from a system volume of 0.0623 m^3 (2.2 ft^3) at 0.67 MPa (gauge)(100 psig) inlet pressure. The membrane is also of a hollow fiber configuration, and the feed air is supplied to the bore side of the fiber. A schematic flow diagram is given in Figure 10.22. This system can produce 37 or 60% oxygen-enriched air.

The NitroGEN^{TM} system is available from Union Carbide Industrial Gases [323]. It is based on composite hollow fibers, the selective skin layer being on the shell side of the hollow fiber. Capacities of the system ranges from 28.32 Nm^3/h (1000 Ncfh) to 850 Nm^3/h (30,000 Ncfh), with purity from 5% oxygen to 5 ppm.

10.3.4 Helium Recovery from Natural Gas

Because of its low boiling point ($-268.9°C$), helium in its liquid state provides an excellent cryogenic environment needed for freezing biological specimens and for superconductors. Helium is also chemically inert and provides a chemically inert environment for gas-shielded arc welding and for a controlled atmosphere in which semiconductor crystals can grow. Helium is abundant in natural gas resources. According to estimates by Alberta's Energy Resources Conservation Board, helium present in proven reserves of Alberta natural gas amounts

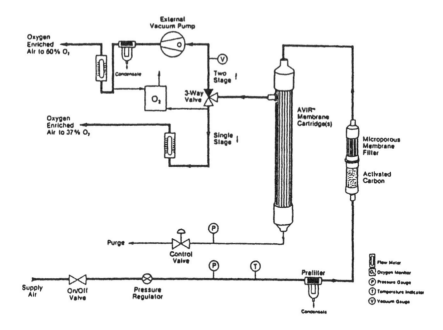

Figure 10.22. Simplified schematic for AVIR^{TM} point-of-use oxygen-enriched air system. (Reproduced from [322] with permission.)

to 900 million m^3. The concentration of helium in natural gas in Alberta ranges from 0.01 to 1.5%, averaging about 0.05%. In the 1970s an attempt was made by the Alberta Research Council to recover and concentrate helium gas by membrane technology [265]. Using these data, Alberta Helium built a pilot plant at the McLeod River Compressor Station of the Alberta Gas Trunk Line Company, to demonstrate the membrane process on a pilot scale [324]. The pilot plant was designed to handle 85,000 m^3/day of natural gas and to separate helium in four consecutive stages. The membranes used were commercially available hollow fiber modules. The hollow fibers are in "water-wet" state, and they are normally used to purify water in the desalination process. The hollow fibers must be dried before being used for the gas separation purpose. The schematic representation of the hollow fiber module is given in Figure 10.23. A typical hollow fiber module contains up to 270,000 individual fibers. Natural gas at normal pipeline pressure is first filtered to remove scale, oil, and other trace liquids and then compressed and passed through the first permeation stage. In the first stage

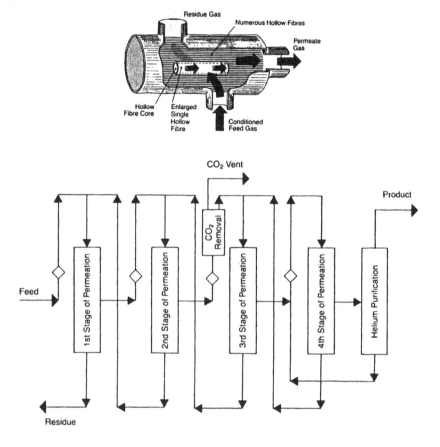

Figure 10.23. Schematic drawing of a typical membrane module for helium recovery. (Reproduced from [324] with permission.)

the helium concentration is increased from 0.05% to 0.6 to 1.4%. The helium-enriched gas is again compressed and passed through a second stage where the helium concentration is increased to 3.3 to 8.1%. After another compression followed by removal of carbon dioxide, the helium-containing gas undergoes two more concentration stages. The helium concentration is increased to 72 to 90%. In the final purification stage, hydrogen and moisture are removed, which brings the concentration of helium in the final product to 99.997%. This process requires only 40% of the energy needed in conventional cryogenic plants. Since 1982 an overcapacity situation has prevailed in helium markets. This led

Alberta Helium to postpone plans for commercial development of their membrane technology. Meanwhile another Calgary-based company, Delta Projects Inc., and several natural gas transmission companies, began research and development programs in membrane technology to separate carbon dioxide and hydrogen sulfide from sour natural gas. In 1985 Alberta Helium and Delta Projects created International Permeation Inc., a Calgary-based firm that provides custom-designed gas separation systems. This service employs hollow fiber membranes that remove or recover such gases as carbon dioxide, hydrogen sulfide, hydrogen, and helium. Helium recovery is also important in producing pure helium from contaminated diving gas [325].

10.3.5   Dehydration of Air

Permea has commercialized a new membrane gas dehydration technology, called Prism Cactus dryer systems, on the basis of hollow fiber membranes [326]. Membranes are chemically or physically posttreated to adjust the pore size and the pore size distribution [327]. Posttreatments consist of gentle chemical annealing with organic liquids, which have swelling effects on, but not strong solvents for, the membrane polymer (for example methyl alcohol for polysulfone polymer), and some thermal annealing. Optimal balance of the transport properties of water vapor and air are depicted in Figure 10.24, where the range of water and air permeabilities suitable for the dehydration system is shown schematically. Outside this region is less suitable for system design of the membrane dehydration system. The central region is bounded by the following combinations of water vapor and air permeabilities: water vapor permeability, 1.005 to 5.024 $\times$ 10^{-6} mol/m^2 s Pa (300 to 1500 $\times$ 10^{-6} cm^3(STP)/cm^2 s cmHg); air permeability, 0.033 to 0.335 $\times$ 10^{-6} mol/m^2 s Pa (10 to 100 $\times$ 10^{-6} cm^3(STP)/cm^2 s cmHg).

The ratio of the respective permeabilities should be in the range of 10 to 50. The membranes are obviously porous, since the O_2/N_2 permeability ratio of the membrane is from 1.5 to 2.0, while the intrinsic permeability of the membrane polymer is greater than 4. Figure 10.25 shows schematically the flow diagram of the dehydration system. Such systems are operated under the pressure of 2069 kPa (300 psig) and produce dry air of 0°C dew point from the feed air of 40°C dew point. The module capacity depends on the size of the module. For example, a module with a 12.7-cm (5 in.) diameter and 63.5-cm (25 in.) length can produce 1.1 Nm^3/min (40SCFM) dried air. The pressure drop across the

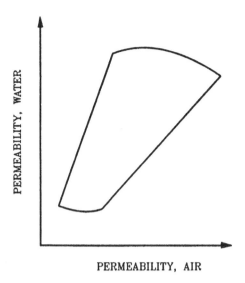

Figure 10.24.   Optimum balance of transport properties for membrane dehydration of compressed air. (Reproduced from [326] with permission.)

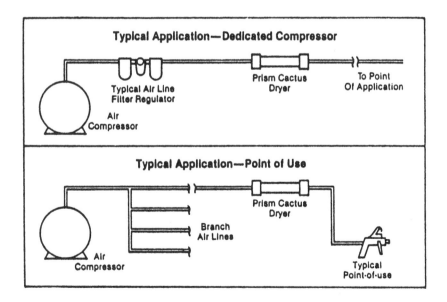

Figure 10.25.   Typical application modes for membrane dehydration of compressed air. (Reproduced from [326] with permission.)

module is less than 0.0138 MPa (2psi). Therefore, the pressure of the dry gas product is essentially the same as that of the feed.

10.3.6   Separation of Hydrocarbons from Air

As regulations regarding the emission of volatile organic compounds (VOC) into air become tighter, the demand for technology to remove VOC from air becomes stronger. The membrane separation process is one of such technologies [328], [329].

According to the solution-diffusion model, the flux of gas through a membrane is given by

$$J_g = -DS\frac{\Delta p}{\delta} \qquad (10.1)$$

The above equation implies that the product of the diffusion coefficient, D, and the solubility, S, governs the permeation flux of the gas. Generally speaking, the diffusion coefficient becomes smaller, whereas the solubility increases with an increase in the molecular size of the gas. Glassy polymers with stiffer polymer backbones let smaller molecules, such as hydrogen and helium, pass faster, while molecules of larger sizes, such as hydrocarbons, permeate the membrane more slowly. The permeation flux is dominated by the diffusion coefficient. Gas and vapor permeation through rubbery polymers, on the other hand, is governed more by solubility. The permeabilities of hydrocarbons with larger molecular sizes are therefore higher than gas molecules of smaller sizes. The tendency of gas permeation through both glassy and rubbery polymers is depicted in Figure 10.26. The upper curve illustrates the permeability of oxygen and nitrogen gases and hydrocarbon vapors, through a polyetherimide/silicone rubber composite membrane that is believed to be governed by the permeability of silicone rubber. As expected for the rubbery polymer, the permeability increases with an increase of the molecular volume of the permeant. The lower curve, on the other hand, shows the permeability of a polyetherimide membrane. Polyetherimide polymer is one of the glassy polymers, and the permeability decreases with an increase in the molecular volume of the permeant. Therefore, the choice of the membrane material should be a rubbery polymer in order to recover highly concentrated hydrocarbon vapors to the permeate side of the membrane, from air contaminated by such hydrocarbon vapors.

GKSS, a German company, developed a composite membrane on the basis of the above principle, a composite membrane that is illustrated schematically in Figure 10.27 [328]. The membrane consists of three layers. The first layer is a nonwoven polyester backing material. A porous membrane of polyetherimide (Ultem, General Electric) material is cast on the backing material by the phase-inversion technique. The porous membrane is further coated with

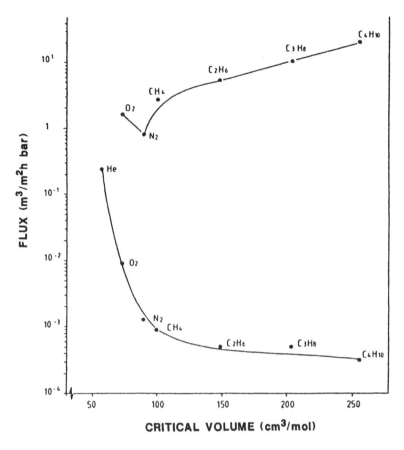

Figure 10.26. Tendencies in flux vs. critical volume curve for rubbery and glassy polymers. (Reproduced from [328] with permission.)

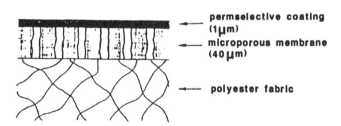

Figure 10.27. Schematic representation of GKSS composite membranes for vapor permeation. (Reproduced from [328] with permission.)

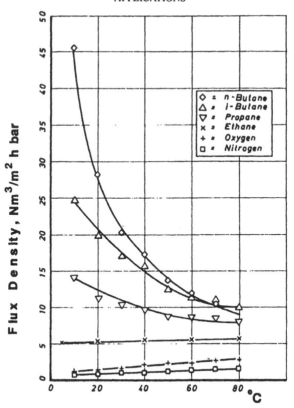

Figure 10.28. Flux density vs. operating temperature. (Operating pressure, 1230 mbar.) (Reproduced from [328] with permission.)

a silicone barrier layer, the thickness of which ranges from 0.5 to 2 μm. As mentioned above, the silicone layer is believed to govern the selectivity of the composite membrane. Polyetherimide material was used instead of popular polysulfone material, because of the higher durability of polyetherimide in an organic environment. When the process variables such as temperature and the vapor pressure of the organic vapor are changed, the permeability changes, as shown in Figures 10.28 and 10.29. GKSS constructed a plate-and-frame-type module, the design of which is shown in Figure 10.30. Volatile organic compounds in tank off-gas from fuel depots vary in concentration from 100 to 1500 g/m^3. The primary components are butane/butene and pentane/pentene, and the minor components are methane, ethane, propane, hexane, heptane, benzene, and toluene. In lead-free grades, methanol, ethanol, TBA, MTBE, and TAME are used as octane number enhancers and may be found in the off-gas as low concentration components. Figure 10.31 is a schematic representation of the process diagram based on the feed hydrocarbon concentration of 37% and the

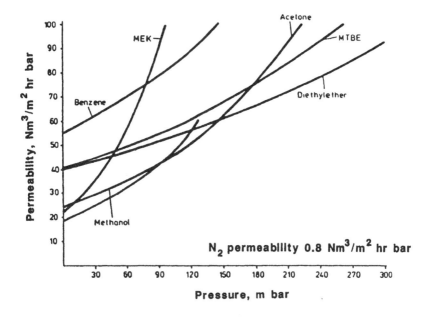

Figure 10.29. Flux density vs. operating pressure. (Reproduced from [328] with permission.)

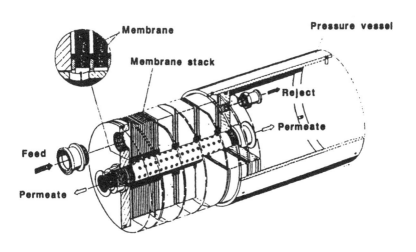

Figure 10.30. Flat sheet membrane module GS5 (GKSS). (Reproduced from [328] with permission.)

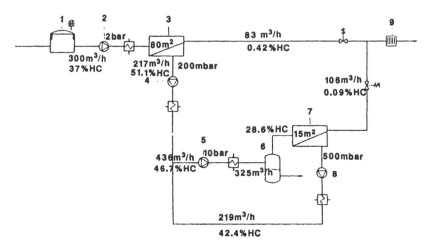

Figure 10.31. Process diagram of a demonstration plant. (Reproduced from [328] with permission.)

feed flow rate of 300 m³/h. The feed air containing hydrocarbon vapors is introduced into a gasometer (1) that also works as a buffer to homogenize the gas content. Then, a rotating compressor (2) pumps the gas and supplies to the first membrane stage. The feed pressure and the permeate pressure of the first membrane stage are 0.202 MPa (2 bar) and 0.020 MPa (0.2 bar), respectively. The membrane area is 80 m². The module inlet flow rate is 300 m³/h, while the module outlet flow rate is 80 m³/h. The rest goes into the permeate. The hydrocarbon concentration at the module inlet and outlet are 37 and 0.42%. A rotary vane vacuum pump (4) pumps the permeate and brings it to a screw compressor (5). The permeate is then compressed to 1.01 MPa (10 bar) to liquefy the hydrocarbons. The gas stream leaving the condenser (6) contains a residual concentration of hydrocarbons, depending on the pressure and the temperature, and this stream is supplied to the second membrane stage (7). The feed pressure and the permeate pressure at the second stage are 10 and 0.5 bar, respectively. The membrane area is 15 m². The hydrocarbon concentration at the module outlet is 0.09%. The hydrocarbon-enriched permeate stream joins the permeate stream from the first membrane stage and is pumped into the condensation unit. The retentates from the outlets of both the first and second stage membrane modules are combined, resulting in a total volume flow of about 190 m³/h, with hydrocarbon concentration of 0.23%. In order to meet the required emission standard, a posttreatment of the retentate is necessary. Membrane Technology and Research Inc., CA, developed a similar membrane system for the recovery of organic vapors from gas streams [329]. Although the membrane is not clearly described, it is believed to have a composite structure similar to that of GKSS. The module, however, is of a spiral-wound design, as illustrated schematically

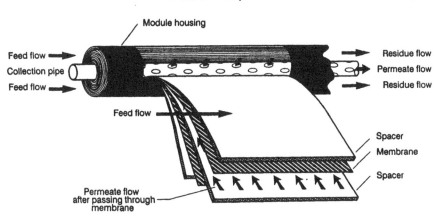

Figure 10.32. Spiral-wound module design (MTR). (Reproduced from [329] with permission.)

in Figure 10.32. The feed air is circulated laterally through the module. The organic vapor passes through the membrane preferentially, enters the permeate channel of the membrane envelope, and spirals inward to the central permeate collection pipe. Various system designs are possible for different applications. The simplest system is the single-stage unit shown in Figure 10.33. In this system the feed gas stream is compressed to 1 to 2 atmospheric pressure and passed through the membrane modules. The treated air is discharged to the atmosphere or recycled to the process. The permeate vapor enriched in the volatile organic compounds is compressed and supplied to a condenser. The condensed solvent is transferred to a solvent holding tank. A single-stage system is generally able to remove 80 to 90% of solvent from the feed air and produce a permeate with a concentration five to ten times higher than in the feed. This degree of vapor recovery is normally sufficient for many applications. An example is the concentration of butane produced as an off-gas in the production of polystyrene from packaging materials. Only 80% recovery of butane is required in this application to meet the requirements of the Environmental Protection Agency. Another goal is to concentrate butane to a sufficient level to allow for its use as a supplemental fuel in an existing boiler. The description of the unit, and the estimated capital and operating costs, are shown in Figure 10.33.

10.3.7 Membrane Reactor

A reversible reaction cannot be completed because the conversion attainable is thermodynamically limited. If the product can be separated from the reacting mixtures, by employing a gas separation membrane, however, the reaction may proceed further in the forward direction and will end in completion.

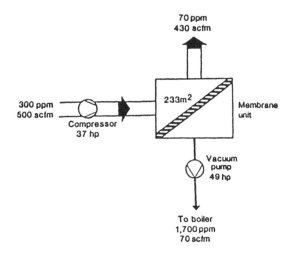

	FEED	PERMEATE	RESIDUE
Flow (scfm)	500	70	430
Concentration (ppm)	300	1,700	70

Membrane Selectivity	30
Membrane Area	233 m²
Vacuum Pumps	49 hp
Compressors	37 hp
CAPITAL COSTS	$282,000
	$560/scfm feed
OPERATING COSTS	
Depreciation + interest	$ 33,200
Miscellaneous	10,400
Module replacement	
(3-year lifetime)	46,600
Energy	22,800
	$113,000/yr
OPERATING COST	$0.52/1000 scf feed
	0.68/lb solvent recovered

Figure 10.33. Schematic design and approximate cost of a single-stage system designed to remove 80% butane from an industrial off-gas for pollution control. (Reproduced from [329] with permission.)

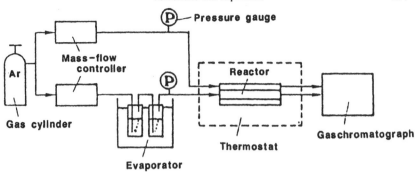

Figure 10.34. Experimental apparatus. (Reproduced from [330] with permission.)

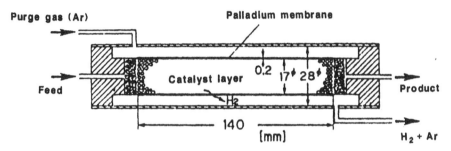

Figure 10.35. Details of the membrane reactor. (Reproduced from [330] with permission.)

A study was made by Shindo et al. to examine whether the expected result can be obtained [330]. Dehydrogenation of cyclohexane to benzene, as a model reaction, was coupled with a palladium membrane that is preferentially permeable to the product hydrogen. Figure 10.34 shows the schematic diagram of the experimental apparatus. Figure 10.35 shows the details of the membrane reactor constructed by Shindo et al. A thin palladium alloy (Pd 77% and Ag 23%) of 0.2-mm thickness was used in a tubular form, the outer diameter and the length of which were 17.0 mm and 140 mm, respectively. The membrane tube was packed uniformly with cylindrical alumina of 3.3-mm outer diameter and 3.6-mm height, which were supporting a platinum catalyst (0.5 wt%). Argon feed gas saturated with cyclohexane vapor (19.7%) was fed into the center of the membrane reactor. An argon stream that was fed into the shell side of the membrane reactor swept hydrogen that permeated through the palladium membrane. The reaction temperature was kept at 473K, and the pressure was atmospheric. Figure 10.36 illustrates some experimental results. The figure shows that the conversion of cyclohexane into benzene is the function of the rate at which cyclohexane is provided to the reactor, u_c^0, and the flow rate of argon sweep gas, v_A^0. When u_c^0 is 0.29, conversion of 99.7% was attained at v_A^0 values above 5

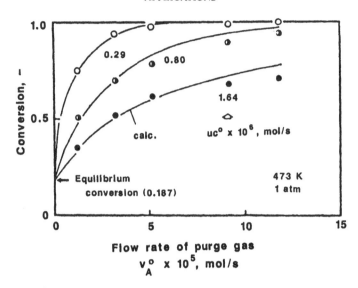

Figure 10.36. Some experimental results of the membrane reactor. (Reproduced from [330] with permission.)

$\times\ 10^{-5}$ mol/s. Considering the thermodynamic limit of 18.7%, the conversion achieved by the membrane reactor is extremely high, testifying to the validity of the membrane reactor concept. It should be noted that benzene leaving the membrane reactor is almost pure, and therefore a cyclohexane/benzene separator is not necessary. Furthermore, if a vacuum is applied on the permeate side, instead of an argon sweep gas stream, pure hydrogen gas is obtained as a reaction product.

10.3.8   Integration of Membrane and PSA Systems for the Purification of Hydrogen and Production of Oxo-alcohol Syngas

Membrane separation processes are always considered in comparison with some other more conventional separation processes, when their economic feasibility is examined. Often, membrane separation alone is not necessarily economically advantageous, but a synergetic effect can be expected when membrane separation is combined with another separation process, in a hybrid system. One such system was proposed by UOP Engineering Products to be incorporated into the oxo-alcohol synthesis process [331]. The oxo-alcohol plant has the following three main operations:

1. Hydrogen/carbon monoxide synthesis gas production by either steam gas reforming or by partial oxidation of hydrocarbons. Steam reforming of

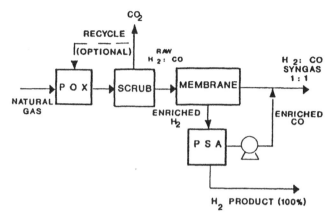

Figure 10.37. Integrated membrane plus PSA system for H_2/CO ratio adjustment. (Reproduced from [331] with permission.)

natural gas produces a product stream with an H_2:CO ratio of approximately 3:1, whereas partial oxidation of hydrocarbons gives a product stream with an H_2:CO ratio of 2:1

2. Reaction of syngas (H_2:CO ratio, 1:1) into formaldehyde

3. Hydrogenation of formaldehyde with pure hydrogen, forming oxo-alcohol product

It is desirable, therefore, to adjust the H_2:CO ratio of the synthesis gas produced in the first stage to 1:1 and simultaneously recover pure hydrogen that will be used in the last stage to hydrogenate formaldehyde to oxo-alcohol product. It is reported that an integrated membrane/PSA system, reduces the capital cost significantly from individual membrane or PSA system by its synergetic effect. A flow diagram of the proposed system is presented schematically in Figure 10.37. The synthesis gas produced at either the steam reformer or at the partial oxidation unit enters a scrubber where CO_2 is removed. The carbon dioxide so produced can be either recycled as feed to the reactor or removed from the system. The carbon-dioxide-free stream enters the membrane module where hydrogen permeates preferentially, leaving on the high-pressure-side carbon-monoxide-enriched stream. Note that the H_2/CO mixture is obtained as a retentate at essentially the feed inlet pressure. Hydrogen-rich permeate is introduced to the PSA unit where CO is adsorbed and a high-purity hydrogen stream is produced for the hydrogenation of formaldehyde. The tail gas from the PSA unit contains CO and unrecovered hydrogen, which is combined with the retentate of the membrane module, thereby adjusting the H_2:CO ratio to 1:1. The unique features of the integrated system, considered as benefits, are as follows:

Table 8. Comparative Economics [331]

	PSA only	Membrane plus PSA
Compression required, BHP	1080	415
Separation equipment cost, MM USD	1.429	1.375
Installation cost, MM USD	0.175	0.225
Installed compressor cost	0.864	0.332
Compressor operating cost, MM USD (3 years-8000 h/year- 5 cents/kW)	0.966	0.371
Total capital cost + 3 years operation, (MM USD)	3.430	2.303

1. The retentate of the membrane module, which constitutes the major portion of the synthesis gas produced at the synthesis stage, is obtained at essentially the feed pressure. Thus, the syngas supplied to the formaldehyde synthesis stage does not require further compression.

2. The membrane permeate introduced to the PSA unit contains a small amount of CO. Therefore, a relatively small size of PSA unit is sufficient to adsorb the residual CO. The hydrogen obtained at the PSA unit is very pure, requiring no further purification.

3. The amount of CO-rich tail gas from the PSA unit is very small, particularly compared to the amount of tail gas that would be produced if a stand-alone PSA unit were used without incorporating a membrane separator. The resulting savings of compressor size and power is the primary advantage of the integrated system. As a result, the integrated system requires substantially smaller capital and operating cost, compared with the PSA stand-alone system. A comparison is made in Table 8.

10.4   Pervaporation

10.4.1   Dehydration of Ethyl Alcohol

The most important application of pervaporation is the dehydration of ethyl alcohol. The ethyl alcohol concentration in the fermentation broth is from 5 to 10%. Ethyl alcohol is concentrated and dehydrated by a distillation process. However, ethyl alcohol forms an azeotropic solution with water at ethyl alcohol concentration of 95.6 wt%, and distillation becomes ineffective at removing the trace amount of water. Azeotropic distillation with an additive, normally either cyclohexane or benzene, needs to be applied. The energy consumption of this dehydration process is very high, and a new process to reduce the dehydration cost has been sought for. Some attempts were made to concentrate ethyl alcohol by the reverse osmosis process [332]–[335]. The maximum concentration of ethyl alcohol achievable by reverse osmosis is limited, however, due to the osmotic pressure of ethyl alcohol, which increases rapidly when ethyl alcohol concentration is higher than 10 wt%. The pervaporation process has been considered as an alternative membrane separation process. So far, the one-step membrane concentration of ethyl alcohol is not considered economically feasible, even by the pervaporation process. The vapor phase-liquid phase ethyl alcohol concentration diagram given for a commercial polyvinyl alcohol/polyacrylonitrile composite membrane, in Figure 10.38, shows that distillation is more favorable than pervaporation for the concentration of ethyl alcohol from a low concentration (for example, 6 wt%) to 85 wt%, while for the concentration of ethyl alcohol from 85 wt% to more than 99 wt%, pervaporation can be used. Figure 10.39 shows the process diagram of the combined distillation-pervaporation process. Tusel claims that the steam consumption of the above arrangement is 2.7 kg/kg of dehydrated ethyl alcohol. When a multipressure three-column system is combined with a pervaporation system, the steam consumption is reduced to 0.8kg/kg. On the other hand, the conventional distillation column results in a steam consumption of 4.0 to 6.0 kg (110°C)/kg or, for a complicated plant operation, 1.0 kg of saturated steam at 150°C per kg of dehydrated alcohol [336]. The process flow sheet of a pervaporation unit by the GFT Company is given in Figure 10.40 GFT systems are used not only in the dehydration of ethanol, but also in the dehydration of other organic solvents. A large-scale pervaporation unit for the dehydration of chlorinated hydrocarbons is shown in Figure 10.41.

While the GFT pervaporation process for alcohol dehydration is based on a polyvinyl alcohol membrane of high hydrophilicity, polyion complex membranes have been recently developed to replace polyvinyl alcohol. The membrane material consists of polyanion (polyacrylic acid, molecular weight 500,000, cross-linked with diepoxide) and polycation (PCA-107, the structure

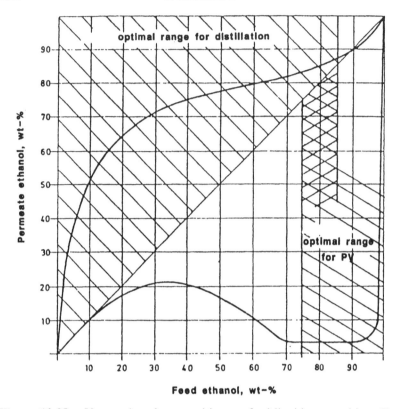

Figure 10.38. Vapor ethanol composition vs. feed liquid composition. (Reproduced from [336] with permission.)

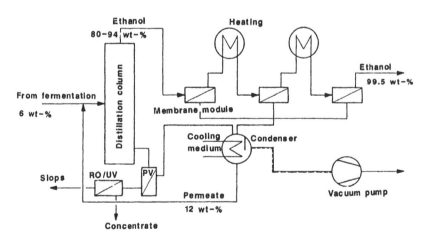

Figure 10.39. Process diagram of combined distillation/pervaporation. (Reproduced from [336] with permission.)

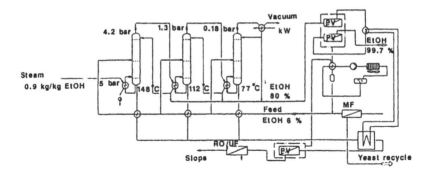

Figure 10.40. Process diagram of combined distillation/pervaporation plant. (Reproduced from [336] with permission.)

▲ pervaporation unit for dehydration of chlorinated hydrocarbons

Figure 10.41. Pervaporation unit for dehydration of chlorinated hydrocarbons. (From Technical Bulletin of GFT gmbH, with permission.)

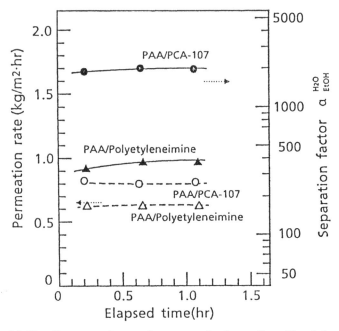

Polyallyl amine(PAAm)

Figure 10.42. Structure of polycation PCA-107. (Reproduced with permission of authors.)

Figure 10.43. Pervaporation performance of polyacrylic acid polyion complex membrane. (Polyacrylic acid molecular weight, 500,000; feed ethanol concentration, 95 wt%; operating temperature, 70°C.) (Reproduced with permission of authors.)

of which is given in Figure 10.42) [337], [338]. The polyacrylic acid was coated and cross-linked on top of the porous polysulfone substrate membrane. The selective skin layer thus formed was found to be 0.5 μ. The polyacrylic film was further reacted with polycation. The membrane so prepared showed an excellent performance and durability for the dehydration of ethyl alcohol, as indicated by the experimental data illustrated in Figure 10.43. In order to prepare composite polyion complex hollow fibers, polyacrylonitrile hollow fibers were spun. Then the inner surface of the hollow fiber was brought into contact with an alkaline solution to partially hydrolyze the nitrile group, which was followed by the immersion of hollow fibers into the polycation solution. Thus, the selective skin layer was formed only inside the hollow fiber. The polyion complex hollow fibers have shown better performance both in selectivity and flux than have the

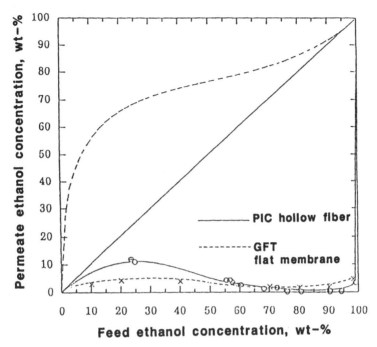

Figure 10.44. Comparison of pervaporation performance data of polyion complex membrane and GFT flat membrane-permeate vapor composition vs. feed liquid composition.

reference GFT pervaporation membranes, as illustrated in Figures 10.44 and 10.45. A pilot plant with a capacity of 3 kl/day is currently being successfully operated in Izumi, Japan.

10.4.2 Methanol Recovery from Methanol/MTBE/C₄ Mixtures

MTBE (methyl tertiary butyl ether) is produced by reacting methanol with isobutene over an acidic ion-exchange catalyst (Figure 10.46). The conversion is limited by thermodynamic equilibrium. In commercial practice the isobutene conversion of 87 to 94% per pass is set with a molar ratio of methanol to isobutene slightly higher than 1.0. When an excess amount of methanol is fed to the reactor, an increase in the isobutene conversion is expected, and 5 to 6% additional conversion is possible. Adding extra methanol, however, imposes an additional separation problem, particularly because of the formation of azeotropes in the reaction product. Air Products proposed the TRIM process in which the pervaporation process is incorporated to remove excessive amounts of methanol, since it is able to break the methanol azeotropes [339]. Two versions of TRIM processes are proposed. In the first version, a pervaporation

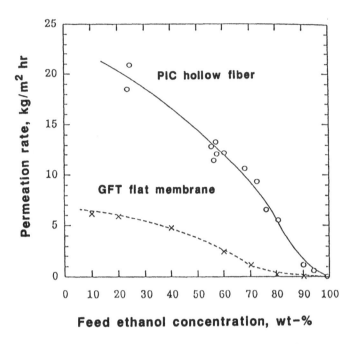

Figure 10.45. Comparison of pervaporation performance data of polyion complex membrane and GFT flat membrane-flux vs. feed liquid composition.

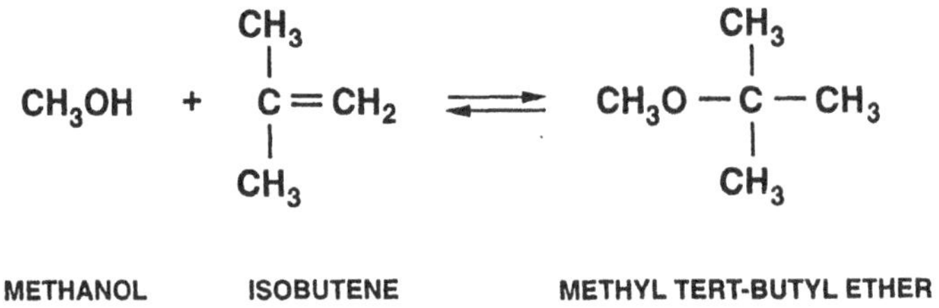

Figure 10.46. MTBE reaction chemistry. (Reproduced from [339] with permission.)

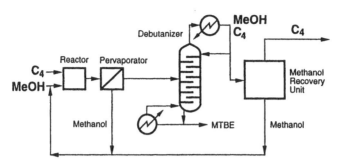

Figure 10.47. Trim process; membrane before debutanizer. (Reproduced from [339] with permission.)

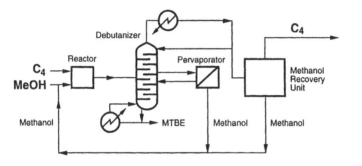

Figure 10.48. Trim process; debutanizer side draw. (Reproduced from [339] with permission.)

unit is added immediately after the reactor (Figure 10.47). methanol:isobutene ratio in the reactor is increased from 1.0 to 1.2, thus increasing the isobutene conversion about 5% per pass. The pervaporation unit removes methanol from the reactor effluent, reducing the methanol concentration from 5 to 2%, which can be fed to the next distillation (debutanizer) column without causing the formation of azeotropes. The MTBE production, therefore, can be increased without touching other parts of the process, simply by attaching a pervaporation unit. The investment required for the additional pervaporation unit is paid back in 18 months or less. The second proposed version of the TRIM system is to take a sidedraw from the distillation column to a pervaporation unit, to remove methanol, Figure 10.48. A composition profile of the distillation column shows two peaks in methanol composition: one above and the other below the feed tray. By taking a sidedraw, removing methanol, and returning the residual, the column operation can be more efficient. In one extreme, the methanol concentration at the top of the column is minimized so that the cost involved in methanol recovery is significantly reduced. Another advantage of this scheme over the first version of the TRIM system is that the methanol concentration in

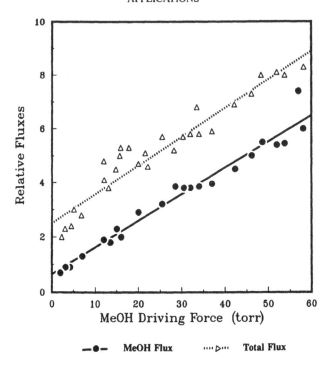

Figure 10.49. Data on relative flux of methanol and total flux of permeant mixture from demonstration test element. (Reproduced from [339] with permission.)

the feed stream of the pervaporation unit is high, resulting in a higher pervaporation flux and a smaller membrane area. The sidedraw scheme is attractive for new plants. It is claimed that as much as 20% of the additional capital investment required for the higher conversion plant can be saved. Seperex cellulose acetate gas membranes (from Air Products) were used for the evaluation of the process. The field demonstration test was conducted at a major MTBE producer's plant on the Gulf coast, and the test was run for 5 months. Two parallel tubes, each containing four spiral-wound elements of 5.08-cm (2 in.) diameter and 101.6-cm (40 in.) length, were used for the test. A module that houses two 20.32-cm (8 in.)-diameter, 101.6-cm (40 in.)-length elements was also tested. The temperature-controlled feed mixture was pumped to the membrane tubes. The permeate was compressed in a 10-HP vacuum pump and then condensed against cooling water and collected in a reservoir. The noncondensables were simply vented from the system. Figures 10.49 and 10.50 illustrate some of the performance data. In Figure 10.49 the relative fluxes of both methanol and the total permeant mixture are plotted vs. the methanol driving force, expressed as the partial vapor pressure difference of methanol between the upstream and the

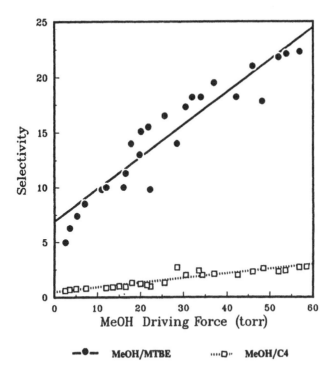

Figure 10.50. Selectivity data from demonstration test element. (Reproduced from [339] with permission.)

downstream side of the membrane. The unit for the flux is unknown. Both fluxes are increased linearly by an increase in the partial vapor pressure difference. Figure 10.50 illustrates the selectivity of the membrane, as shown by flux ratios of C_4/methanol and MTBE/methanol. Both selectivities increase with an increase in the partial vapor pressure difference. In particular, the MTBE/methanol ratio is extremely low, which is important, since very little MTBE is recycled to the reactor. The typical composition of the permeate is 65 to 90 wt% methanol, the rest being mostly C_4s, when the feed methanol concentration is 1 to 10 wt%.

10.4.3 Separation of Azeotropic Mixture of Dimethyl Carbonate (DMC) and Methanol (MeOH)

Dimethyl carbonate (DMC) or DMC blend with alcohols or MTBE are considered as gasoline enhancers alternate to MTBE. Both DMC and MTBE can also serve as phase enhancers in alcohol-containing gasoline. In the process to produce DMC, a DMC/MeOH azeotrope (MeOH 69 wt%) is formed at atmospheric pressure. Therefore, it is necessary to separate the azeotrope into high-purity (MeOH 1 wt%) DMC and a DMC/MeOH mixture (MeOH >95 wt%). The latter mixture is recycled to the reactor, and for the optimum reaction efficiency, no less than 95 wt% of MeOH content is desirable. There are two alternatives to achieve this goal. At a pressure of 1172 kPa (170 psi), the azeotrope shifts to a MeOH concentration of 96 wt%. Therefore, high-pressure distillation can satisfy the above requirement. Another alternative is a membrane-assisted distillation process. A methanol-selective membrane is used just to break the azeotrope. While the permeate that contains 95 wt% methanol is recycled to the reactor, the retentate with a MeOH content of 55 wt% is subjected to distillation at 310 kPa (45 psi). An azeotropic mixture is obtained at the top of the distillation column, whereas high-purity (MeOH 1 wt%) DMC is obtained at the bottom of the distillation column. Flow diagrams of both processes are illustrated in Figure 10.51.

Shah et al. made an economic comparison of both processes, assuming a DMC production rate of 907×10^3 kg (2 million lbs) per year [340]. The high-pressure distillation required 30 theoretical trays, and the reboiler duty was 0.82 MMW (2.8 MMBtu/h). High-pressure steam was used as a heating medium in the reboiler. The basis for the calculation of membrane-assisted distillation was that the membrane flux was 0.6 kg/m^2 h when operated at about 75 to 80°C. A membrane area of 540 m^2 was required. The permeate vapors were recovered by condensing and cooling them to a temperature of 0°C. Separation of the retentate required a distillation column with nine theoretical trays and a reboiler duty of 0.2 MMW (0.68 MMBtu/h). Medium-pressure steam was used as a heating medium in the reboiler. Both capital and utility costs are listed in Table 9. It is obvious that the membrane-assisted distillation is more advantageous.

10.4.4 Removal of Organic Solvent Contaminants from Industrial Effluent Streams

When hydrophobic membranes, such as those made of silicone rubber material, are used for the separation of an aqueous solution, organic components of the mixture permeate preferentially through the membrane [341]. Thus, it is possible to remove and concentrate organic contaminants in the industrial waste water. This principle was applied by Membrane Technology Research Inc., CA, using their membranes and membrane modules similar to those developed for the separation of volatile organic compounds from air (Figure 10.32). One

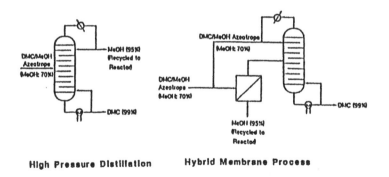

High Pressure Distillation   **Hybrid Membrane Process**

Figure 10.51. DMC/MeOH separation processes by Texaco. (Reproduced from [340] with permission.)

example of such applications is the pervaporation treatment of effluent samples supplied from a chemical producer. The sample was taken from the scrubber blowdown stream exiting the decanter of a diisopropyl benzene plant. The volume of the stream was 15,000 gal/day and the composition was 900 to 17,000 ppm benzene, 10 ppm cumene, 0.08 to 0.13 wt% chlorides, 0.42 mg/l dissolved solid, and a trace of aluminum. The temperature of the stream was 15 to 55°C, and the pH ranged from 2 to 4. Table 10 shows the permeate fluxes. Note that the permeate was separated into the aqueous and organic phase, and the latter phase consisted primarily of benzene. It is shown in the table that most of the benzene was removed in the first hour of the batchwise operation. The table also shows that the concentration of benzene in the permeate vapor phase took place before condensation of the liquid. It is obvious that benzene concentration was dramatically increased in the permeate by both MTR-100 and MTR-200 membranes, producing permeate streams 100 to 500 times more concentrated than the feed. Nitrogen and oxygen fluxes were also measured before and after the pervaporation experiment. No significant change in the permeation rate of both gases was noticed, and therefore it was concluded that neither deposition of foulants on the membrane surface, nor notable change in the membrane structure, took place during the pervaporation experiment. The capital and operating costs to remove 90% of benzene in a feed stream containing 1,300 ppm benzene are summarized in Table 11 for both MTR-100 and MTR-200 membranes. The operating cost for the MTR-200 membrane is $10/1000 gallon feed or $7.60/gal of recovered benzene. The cost for the MTR-100 membrane is slightly higher. If the price of benzene is included, the operating cost is reduced to $6 to $8/1000 gal of feed.

Table 9. Cost Comparison Between
High-Pressure Distillation and Hybrid Membrane Process
for DMC/MeOH Azeotrope Separation [340]

	High pressure distillation	Membrane assisted distillation
Capital cost $\times 10^{-6}$, $	1.5	1.0
Annual utility requirement		
1. Steam $\times 10^{-6}$, kg		
a. Memb. unit (138 kPa)	—	1.1
b. Distillation column		
i. 1.38 MPa	—	2.9
ii. 4.14 MPa	14	—
2. Electricity $\times 10^{-9}$, kJ	—	1.1
for refrigeration		
3. Cooling water $\times 10^{-6}$, m^3	0.3	0.1
Utility cost $\times 10^{-6}$, $	0.171	0.045

Note: DMC production rate = 907 $\times 10^3$ kg/year.

Table 10. Permeate Composition and Permeate Flux Data
of Pervaporation[a] for Benzene Separation [341]

Time (h)	Total permeate benzene conc. (%)	Permeate benzene phase (ml)	Permeate aqueous phase (ml)	Total weight of permeate sample (g)	Permeate flux (kg/m^2 h)
MTR–100					
1	17.2	26	11135	0.75	
2	4.3	5	102	104	0.58
3	1.0	2	101	102	0.57
4	0.2	—[b]	102	101	0.56
MTR–200					
2	45.4	27	29	53	0.15
4	6.7	1.8	23	24	0.068
6	2.7	0.5	23	23	0.063

[a] Operating conditions: temperature, 40°C; downstream pressure, 2000 Pa; feed flow, 6 l/min; feed volume, 22.7 l; total benzene content in feed, 30–32 ml.
[b] This sample contained only one phase.

Table 11. Capital and Operating Cost of
Pervaporation Systems for Benzene Recovery from
a Specialty Chemical Producer's Effluent Stream [341]

	MTR–100	MTR–200
Capital cost	$150,000[a]	$160,000[a]
Operating cost		
Depreciation + interest at 20%		
(excluding modules)	24,000	24,000
Module replacement (3–year life)	12,000	15,000
Maintenance (5% of capital cost)	7,500	8,000
Energy	9,000	3,000
Total	$52,500/year	$50,000/year
Operating cost	$11/1000 gal feed	$10/1000 gal feed

[a] System includes chiller and heater. If cooling water is available on-site,
the capital cost can be decreased by $10,000.

10.4.5 Concentration and Recovery of Flavor Components in Apple Essence

Another interesting application of pervaporation using hydrophobic polydi-methylsiloxane membranes is to recover and concentrate natural flavor compo-nents for fruit juices, into the permeate. An attempt was made to concentrate the flavor components of apple essence ($\times$ 500) [342]. The thickness of the polydimethylsiloxane membrane used was 25.4 μm. At room temperature, the permeation flux of 3.76×10^{-5} kg/m^2 s was obtained. The concentration of the flavor components in the feed and the permeate was determined by the GC-MS method. The results are summarized in Table 12. The ratio of the concentration in the permeate to that in the feed is strongly correlated to the boiling point of the flavor compound. In particular, concentration ratios above 20 were achieved for flavor components with boiling points lower than 100°C.

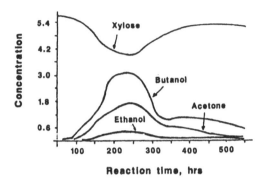

Figure 10.52. Composition of BAE fermentation product. (Reproduced from [344] with permission.)

10.4.6 Pervaporation Combined with Fermentation

The pervaporation process has progressed, undoubtedly, together with the progress in biotechnology. Although, in its early stage, pervaporation has been used for the dehydration of ethanol product from fermenters, a unique application of pervaporation was proposed by Sodeck et al. [343], in which the pervaporation process is combined with a bioreactor. The purpose of incorporating the pervaporation process is to remove ethanol product from fermentation broth, thus preventing product inhibition. In order to achieve this goal, membranes that are preferentially permeable to ethanol have to be used. Polydimethylsiloxane, or silicone rubber, is known as such a membrane material. The effect of product removal is clearly shown in Figures 10.52 and 10.53. Figure 10.52 shows the result of ABE (acetone, butanol, and ethanol) production from xylose substrate, using C-acetobutylicum in a conventional fermenter. The figure indicates that butanol production reached a maximum value after 90 h. But the production dropped after 260 h. Figure 10.53 shows a bioreactor combined with a pervaporation process. It shows that butanol production reached a maximum after 300 h, and the high value was maintained due to the removal of product.

Table 12. GC-MS Analytical Results [342]

| | Feed[a] | | | Boiling point |
	Sample 1	Sample 2	Permeate[a]	(°C)
1. Acetic acid, Ethyl Ester	0.2		4.5	77.1
2. 1-Propanol, 2-Methyl-			6.8	108.0
3. 1-Butanol	14.6	15.0	228.5	117.2
4. Propionic acid, Ethyl Ester	0.25	0.25	11.0	99.1
5. Ethane, 1,1-Diethoxy-			53.3	
6. 1-Hexene, 4-Methyl-	1.4	0.7	34.0	87.5
7. Propionic acid	2.3	5.1	2.1	141
8. Propionic acid, 2-Methyl,				
Ethyl Ester	0.2		15.3	109–111
9. Hexanal	2.3	1.5	33.1	128
10. 2-Furfural	2.5	1.4	0.1	161.7
11. Propanoic acid, 2-Methyl-	5.0	7.2	11.2	153.2
12. 2-Hexanal	9.8	9.2	224.7	146–147
13. Butanoic acid, 3-Methyl-,				
Ethyl Ester	0.5	0.5	4.0	
14. 1-Hexanol	7.4	6.3	91.2	158
15. 1,3-Dioxolane, 2-Ethyl-			5.9	
16. 1,3-Dioxolane, 2-Propyl-			0.6	
17. 3-Heptanol, 3-Methyl-	0.6	0.5	5.3	163
18. Ethanone, 1-Phenyl-	100.0	100.0	100.0	
19. Hexane, 1,1-Diethoxy-			4.7	
20. Hexane, 1-(1-Ethoxyethoxy)-			1.2	
21. 2-Furfural, 5-(Hydroxymethyl)-	28.4	33.2		114–116

[a] Peak ratio (%) of flavor component to the internal standard (Ethanone,
1-Phenyl-).

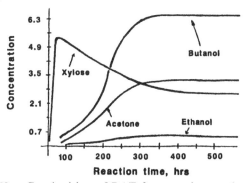

Figure 10.53. Composition of BAE fermentation product with product removal by pervaporation. (Reproduced from [344] with permission.)

References

[1] Hildebrand, J.H., Scott, R.L., *The Solubility of Non-electrolyte*, 3rd ed.; Reinhold: New York, 1950; pp. 123–124.

[2] Small, P.A., *J. Appl. Chem.* 1953, 3, 71.

[3] Hansen, C.M., Beerbower, A., in *Encyclopedia of Chemical Technology, Supplement Volume*; John Wiley & Sons: New York, 1971; pp. 889–910.

[4] Van Krevelen, D.W., *Properties of Polymers*; Elsevier: Amsterdam, 1976.

[5] Sourirajan, S., Matsuura, T., *Reverse Osmosis/Ultrafiltration Process Principles*; National Research Council of Canada: Ottawa, 1985.

[6] Mulder, M.H.V., Fruitz, F., Smolders, C.A., *J. Membrane Sci.* 1982, 11, 349.

[7] Lee, Y.M., Bourgeois, D., Belfort, G., in *Proc. Second Int. Conf. on Pervaporation Processes in the Chemical Industry*; R. Bakish, Ed.; Bakish Materials Corp.: NJ, 1987; pp. 249–265.

[8] Feng, X, Sourirajan, S., Tezel, H., Matsuura, T., *J. Appl. Polym. Sci.* 1991, 43, 1071.

[9] Matsuura, T., Sourirajan, S., *J. Colloid Interface Sci.* 1978, 66, 589.

[10] Blair, H.S., McElroy, M.J., *J. Appl. Polym. Sci.* 1975, 19, 3161.

[11] Stamm, A.J., Millet, M.A., *J. Phys. Chem.* 1941, 45, 43.

[12] Barrie, J.A., in *Diffusion in Polymers*; J. Crank and G.S. Park, Eds.; Academic Press: New York, 1968; pp. 274–276.

[13] Okita, K., in *Recent Advances in Membrane Treatment Technology and Its Applications*; H. Shimizu, M. Nishimura, Eds.; Fuji Technosystem: Tokyo, 1984; p. 271.

[14] Matsuura, T., Taketani, Y., Sourirajan, S., *Desalination* 1981, 38, 319.

[15] Taketani, Y., Matsuura, T., Sourirajan, S., *Sep. Sci. Technol.* 1982, 17, 821.

[16] Taketani, Y., Matsuura, T., Sourirajan, S., *J. Electrochem. Soc.* 1982, 129, 1485.

[17] Matsuura, T., Taketani, Y., Sourirajan, S., *J. Colloid Interface Sci.* 1983, 95, 10.

[18] Lloyd, D.R., Meluch, T.B., Selection and evaluation of membrane materials for liquid separations, in *Materials Science of Synthetic Membranes*; D.R. Lloyd, Ed.; American Chemical Society: Washington, D.C., 1985; pp. 47–79.

[19] Kiso, Y., Kitao, T., *Chromatographia* 1986, 22, 341.

[20] Kiso, Y., Kitao, T., Yong-Sheng, G., Jinno, K. *Chromatographia* 1989, 28, 279.

[21] Gao, S., Bao, Q., *J. Chromatogr.* 1989, 12, 2083.

[22] Sun, X., Gao, K., Lang, K., Matsuura, T., *Technol. Water Treat.* 1987, 13, 247.

[23] Tam, C.M., Kutowy, O., Talbot, F.D.F., *J. Liq. Chromatogr.* 1991, 14, 45.

[24] Tam, C.M., Kutowy, O., Talbot, F.D.F., *Chromatographia* 1991, 32, 224.

[25] Long, V.T., Minhas, B.S., Matsuura, T., Sourirajan, S., *J. Colloid Interface Sci.* 1988, 125, 478.

[26] Bao, S., Sourirajan, S., Talbot, F.D.F., Matsuura, T., Gas and vapor adsorption on polymeric materials by inverse gas chromatography, in *Inverse Gas Chromatography, Characterization of Polymer and Other Materials*; D.R. Lloyd, T.C. Ward, H.P. Schreiber, C.C. Pizana, Eds.; American Chemical Society: Washington, D.C., 1989; pp. 59–76.

[27] Jiang, J., Minghi, S., Minling, F., Jiayan, C., *Desalination* 1989, 71, 107.

[28] Francis, P.S., Luzio, F.C.D., Gillam, W.S., Kotch, A., Fabrication and evaluation of new ultrathin reverse osmosis membranes, *Research and Development Progress Report No. 177*; Office of Saline Water, U.S. Department of Interior: Washington, D.C., 1966.

[29] Riley, R.L., Hightower, G., Lyons, C.R., Composite reverse osmosis membranes, in *Reverse Osmosis Membrane Research*; H.K. Lonsdale, H.E. Podall, Eds.; Plenum Press: New York, 1972; p. 437.

[30] Rozelle, L.T., Cadotte, J.E., Cobian, K.E., Kopp, C.V., Jr., Nonpolysaccharide membranes for reverse osmosis: NS- 100 membranes, in *Reverse Osmosis and Synthetic Membranes, Theory-Technology-Engineering*; S. Sourirajan, Ed.; National Research Council Canada: Ottawa, 1977; p. 249.

[31] Chen, Y., Miyano, T., Fouda, A., Matsuura, T., *J. Membrane Sci.* 1990, 48, 203.

[32] Hirotsu, T., *Ind. Eng. Chem. Res.* 1987, 26, 1287.

448

[33] Johnson, J.S., Jr., Polyelectrolytes in aqueous solutions-filtration, hyperfiltration and dynamic membranes, in *Reverse Osmosis Membrane Research*; H.K. Lonsdale, H.E. Podall, Eds.; Plenum Press: New York, 1972; p. 379.

[34] Leenaars, A.F.M., Keizer, K., Burggraaf, A.J., Structure, permeability and separation characteristics of porous alumina membranes, in *Reverse Osmosis and Ultrafiltration*; S. Sourirajan, T. Matsuura, Eds.; ACS Symp. Ser. 281; American Chemical Society: Washington, D.C., 1985; p. 57.

[35] Liu, T., Xu, S., Zhang, D., Matsuura, T., *Desalination* 1991, 85, 1.

[36] Tompa, H., *Polymer Solutions*; Butterworths: London, 1956.

[37] Zeman, L., Tkacik, G., *J. Membrane Sci.* 1988, 36, 119.

[38] Swinyard, B.T., Barrie, J.A., *Br. Polym. J.* 1988, 20, 317.

[39] Wijmans, J.G., Kant, J., Mulder, M.H.V., Smolders, C.A., *Polymer* 1988, 26, 1539.

[40] Lau, W.W.Y., Guiver, M.D., Matsuura, T., *J. Appl. Polym. Sci.* 1991, 42, 3215.

[41] Lau, W.W.Y., Guiver, M.D., Matsuura, T., *J. Membrane Sci.* 1991, 59, 219.

[42] Yilmaz, L., McHugh, A.J., *J. Appl. Polym. Sci.* 1986, 31, 997.

[43] Yilmaz, L., McHugh, A.J., *J. Membrane Sci.* 1986, 28, 287.

[44] Paul, D.R., *J. Appl. Polym. Sci.* 1968, 12, 383.

[45] Strathmann, H., Scheible, P., Baker, R.W., *J. Appl. Polym. Sci.* 1971, 15, 811.

[46] Strathmann, H., Kock, K., Amar, P., Baker, R.W., *Desalination* 1975, 16, 179.

[47] So, M.T., Eirich, F.R., Strathmann, H., Baker, R.W., *Polymer Lett. Ed.* 1973, 11, 201.

[48] Frommer, M.A., Matz, R., Rosenthal, U., *Ind. Eng. Chem. Prod. Res. Develop.* 1971, 10, 193.

[49] Yilmaz, L., McHugh, A.J., *J. Appl. Polym. Sci.* 1988, 35, 1967.

[50] Frommer, M.A., Lancet, D., *Polymer Preprints* 1971, 12, 245.

[51] Sourirajan, S., *Reverse Osmosis*; Academic Press: New York, 1970.

[52] Keilin, B., *Research and Development Progress Report No. 84*; Office of Saline Water, U.S. Department of Interior: Washington, D.C., 1963; p. VII-2.

[53] Hermans, P., *J. Text. Res.* 1950, 20, 553.

[54] Riley, R.L., Gardner, J.O., Merten, U., *Science* 1964, 143, 801.

[55] Riley, R.L., Merten, U., Gardner, J.O., *Desalination* 1966, 1, 30.

[56] Schultz, R.D., Asunmaa, S.K., Ordered water and the ultrastructure of the cellular plasma membrane, in *Recent Progress in Surface Science, Vol. 3*; J.F. Danielli, A.C. Riddiford, M. Rosenberg, Eds.; Academic Press: New York, 1970; pp. 291–332.

[57] Manjikian, S., Loeb, S., McCutchan, J.W., in *Proc. 1st Int. Symp. on Water Desalination, Vol. 2*; Office of Saline Water, U.S. Department of Interior: Washington, D.C., 1965; p. 159.

[58] Carman, P.C., *Disc. Faraday Soc.* 1948, 3, 72.

[59] Némethy, G., Scheraga, H.A., *J. Chem. Phys.* 1964, 41, 680.

[60] Némethy, G., Scheraga, H.A., *J. Chem. Phys.* 1962, 39, 3382.

[61] Miller, A.A., *J. Chem. Phys.* 1961, 38, 1563.

[62] Glasstone, S., Laidler, K.J., Eyring, H., *Theory of Rate Processes*; McGraw-Hill: New York, 1941; p. 977.

[63] Panar, M., Hoehn, H.H., Hebert, R.R., *Macromolecules* 1973, 6, 777.

[64] Kesting, R.E., *Synthetic Polymeric Membranes*; John Wiley & Sons: New York, 1985.

[65] Chan, K., Matsuura, T., Sourirajan, S., *J. Polym. Sci. Polym. Let. Ed.* 1983, 21, 417.

[66] Smolders, C.A., Vegteveen, E., New characterization methods for asymmetric ultrafiltration membranes, in *Materials Science of Synthetic Membranes*; D.R. Lloyd, Ed.; American Chemical Society: Washington, D.C., 1985; pp. 327–338.

[67] Zeman, L., Tkacik, G., Pore volume distribution in ultrafiltration membranes, in *Materials Science of Synthetic Membranes*; D.R. Lloyd, Ed.; American Chemical Society: Washington, D.C., 1985; pp. 339–350.

[68] Ohya, H., Imura, Y., Moriyama, T., Kitaoka, M., *J. Appl. Polym. Sci.* 1974, 18, 1855.

[69] Ohya, H., Konuma, J., Negishi, Y., *J. Appl. Polym. Sci.* 1977, 21, 2515.

[70] Broens, L., Bargeman, D., Smolders, C.A., On the mechanism of formation of P.P.O. ultrafiltration membranes, in *Proc. Sixth Int. Symp. on Fresh Water from the Sea, Vol. 3*; A. Delyannis and E. Delyannis, Eds.: Athens, 1978; pp. 165–171.

[71] Han, M.J., Bhattacharyya, D., *J. Membrane Sci.* 1991, 62, 325.

[72] Brun, M., Lallemand, A., Quinson, J.-F., Eyraud, C., *Thermochimica Acta* 1977, 21, 59.

[73] Zeman, L., Tkacik, G., *J. Membrane Sci.* 1987, 32, 329.

[74] Cuperus, F.P., Bargeman, D., Smolders, C.A., *J. Membrane Sci.*, 1992, 66, 45.

[75] Michaels, A.S., *Sep. Sci. Technol.* 1980, 15, 1980.

[76] Zeman, L., Wales, M., *Sep. Sci. Technol.* 1981, 16, 275.

[77] Dietz, P., Hansma, P.K., Inacker, O., Lehmann, H.-D., Herrmann, K.-H., *J. Membrane Sci.* 1992, 65, 101.

[78] Flory, P.J., *Principles of Polymer Chemistry*; Cornell University Press: New York, 1953.

[79] Rudiń, A., Johnston, H.K., *J. Paint Technol.* 1971, 43 39.

[80] Maron, S.H., Chiu, T.T., *J. Polym. Sci. A.* 1963, 1, 2641.

[81] Nguyen, T.D., Chan, K., Matsuura, T., Sourirajan, S., *Ind. Eng. Chem. Prod. Res. Dev.* 1985, 24, 655.

[82] Kesting, R.E., The nature of pores in integrally skinned phase inversion membranes, in *Advances in Reverse Osmosis and Ultrafiltration*; T. Matsuura and S. Sourirajan, Eds.; National Research Council of Canada: Ottawa, 1989; p. 3.

[83] Zhu, Z., Matsuura, T., *J. Colloid Interface Sci.* 1991, 147, 307.

[84] Kamide, K., *Thermodynamics of Polymer Solutions, Phase Equilibrium and Critical Phenomena*; Elsevier: New York, 1990; pp. 442–618.

[85] Tam, C.M., Matsuura, T., Tremblay, A.Y., *J. Colloid Interface Sci.* 1991, 147, 206.

[86] Kazama, S., Kaneta, T., Sakashita, M., Gas separation membranes of Cardo type polyamides, paper read at the Int. Symp. on Gas Separation Technology, Sept. 10–15, 1989, Antwerp, Belgium.

[87] Miyano, T., Matsuura, T., Sourirajan, S., *Chem. Eng. Comm.*, 1992.

[88] Miyano, T., Matsuura, T., Carlsson, D.J., Sourirajan, S., *J. Appl. Polym. Sci.* 1990, 41, 407.

[89] Nguyen, T.D., Matsuura, T., Sourirajan, S., *Chem. Eng. Comm.* 1987, 54, 17.

[90] Nguyen, T.D., Matsuura, T., Sourirajan, S., *Chem. Eng. Comm.* 1987, 57, 351.

[91] Nguyen, T.D., Matsuura, T., Sourirajan, S., *Chem. Eng. Comm.* 1990, 88, 91.

[92] Lafreniere, L.Y., Talbot, F.D.F., Matsuura, T., Sourirajan, S., *Ind. Eng. Chem. Res.* 1987, 26, 2385.

[93] Miyano, T., Matsuura, T., Sourirajan, S., *Chem. Eng. Comm.* 1990, 95, 11.

[94] Frommer, M.A., Lancet, D., *J. Appl. Polym. Sci.* 1972, 16, 1295.

[95] Taniguchi, Y., Horigome, S., *J. Appl. Polym. Sci.* 1975, 19, 2743.

[96] Yoshikawa, M., Matsuura, T., Study on the physicochemical property of solution components in the membrane and its effect on the pervaporation transport, in *Proc. Fourth Int. Conf. on Pervaporation Processes in the Chemical Industry*; R. Bakish, Ed.; Bakish Materials Corp.: Englewood, NJ, 1990; pp. 406–411.

[97] Yoshikawa, M., Handa, Y.P., Cooney, D., Matsuura, T., *Makromol. Chem. Rapid Comm.* 1990, 11, 387.

[98] Yoshikawa, M., Matsuura, T., Cooney, D., *J. Appl. Polym. Sci.* 1991, 42, 1417.

[99] Yoshikawa, M., Matsuura, T., *Polymer J.* 1991, 23, 1025.

[100] Crank, J., Park, G.S., *Diffusion in Polymers*; Academic Press: London, 1968.

[101] Lonsdale, H.K., Properties of cellulose acetate membranes, in *Desalination by Reverse Osmosis*; U. Merten, Ed.; M.I.T. Press: Cambridge, 1966; chap. 4.

[102] Cheryan, M., *Ultrafiltration Handbook*; Technomic: Lancaster, 1986.

[103] Kimura, S., Sourirajan, S., *A.I.Ch.E.J.* 1967, 13, 497.

[104] Sourirajan, S., Kimura, S., *Ind. Eng. Chem. Process Des. Dev.* 1967, 6, 504.

[105] Kimura, S., Sourirajan, S., *Ind. Eng. Chem. Process Des. Dev.* 1968, 7, 41.

[106] Kimura, S., Sourirajan, S., *Ind. Eng. Chem. Process Des. Dev.* 1968, 7, 197.

[107] Kimura, S., Sourirajan, S., *Ind. Eng. Chem. Process Des. Dev.* 1968, 7, 539.

[108] Kimura, S., Sourirajan, S., *Ind. Eng. Chem. Process Des. Dev.* 1968, 7, 548.

450

[109] Agrawal, J.P., Sourirajan, S., *Ind. Eng. Chem. Process Des. Dev.* 1969, 8, 439.
[110] Matsuura, T., Sourirajan, S., *J. Appl. Polym. Sci.* 1973, 17, 1043.
[111] Matsuura, T., Bednas, M.E., Sourirajan, S., *J. Appl. Polym. Sci.* 1974, 18, 567.
[112] Matsuura, T., Blais, P., Dickson, J.M., Sourirajan, S., *J. Appl. Polym. Sci.* 1974, 18, 3671.
[113] Matsuura, T., Pageau, L., Sourirajan, S., *J. Appl. Polym. Sci.* 1975, 19, 179.
[114] Matsuura, T., Bednas, M.E., Dickson, J.M., Sourirajan, S., *J. Appl. Polym. Sci.* 1975, 19, 2473.
[115] Matsuura, T., Bednas, M.E., Dickson, J.M., Sourirajan, S., *J. Appl. Polym. Sci.* 1974, 18, 2829.
[116] Dickson, J.M., Matsuura, T., Blais, P., Sourirajan, S., *J. Appl. Polym. Sci.* 1975, 19, 801.
[117] Dickson, J.M., Matsuura,T., Blais, P., Sourirajan, S., *J. Appl. Polym. Sci.* 1976, 20, 1491.
[118] Matsuura, T., Dickson, J.M., Sourirajan, S., *Ind. Eng. Chem. Process Des. Dev.* 1976, 15, 149.
[119] Matsuura, T., Dickson, J.M., Sourirajan, S., *Ind. Eng. Chem. Process Des. Dev.* 1976, 15, 350.
[120] Rangarajan, R., Matsuura, T., Goodhue, E.C., Sourirajan, S., *Ind. Eng. Chem. Process Des. Dev.*, 1976, 15, 529.
[121] Kutowy, O., Matsuura, T, Sourirajan, S., *J. Appl. Polym. Sci.* 1977, 21, 2051.
[122] Matsuura, T., Baxter, A.G., Sourirajan, S., *Ind. Eng. Chem. Process Des. Dev.* 1977, 16, 82.
[123] Matsuura, T., Blais, P., Pageau, L., Sourirajan, S., *Ind. Eng. Chem. Process Des. Dev.* 1977, 16, 510.
[124] Rangarajan, R., Matsuura, T., Goodhue, E.C., Sourirajan, S., *Ind. Eng. Chem. Process Des. Dev.* 1978, 17, 71.
[125] Malaiyandi, P., Matsuura, T., Sourirajan, S., *Ind. Eng. Chem. Process Des. Dev.* 1982, 21, 277.
[126] Agrawal, J.P., Sourirajan, S., *Ind. Eng. Chem. Process Des. Dev.* 1970, 9, 12.
[127] Rangarajan, R., Matsuura, T., Goodhue, E.C., Sourirajan, S., *Ind. Eng. Chem. Process Des. Dev.* 1978, 17, 46.
[128] Rangarajan, R., Matsuura, T., Goodhue, E.C., Sourirajan, S., *Ind. Eng. Chem. Process Des. Dev.* 1979, 18, 278.
[129] Rangarajan, R., Baxter, A.G., Matsuura, T., Sourirajan, S., *Ind. Eng. Chem. Process Des. Dev.* 1984, 23, 367.
[130] Matsuura, T., Sourirajan, S., *Ind. Eng. Chem. Process Des. Dev.* 1985, 24, 297.
[131] Rangarajan, R., Majid, M.A., Matsuura, T., Sourirajan, S., *Ind. Eng. Chem. Process Des. Dev.* 1985, 24, 977.
[132] Rangarajan, R., Matsuura, T., Sourirajan, S., Predictability of membrane performance in RO systems involving mixed ionized solutes in aqueous solutions — a general approach, in *Reverse Osmosis and Ultrafiltration*; S. Sourirajan, T. Matsuura, Eds.; American Chemical Society: Washington, D.C., 1985.
[133] Parsons, R., *Handbook of Electrochemical Constants*; Butterworths: London, 1959.
[134] Stannett, V., *J. Membrane Sci.* 1978, 3, 97.
[135] Barrer, R.M., Surface and volume flow in porous media, in *The Solid-Gas Interface*; E.A. Flood, Ed.; Marcel Dekker: New York, 1967; chap. 19.
[136] Hopfenberg, H.B., Ed., *Permeability of Plastic Films and Coatings to Gases, Vapors and Liquids*; Plenum Press: New York, 1974.
[137] Hwang, S.-T., Kammermeyer, K., *Membranes in Separations Techniques of Chemistry, Vol. VII*; Wiley-Interscience: New York, 1975; chap. 5.
[138] Stern, S.A., in *Membrane Separation Processes*; P. Meares, Ed.; Elsevier: New York, 1976; chap. 8.
[139] Stannett, V., Koros, W.J., Paul, D.R., Lonsdale, H.K., Baker, R.W., *Adv. Polym. Sci.* 1979, 32, 69.
[140] Stern, S.A., Frisch, H.L., *Ann. Rev. Mater. Sci.* 1981, 11, 523.
[141] Cohen, M.H., Turnbull, D., *J. Chem. Phys.* 1959, 31, 1164.

[142] Fujita, H., Kishimoto, A., Matsumoto, K., *Trans. Faraday Soc.* 1960, 56, 424.

[143] Frish, H.L., *J. Elastoplast.* 1970, 2, 130.

[144] Stern, S.A., Fang, S.-M., Frisch, H.L., *J. Polym. Sci.* 1972, A210, 201.

[145] Fang, S.-M., Stern, S.A., Frisch, H.L., *Chem. Eng. Sci.* 1975, 30, 773.

[146] Koros, W.J., Chan, A.H., Paul, D.R., *J. Membrane Sci.* 1977, 2, 165.

[147] Chan, A.H., Koros, W.J., Paul, D.R., *J. Membrane Sci.* 1978, 3, 117.

[148] Erb, A.J., Paul, D.R., *J. Membrane Sci.* 1981, 8, 11.

[149] Petropoulos, J.H., *J. Polym. Sci. Part A2.* 1970, 8, 1797.

[150] Vieth, W.R., Howell, J.M., Hsieh, J.H., *J. Membrane Sci.* 1976, 1, 177.

[151] Paul, D.R., *Ber. Bunsenges. Phys. Chem.* 1979, 83, 294.

[152] Vieth, W.R., *Diffusion In and Through Polymers*; Carl Hanser: Munich, 1991; chap. 4.

[153] Stern, S.A., Sen, S.K., Rao, A.K., *J. Macromol. Sci. Phys.* 1974, B10, 507.

[154] Binning, R.C., James, F.E., *Petroleum Refiner.* 1958, 37, 214.

[155] Binning, R.C., Lee, R.J., Jennings, J.F., Martin, E.C., *Ind. Eng. Chem.* 1961, 53, 45.

[156] Lee, C.H., *J. Appl. Polym. Sci.* 1975, 19, 83.

[157] Paul, D.R. Ebra-Lima, O.M., *J. Appl. Polym. Sci.* 1971, 15, 2199.

[158] Greenlaw, F.W., Prince, W.D., Shelden, R.A., Thompson, E.V., *J. Membrane Sci.* 1977, 2, 141.

[159] Greenlaw, F.W., Shelden, R.A., Thompson, E.V., *J. Membrane Sci.* 1977, 2, 333.

[160] Shelden, R.A., Thompson, E.V., *J. Membrane Sci.* 1978, 4, 115.

[161] Rautenbach, R., Albrecht, R., *J. Membrane Sci.* 1984, 19, 1.

[162] Mulder, M.H.V., Smolders, C.A., *J. Membrane Sci.* 1984, 17, 289.

[163] Hauser, J., Heintz, A., Reinhardt, G.A., Schmittecker, B., Wesslein, M., Lichtenthaler, R.N., Sorption, diffusion and pervaporation of water/alcohol mixtures in PVA-membranes. Experimental results and theoretical treatment, in *Proc. Second Int. Conf. on Pervaporation Processes in the Chemical Industry*; R. Bakish, Ed.; Bakish Materials Corp.: Englewood, NJ, 1987; pp. 15–34.

[164] Rhim, J.-W., Huang, R.Y.M., *J. Membrane Sci.* 1989, 46 335.

[165] Huang, R.Y.M., Rhim, J.-W., Separation characteristics of pervaporation membrane separation processes, in *Pervaporation Membrane Separation Processes*; R.Y.M. Huang, Ed.; Elsevier: Amsterdam, 1991; chap. 2.

[166] Bitter, J.G.A., *Transport Mechanisms in Membrane Separation Processes*; Plenum Press: New York, 1991.

[167] Iwatsubo, T., Yamanaka, T., Yamamoto, S., Mizoguchi, K., Suda, Y., *Sen-i Gakkaishi* 1988, 44, 367.

[168] Bode, E., Transport characteristics for permeants in a membrane as derived from steady state pervaporation experiments, in *Proc. Fourth Int. Conf. on Pervaporation Processes in the Chemical Industry*; R. Bakish, Ed.; Bakish Materials Corp.: Englewood, NJ, 1989; pp. 103–113.

[169] Bode, E., *J. Membrane Sci.* 1990, 50, 1.

[170] Rogers, C.E., Stannett, V., Szwarz, M., *J. Polym. Sci.* 1960, 45, 61.

[171] Henis, J.M.S., Tripodi, M.K., *Multicomponent Membranes for Gas Separations*, U.S. Patent 4,230,463.

[172] Fouda, A., Chen, Y., Bai, J., Matsuura, T., *J. Membrane Sci.* 1991, 64, 263.

[173] Henis, J.M.S., Tripodi, M.K., *Sep. Sci. Technol.* 1980, 15, 1059.

[174] Henis, J.M.S., Tripodi, M.K., *J. Membrane Sci.* 1981, 8, 233.

[175] Pinnau, I., Wijmans, J.G., Blume, I., Kuroda, T., Peinemann, K.-V., *J. Membrane Sci.* 1988, 37, 81.

[176] te Hennepe, H.J.C., Smolders, C.A., Bargeman, D., Mulder, M.H.V., *Sep. Sci. Technol.* 1991, 26, 585.

[177] Kedem, O., Katchalsky, A., *J. Gen. Physiol.* 1961, 45, 143.

[178] Spiegler, K.S., Kedem, O., *Desalination* 1966, 1, 311.

[179] Spiegler, K.S., *Salt-Water Purification, 2nd Ed.*; Plenum Press: New York, 1977.

[180] Pusch, W., *Ber. Bunsenges. Phys. Chem.* 1977, 81, 269.

452

[181] Pusch, W., *Ber. Bunsenges. Phys. Chem.* 1977, 81, 854.
[182] Merten, U., in *Desalination by Reverse Osmosis*; U. Merten, Ed.; MIT Press: Cambridge, 1966; pp. 15–54.
[183] Jonsson, G., Boesen, C.E., *Desalination* 1975, 17, 145.
[184] Glückauf, E., On the mechanism of osmotic desalting with porous membranes, in *Proc. First Int. Conf. on Water Desalination*, Vol. 1; Office of Saline Water, U.S. Department of Interior: Washington, D.C., 1965; pp. 143–156.
[185] Bean, C.P., in *Research and Development Report No. 465*; Office of Saline Water, U.S. Department of Interior: Washington, D.C., 1969.
[186] Jacazio, G., Probstein, R.F., Sonin, A.A., Yung, D., Porous materials for reverse osmosis membranes, theory and experiment, paper presented at Third Office of Saline Water Conference on Reverse Osmosis, May 7–11, 1972, Las Vegas, NV.
[187] Dresner, L., *Desalination* 1974, 15, 371.
[188] Matsuura, T, Sourirajan, S., *Ind. Eng. Chem. Process Des. Dev.* 1981, 20, 273.
[189] Matsuura, T., Taketani, Y., Sourirajan, S., Estimation of interfacial forces governing the reverse osmosis system: nonionized polar organic solute-water-cellulose acetate membrane, in *Synthetic Membranes*, Vol. II; A.F. Turbak, Ed.; American Chemical Society: Washington, D.C., 1981; p. 315.
[190] Chan, K., Matsuura, T., Sourirajan, S., *Ind. Eng. Chem. Prod. Res. Dev.* 1982, 21, 605.
[191] Matsuura, T., Tweddle, T.A., Sourirajan, S., *Ind. Eng. Chem. Process Des. Dev.* 1984, 23, 674.
[192] Liu, T., Matsuura, T., Sourirajan, S., *Ind. Eng. Chem. Prod. Res. Dev.* 1983, 22, 77.
[193] Taketani, Y., Matsuura, T., Sourirajan, S., *Desalination* 1983, 46, 455.
[194] Liu, T., Chan, K., Matsuura, T., Sourairajan, S., *Ind. Eng. Chem. Prod. Res. Dev.* 1984, 23, 116.
[195] Chan, K., Liu, T., Matsuura, T., Sourirajan, S., *Ind. Eng. Chem. Prod. Res. Dev.* 1984, 23, 124.
[196] Chan, K., Matsuura, T., Sourirajan, S., *Ind. Eng. Chem. Prod. Res. Dev.* 1984, 23, 492.
[197] Nguyen, T.D., Chan, K., Matsuura, T., Sourirajan, S., *Ind. Eng. Chem. Prod. Res. Dev.* 1984, 23, 501.
[198] de Pinho, M.N., Matsuura, T., Nguyen, T.D., Sourirajan, S., *Chem. Eng. Comm.* 1988, 64, 113.
[199] Bao, S., Talbot, F.D.F., Nguyen, T.D., Matsuura, T., Sourirajan, S., *Sep. Sci. Technol.* 1988, 23, 77.
[200] Farnand, B.A., Talbot, F.D.F., Matsuura, T., Sourirajan, S., *Ind. Eng. Chem. Proc. Des. Dev.* 1983, 22, 179.
[201] Farnand, B.A., Talbot, F.D.F., Matsuura, T., Sourirajan, S., *Sep. Sci. Technol.* 1984, 19, 33.
[202] Farnand, B.A., Talbot, F.D.F., Matsuura, T., Sourirajan, S., *Ind. Eng. Chem. Research.* 1987, 26, 1080.
[203] Mehdizadeh, H., Dickson, J.M., *J. Membrane Sci.* 1989, 42, 119.
[204] Tremblay, A.Y., The Role of Structural Forces in Membrane Transport: Cellulose Membranes, Ph.D. thesis, University of Ottawa, 1989.
[205] Bouchard, C., Separation partielle en ultrafiltration: Etude experimentale et simulation, Ph.D. thesis, Ecole Polytechnique, Universite de Montreal, 1990.
[206] Lebrun, R., Etude de la porometrie des membranes de separation en acetate de cellulose a l'aide de la morphologie des solutions de coulage, Ph.D. thesis, Ecole Polytechnique, Universite de Montreal, 1990.
[207] Bhattacharyya, D., Jevtitch, M., Schrodt, J.T., Fairweather, G., *Chem. Eng. Comm.* 1986, 42, 111.
[208] Onsager, L. Samaras, N.N.T., *J. Chem. Phys.* 1934, 2, 528.
[209] Matsuura, T., Sourirajan, S., Preferential sorption-capillary flow mechanism and surface force-pore flow model, applicability to different membrane separation processes, in *Advances in Reverse Osmosis and Ultrafiltration*; T. Matsuura and S. Sourirajan, Eds.; National Research Council of Canada: Ottawa, 1989; p. 139.

[210]   Faxen, H., *Kolloid Z.* 1959, 167, 146.
[211]   Satterfield, C.N., Colton, C.K., Pitcher, W.R., Jr., *A.I.Ch.E.J.* 1973, 19, 628.
[212]   Lane, J.A., Riggle, J.W., *Chem. Eng. Progr. Symp. Ser.* 1959, 55, 127.
[213]   Deen, W.M., Bohrer, M.P., Epstein, N.B., *A.I.Ch.E.J.* 1981, 27, 952.
[214]   Tomlinson, R.H., Flood, E.A., *Can. J. Res.* 1948, B26, 38.
[215]   Wicke, E., Vollmer, W., *Chem. Eng. Sci.* 1952, I(6), 282.
[216]   Weber, S., *Kgl. Danske Videnskab. Selskab. Mat. Fys. Medd.* 1954, 28, 1.
[217]   Frisch, H.L., *J. Phys. Chem.* 1956, 60, 1177.
[218]   Gilliland, E.R., Baddour, R.F., Russell, J.L., *A.I.Ch.E.J.* 1958, 4, 90.
[219]   Kammermyer, K., *Chem. Eng. Progr. Symp. Ser.* 1959, 55, 115.
[220]   Kammermyer, K., Rutz, L.O., *Chem. Eng. Progr. Symp. Ser.* 1959, 55, 163.
[221]   Higashi, K., Ito, H., Oishi, J., *J. At. Energy Soc. Jpn.* 1963, 5, 846.
[222]   Barrer, R.M., *Proc. Br. Ceram. Soc.* 1965, 5, 21.
[223]   Bartholemew, R.F., Flood, E.A., *Can. J. Chem.* 1965, 43, 1968.
[224]   Weaver, J.A., Metzner, A.B., *A.I.Ch.E.J.* 1966, 12, 655.
[225]   Agrawal. J.P., Sourirajan, S., *J. Appl. Polym. Sci.* 1970, 14, 1303.
[226]   Gantzel, P.K., Merten, U., *Ind. Eng. Chem. Process Des. Dev.* 1970, 9, 336.
[227]   Nohmi, T., Manabe, S., Kamide, K., Kawai, T., *Kobunshi Ronbunshu.* 1977, 34, 737.
[228]   Kakuta, A., Ozaki, O., Ohno, M., *J. Polym. Sci. Polym. Chem. Ed.* 1978, 16, 3249.
[229]   Kakuta, A., Kuramoto, M., Ohno, M., Tanioka, A., Ishikawa, K., *J. Polym. Sci. Polym. Chem. Ed.* 1980, 18, 3229.
[230]   Rangarajan, R., Mazid, M.A., Matsuura, T., Sourirajan, S., *Ind. Eng. Chem. Process Des. Dev.* 1984, 23, 79.
[231]   Majid, M.A., Rangarajan, R., Matsuura, T., Sourirajan, S., *Ind. Eng. Chem. Prod. Res. Dev.* 1985, 24, 907.
[232]   Minhas, B.S., Matsuura, T., Sourirajan, S., Characterization of cellulose acetate membranes and predictability of their performance for the separation of hydrogen-methane gas mixtures, in *Reverse Osmosis and Ultrafiltration*; S. Sourirajan, T. Matsuura, Eds.; American Chemical Society: Washington, D.C., 1985; pp. 451–466.
[233]   Fouda, A.E., Matsuura, T., Lui, A., Talbot, F.D.F., Sourirajan, S., *Sep. Sci. Technol.* 1988, 23, 1839.
[234]   Tremblay, A.Y., Fouda, A.E., Lui, A., Matsuura, T., Sourirajan, S., *Can. J. Chem. Eng.* 1988, 66, 1027.
[235]   Chen, Y., Fouda, A.E., Matsuura, T., A study on dry cellulose acetate membrane for the separation of carbon dioxide/methane gas mixtures, in *Advances in Reverse Osmosis and Ultrafiltration*; T. Matsuura, S. Sourirajan, Eds.; National Research Council of Canada: Ottawa, 1989; pp. 259–274.
[236]   Chen, Y., Fouda, A.E., Matsuura, T., *A.I.Ch.E. Symp. Ser.* 272. 1989, 85, 18.
[237]   Moore, W.J., *Physical Chemistry*; Longmans Green: New York, 1958; p. 175.
[238]   Deng, S., Bao, S., Sourirajan, S., Matsuura, T., *J. Colloid Interface Sci.* 1990, 136, 283.
[239]   Okada, T., Matsuura, T., *J. Membrane Sci.* 1991, 59, 133.
[240]   Okada, T., Yoshikawa, M., Matsuura, T., *J. Membrane Sci.* 1991, 59, 151.
[241]   Okada, T., Matsuura, T., *J. Membrane Sci.* 1992, 70, 163.
[242]   Zhang, S.Q., Matsuura, T., *J. Membrane Sci.* 1992, 70, 249.
[243]   Japanese Chemical Society; *Chemistry Handbook*; Maruzen: Tokyo, 1987; p. 736.
[244]   Neel, J., Nguyen, Q.T., Clement, R., François, R., Separation of water-organic liquid mixtures by pervaporation: an insight into the mechanism of the process, in *Proc. Second Int. Conf. on Pervaporation Processes in the Chemical Industry*; R. Bakish, Ed.; Bakish Materials Corp.: Englewood, NJ, 1987; p. 35.
[245]   Tanigaki, M., Yoshikawa, M., Eguchi, W., Selective separation of alcohol from aqueous solution through polymer membrane, in *Proc. Second Int. Conf. on Pervaporation Processes in the Chemical Industry*; R. Bakish, Ed.; Bakish Materials Corp.: Englewood, NJ, 1987; p. 126.

454

[246] Brüschke, H.E.A., Tusel, G.F., Rautenbach, R., Pervaporation membranes: application in the chemical process industry, in *Reverse Osmosis and Ultrafiltration;* S. Sourirajan, T. Matsuura, Eds.; ACS Symp. Ser. 281, American Chemical Society: Washington, D.C., 1985; p. 467.

[247] Hauser, J., Reinhardt, G.A., Stumm, F., Heintz, A., *J. Membrane Sci.* 1989, 47, 261.

[248] Taketani, Y., Minematsu, H., Dehydration of alcohol-water mixtures through composite membranes by pervaporation, in *Reverse Osmosis and Ultrafiltration;* S. Sourirajan, T. Matsuura, Eds.; ACS Symp. Ser. 281; American Chemical Society: Washington, D.C., 1985; p. 479.

[249] Wesslein, M., Heintz, A., Reinhardt, G.A., Lichtenthaler, R.N., Pervaporation of binary and multicomponent mixtures using PVA-membranes: experiments and model calculations, in *Proc. Third Int. Conf. on Pervaporation Processes in the Chemical Industry*; R. Bakish, Ed.; Bakish Materials Corp.: Englewood, NJ, 1989; p. 172.

[250] Hoover, K.C., Hwang, S.T., *J. Membrane Sci.* 1982, 10, 253.

[251] Tyagi, R., Handa, P., Fouda, A., Matsuura, T., Transport studies in pervaporation, in *Proc. Sixth Int. Conf. on Pervaporation Processes in the Chemical Industry*; R. Bakish, Ed.; Bakish Materials Corp.: Englewood, NJ, 1992.

[252] Gill, W.N., Bansal, B., *A.I.Ch.E.J.* 1973, 19, 823, pp. 137–152.

[253] Dandavati, M.S., Doshi, M.R., Gill, W.N., *Chem. Eng. Sci.* 1975, 30, 877.

[254] Gill, W.N., Matsumoto, M.R., Gill, A.L., Lee Y.-T., *Desalination* 1988, 68 11.

[255] Ohya, H., Sourirajan, S., *A.I.Ch.E.J.* 1969, 15, 829.

[256] Ohya, H., Sourirajan, S., *Reverse Osmosis System Specification and Performance Data for Water Treatment*; Thayer School of Engineering, Dartmouth College: Hanover, NH, 1971.

[257] Ohya, H., Taniguchi, Y., *Desalination* 1975, 16, 359.

[258] Ohya, H., Nakajima, N., Takagi, K., Kagawa, S., Negishi, Y., *Desalination* 1977, 21, 257.

[259] Taniguchi, Y., *Desalination* 1978, 25, 71.

[260] Darwish, B.A.Q., Aly, G.S., Al-Rqobah, H.A., Abdel-Jawad, M., *Desalination* 1989, 75, 55.

[261] Abdel-Jawad, M., Darwish, M.A., *Desalination* 1989, 75, 97.

[262] Rautenbach, R., Dahm, W., *Desalination* 1987, 65, 259.

[263] Rautenbach, R, Albrecht, R., *Membrane Processes*; John Wiley & Sons: New York, 1989; p. 151.

[264] Pan, C.-Y., Habgood, H.W., *Can. J. Chem. Eng.* 1978, 56, 197.

[265] Pan, C.-Y., Habgood, H.W., *Can. J. Chem. Eng.* 1978, 56, 210.

[266] Sherwood, T.K., Wei, J.C., *A.I. Ch.E.J.* 1955, 1, 522.

[267] Merten, U., *Ind. Eng. Chem. Fundam.* 1963, 2, 229.

[268] Sherwood, T.K., Brian, P.L.T., Fisher, R.E., in M.I.T. Desalination Research Laboratory, Report 295-1, August 1963.

[269] Merten, U., Lonsdale, H.K., Riley, R.L., *Ind. Eng. Chem. Fundam.* 1964, 3, 210.

[270] Dresner, L., in Oak Ridge National Laboratory, Report 3621, May 1964.

[271] Sherwood, T.K., Brian, P.L.T., Fisher, R.E., Dresner, L., *Ind. Eng. Chem. Fundam.* 1965, 4, 113.

[272] Brian, P.L.T., *Ind. Eng. Chem. Fundam.* 1965, 4, 439.

[273] Sherwood, T.K., Brian, P.L.T., Sarofin, A.F., in M.I.T. Desalination Research Laboratory, Report 295-8, December 1965.

[274] Brian, P.L.T., Influence of concentration polarization on reverse osmosis system design, in Proc. First Int. Desalination Symp., Washington, D.C., Oct. 3–9, 1965, Paper SWD/79.

[275] Brian, P.L.T., Mass transport in reverse osmosis, in *Desalination by Reverse Osmosis*; U. Merten, Ed.; M.I.T. Press: Cambridge, 1966; pp. 161–202.

[276] Sherwood, T.K., Brian, P.L.T., Fisher, R.E., *Ind. Eng. Chem. Fundamentals.* 1967, 6, 2.

[277] Berman, A.S., *J. Appl. Phys.* 1953, 24, 1232.

[278] Cheryan, M., Membrane bioreactors for high-performance fermentations, in *Reverse Osmosis and Ultrafiltration*; S. Sourirajan and T. Matsuura, Eds.; ACS Symp. Ser. 281, American Chemical Society: Washington, D.C., 1985; pp. 231–245.

[279] Mulder, M.H.V., Smolders, C.A., Pervaporation in continuous alcohol fermentation, in *Proc. First Int. Conf. on Pervaporation Processes in the Chemical Industry*; R. Bakish, Ed.; Bakish Materials Corp.: Englewood, NJ, 1986; p. 187.

[280] Vasdevan, M., Matsuura, T., Chotani, G.K., Vieth, W.R., *Sep. Sci. Technol.* 1987, 22, 1651.

[281] Vasdevan, M., Matsuura, T., Chotani, G.K., Vieth, W.R., *Ann. N.Y. Acad. Sci.* 1988, 506, 345.

[282] Michaels, A.S., *Desalination* 1980, 35, 329.

[283] Jeong, Y.S., Vieth, W.R., Matsuura, T., *Ind. Eng. Chem. Res.* 1989, 28, 231.

[284] Jeong, Y.S., Vieth, W.R., Matsuura, T., *Ann. N.Y. Acad. Sci.* 1991, 589, 214.

[285] Jeong, Y.S., Vieth, W.R., Matsuura, T., *Biotechnol. Prog.* 1991, 7, 130.

[286] Matsuura, T., Sourirajan, S., *Ind. Eng. Chem. Proc. Des. Dev.* 1971, 10, 102.

[287] Kremen, S., Wilf, M., Lange, P., *Desalination* 1991, 82, 15.

[288] Channabasappa, K.C., Strobel, J.J., in *Proc. Fifth Int. Symp. on Fresh Water from the Sea, Alghero, May 16–20, 1976*; A. Delyannis and E. Delyannis, Eds.; Athens, 1976; p. 267.

[289] Redondo, J.A., Frank, K.F., *Desalination* 1991, 82, 31.

[290] Riley, R.L., Fox, R.L., Lyons, C.R., Milstead, C.E., Sorey, M.W., Tagami, M., *Desalination* 1976, 19, 113.

[291] Kurihara, M., Harumiya, N., Kanamaru, N., Tonomura, T., Nakasatomi, M., *Desalination* 1981, 38, 449.

[292] Allegrezza, A.E., Jr., Parekh, B.S., Parize, P.L., Swiniarski, E.J., White, J.L., *Desalination* 1987, 64, 285.

[293] Chu, H.C., Tran, C.N., Light, W.G., in Proc. 1987 Int. Congr. on Membranes and Membrane Processes (ICOM '87), Tokyo, June 8–12, 1987, pp. 358–359.

[294] Koo, J.-Y., Petersen, R.J., Cadotte, J.E., in Proc. 1987 Int. Congr. on Membranes and Membrane Processes (ICOM '87), Tokyo, June 8–12, 1987, p. 350.

[295] Cadotte, J.E., Schaffenberg, R.W., Petersen, R.J., in *Proc. Int. Membrane Conference, Sept. 24–26, 1986*; M. Malaiyandi, O. Kutowy, F. Talbot, Eds.; National Research Council: Ottawa, 1986; p. 203.

[296] Kurihara, M., Uemura, T., Nakagawa, Y., Tonomura, T., *Desalination* 1985, 54, 75.

[297] Kawada, I., Inoue, K., Kazuse, Y., Ito, H., Shintani, T., Kamiyama, Y., *Desalination* 1987, 64, 387.

[298] Ohya, H., *Separation Membranes*; Nihonkikaku Kyokai: Tokyo, 1989; pp. 105–113.

[299] Sourirajan, S. *Nature* 1964, 203, 1348.

[300] Nader, C., Farnand, B., Selection of membrane materials by liquid phase affinity chromatography for naphtha upgrading, in *Energy Research Laboratories Rep. No. ERP/ERL 83-69 (TR)*; CANMET, Energy, Mines and Resources Canada: Ottawa, 1983.

[301] Farnand, B.A., Sawatzky, H., Reverse osmosis fractionation of petroleum and synthetic crude distillates, in *Energy Research Laboratories Rep. No. ERP/ERL 84-54(TR)*; CANMET, Energy, Mines and Resources Canada: Ottawa, 1984.

[302] Alzetta, P., Farnand, B.A., Sawatzky, H., Fractionation of dilute methanol in pentane solution by reverse osmosis, in *Energy Research Laboratories Rep. No. ERP/ERL 84-41(CF)*; CANMET, Energy, Mines and Resources Canada: Ottawa, 1984.

[303] Farnand, B.A., Talbot, F.D.F., Matsuura, T., Sourirajan, S., Reverse osmosis separations of alkali metal halides in methanol solutions using cellulose acetate membranes, in *Synthetic Membranes*, Vol. II; A.F. Turbak, Ed.; American Chemical Society: Washington, D.C., 1981; pp. 339–359.

[304] Bitter, J.G.A., Haan, J.P., Rijkens, H.C., *A.I.Ch.E. Symp. Ser.* 272. 1989, 85, 98.

[305] Hazlett, J.D., Kutowy, O., Tweddle, T.A., *A.I.Ch.E. Symp. Ser.* 272. 1989, 85, 101.

[306] Ishii, K., Higashi, T., Yoshikane, M., Maeda, H., Takahashi, Y., Miyake, M., Ifuku, Y., in Proc. 1987 Int. Congr. on Membranes and Membrane Processes (ICOM '87), Tokyo, June 8–12, 1987.

[307] *Molsep System*; company's catalogue; Daicel: Tokyo.

[308] Ohya, H., *Separation Membranes*; Nihon Kikaku Kyokai: Tokyo, 1989; pp. 115–118.

[309] Sourirajan, S., *Nature* 1963, 199, 590.

456

[310]   Agrawal, J.P., Sourirajan, S., *J. Appl. Polym. Sci.* 1969, 13, 1065.

[311]   Agrawal, J.P., Sourirajan, S., *J. Appl. Polym. Sci.* 1970, 14, 1303.

[312]   U.S. Patent 4,134,742.

[313]   Laverty, B.W., O'Hair J.G., Applications of membrane technology in the gas industry, in *Proc. Fourth BOC Priestley Conf.*; Royal Society of Chemistry: London, 1986; pp. 291–310.

[314]   Backhouse, I.W., in *Proc. Fourth BOC Priestley Conf.* Royal Society of Chemistry: London, 1986; p. 265.

[315]   Spillman, R.W., An overview of gas separation membranes, in *1989 Seventh Annu. Membrane Technology/Planning Conf. Proc.*; Business Communications Co.: Stamford, CT, 1989; pp. 132–143.

[316]   Anon., *Conserving the Global Environment*; Research Institute of Innovative Technology for the Earth: Kyoto, 1991.

[317]   Sourirajan, S., *Lectures on Membrane Separations*; Industrial Membrane Research Institute: University of Ottawa, 1991; pp. 369–443.

[318]   Minhas, B.S., Wang, S.P., Sherwin, M.B., Effect of process variables on cellulose acetate membrane performance for separation of carbon dioxide-methane mixtures, in *Proc. Int. Membrane Conference, Sept. 24-26, 1986*; M. Malaiyandi, O. Kutowy, F. Talbot, Eds.; National Research Council of Canada: Ottawa, 1986; pp. 213–228.

[319]   *DE-3 Membrane Oxygen Enricher Catalogue*; Oxygen Enrichment Co., 1976.

[320]   McReynolds, K.B., Generon* air separation systems-membranes in gas separation and enrichment, in *Proc. Fourth BOC Priestley Conf.*; Royal Society of Chemistry: London, 1986; pp. 342–350.

[321]   Beaver, E.R., PERMEA-gas separation membranes developed into a commercial reality, in *1989 Seventh Annu. Membrane Technology/Planning Conf. Proc.*; Business Communications Co.: Stamford, CT, 1989; pp. 144–154.

[322]   Gollan, A., Point-of-use air separation systems, in *1989 Seventh Annu. Membrane Technology/Planning Conf. Proc.*; Business Communications Co.: Stamford, CT, 1989; pp. 172–185.

[323]   Romano, L., Gottzmann, C.F., Thompson, D.R., Prasad, R., Nitrogen production using membranes, in *1989 Seventh Annu. Membrane Technology/Planning Conf. Proc.*; Business Communications Co.: Stamford, CT, 1989; pp. 163–171.

[324]   Anon., *Helium Recovery From Natural Gas*, Pub. No. 1/172; Alberta Energy Information Center: Edmonton.

[325]   Peinemann, K.-V., Ohlrogge, K., Knauth, H.-D., The Recovery of helium from diving gas with membranes, in *Proc. Fourth BOC Priestley Conf.*; Royal Society of Chemistry: London, 1986; pp. 329–341.

[326]   Murphy, M.K., Beaver, E.R., Rice, A.W., *A.I.Ch.E. Symp. Ser.* 272 1989, 85, pp. 34–40.

[327]   Rice, A.W., Murphy, M.K., U.S. Patent 4,783,201.

[328]   Behling, R.-D., Ohlrogge, K., Peinemann, K.-V., *A.I.Ch.E. Symp. Ser.* 272 1989, 85, 68–73.

[329]   Wijmans, J.G., Helm, V.D., *A.I.Ch.E. Symp. Ser.* 272 1989, 85, pp. 74–79.

[330]   Shindo, Y., Itoh, N., Haraya, K., *A.I.Ch.E. Symp. Ser.* 272 1989, 85, pp. 80–81.

[331]   Doshi, K.J., Werner, R.G., Mitariten, M. J., *A.I.Ch.E. Symp. Ser.* 272 1989, 85, pp. 62–67.

[332]   Taketani, Y., Kunst, B., Matsuura, T., Sourirajan, S., Preliminary studies on reverse osmosis separation of ethyl alcohol present in high concentrations in aqueous solutions, in *Proc. Third Bioenergy R & D Seminar*; National Research Council of Canada: Ottawa, 1981; pp. 145–149.

[333]   Matsuura, T., Taketani, Y., Sourirajan, S., Reverse osmosis of ethyl alcohol in aqueous solutions using different polymeric membranes, in *Proc. Fourth Bioenergy R & D Seminar*; National Research Council of Cananda: Ottawa, 1982; pp. 529–533.

[334]   Mehta, G.D., *J. Membrane Sci.* 1982, 12, 1.

[335]   Lee, E.K.L., Babcock. W.C., Bresnahan, P.A., Countercurrent reverse osmosis for ethanol-water separation, in NTIS, Report No. DOE/ID/12320-T1, March 1983.

[336] Tidball, R.A., Tusel, G.F., in *Proc. First Int. Conf. on Pervaporation Processes in the Chemical Industry*; R. Bakish, Ed.; Bakish Materials Corp.: Englewood, NJ, 1986; pp. 142–153.

[337] Tsuyumoto, M., Karakane, H., Maeda, Y., Tsugaya, H., *Desalination* 1991, 80, 139.

[338] Karakane, H., Tsuyumoto, M., Maeda, Y., Honda, Z., *J. Appl. Polym. Sci.* 1991, 42, 3229.

[339] Chen, M.S.K., Markiewicz, G.S., Venugopal, K.G., *A.I.Ch.E. Symp. Ser.* 272 1989, 85, 82–88.

[340] Shah, V.M., Bartels, C.R., Engineering considerations in pervaporation application, in *Proc. Fifth Int. Conf. on Pervaporation Processes in the Chemical Industry*; R. Bakish, Ed.; Bakish Materials Corp.: Englewood, NJ, 1991; pp. 331–337.

[341] Kaschemekat, J., Wijmans, J.G., Baker, R.W., Removal of organic solvent contaminants from industrial effluent streams by pervaporation, in *Proc. Fourth Int. Conf. on Pervaporation Processes in the Chemical Industry*; R. Bakish, Ed.; Bakish Materials Corp.: Englewood, NJ, 1989; pp.321–331.

[342] Zhang, S.Q., Matsuura, T., *J. Food Proc. Eng.*, 1992, in press.

[343] Sodeck, G., Effenberger, H., Steiner, E., Salzbrunn, W., in *Proc. Second Int. Conf. on Pervaporation in the Chemical Industry*; R. Bakish, Ed.; Bakish Materials Corp.: Englewood, NJ, 1987; p. 157.

[344] Tusel, G., in *Proc. Second Int. Conf. on Pervaporation in the Chemical Industry*; R. Bakish, Ed.; Bakish Materials Corp.: Englewood, NJ, 1987; p. 277.

Appendix

```
      REAL X(100),P3(100),CA3BAR(100),U3(100)          MEM00010
      AL=0.6                                           MEM00020
      PATM=101.3                                       MEM00030
      U1=0.35                                          MEM00040
      CA1=0.1                                          MEM00050
      P2=2758                                          MEM00060
      NH=5.E7                                          MEM00070
      AK=1.24E-6                                        MEM00080
      B=4637.41                                        MEM00090
      A=2.E-8                                          MEM00100
      DAM=7.5E-9                                        MEM00110
      C=55.3                                           MEM00120
      RI=0.0125                                        MEM00130
      RO=0.0486                                        MEM00140
      DI=0.0001                                         MEM00150
      DE=0.000105                                      MEM00160
      VISC=9.6E-7                                       MEM00170
      PI=3.14159                                       MEM00180
                                                       MEM00190
C                                                      MEM00200
      DX=AL/30                                         MEM00210
      R=RI                                             MEM00220
      DR=(RO-RI)/10                                     MEM00230
      SV=PI*NH*DE                                       MEM00240
      EP=1-PI*NH*DE**2/4                                MEM00250
      EPP=(1-EP)**2/EP**3                              MEM00260
      FA=-PI*NH*DI**2/4/AL                             MEM00270
      CA3T=0.0                                         MEM00280
      U1T=0.0                                          MEM00290
                                                       MEM00300
    3 IF(R.GT.RO) GOTO 200                             MEM00310
                                                       MEM00320
      WRITE(2,*)'                                      MEM00330
    5 WRITE(2,6)'R=',R,'U1=',U1,'CA1=',CA   .'P1=',P2  MEM00340
    6 FORMAT(4X,A5,F6.4,2X,A5,F4.2,2X,A5,=6.4,2X,A5,F6.1)  MEM00350
      WRITE(2,*)'----------------------------------------'  MEM00360
      WRITE(2,7)'X','P3','CA3','U3'                    MEM00370
    7 FORMAT(28X,A5,5X,A5,8X,A5,6X,A5)                 MEM00380
      PGUESS=PATM                                      MEM00390
   10 I=1                                              MEM00400
      X(I)=DX                                          MEM00410
      P3(I)=PGUESS                                     MEM00420
   20 IF (X(I).GT.AL) GO TO 100                        MEM00430
      CALL TRANSP(I,AK,DAM,A,P2,P3,100,CA1,B,C,AJ,BJ,VW,CA3,CA2)  MEM00440
      CALL NEWVAL(I,CA3,DE,VW,DI,AJ,VISC,DX,U3,CA3BAR,P3,100)  MEM00450
      PD=P3(I)-PATM                                    MEM00460
      IF (PD.LT.-.01) THEN                             MEM00470
         IF (X(I).LT.AL/2) THEN                        MEM00480
            PGUESS=PGUESS+1.                           MEM00490
         ELSE                                          MEM00500
            PGUESS=PGUESS+.1                           MEM00510
         ENDIF                                         MEM00520
         GO TO 10                                      MEM00530
      ELSE                                             MEM00540
         I=I+1                                         MEM00550
         X(I)=X(I-1)+DX
```

459

```
          GO TO 20                                          MEM00560
          END IF                                            MEM00570
                                                            MEM00580
100       CA3T=CA3T+CA3BAR(I-1)                             MEM00590
          U1T=U1T+U1                                        MEM00600
                                                            MEM00610
          DO 120 J=3,I-1,3                                  MEM00620
             WRITE(2,110)X(J),P3(J),CA3BAR(J),U3(J)         MEM00630
110          FORMAT(30X,F4.2,5X,F6.2,5X,F8.6,5X,F5.3)       MEM00640
120       CONTINUE                                          MEM00650
                                                            MEM00660
          CALL RADIAL(I,FA,R,U3,100,CA3BAR,VISC,SV,EPP,U1,CA1,P2,DR)  MEM00670
          R=R+DR                                            MEM00680
          GO TO 3                                           MEM00690
                                                            MEM00700
200       CA3AVG=CA3T/11                                    MEM00710
          CSP=PI**2*(RO**2-RI**2)*DI**2*NH/4                MEM00720
          FRP=CSP*U3(I-1)*3600                              MEM00730
          CSC=PI*(RO**2-RI**2)-CSP                          MEM00740
          FRC=U1T*CSC*3600/11                               MEM00750
                                                            MEM00760
          WRITE(2,*)' '                                     MEM00770
          WRITE(2,201)'PERMEAT FLOW RATE IS',FRP,'M3/HR'    MEM00780
          WRITE(2,202)'AVERAGE CONC. OF PERMEAT IS',CA3AVG,'KMOL/M3'  MEM00790
          WRITE(2,203)'CONCENTRATE FLOW RATE IS',FRC,'M3/HR'  MEM00800
          WRITE(2,204)'AVERAGE CONC. OF CONCENTRATE IS',CA1,'KMOL/M3'  MEM00810
201       FORMAT(A40,F8.5,A6)                               MEM00820
202       FORMAT(A40,F8.5,A8)                               MEM00830
203       FORMAT(A40,F8.5,A6)                               MEM00840
204       FORMAT(A40,F8.5,A8)                               MEM00850
                                                            MEM00860
          STOP                                              MEM00870
          END                                               MEM00880
C                                                           MEM00890
C                                                           MEM00900
C         *************************************************************  MEM00900
C                                                           MEM00910
          SUBROUTINE TRANSP(I,AK,DAM,A,P2,P3,N,CA1,B,C,AJ,BJ,VW,CA3,CA2)  MEM00920
          REAL P3(N)                                        MEM00930
C                                                           MEM00940
          CALL CA2CA3(I,B,P2,P3,N,CA1,DAM,C,A,AK,S)         MEM00950
          AJ=DAM*S                                          MEM00960
          BJ=A*(P2-P3(I))-A*B*S                             MEM00970
          VW=BJ/C                                           MEM00980
          CA3=AJ/VW                                         MEM00990
          CA2=CA3+S                                         MEM01000
          RETURN                                            MEM01010
          END                                               MEM01020
C                                                           MEM01030
C                                                           MEM01040
C         *************************************************************  MEM01040
C                                                           MEM01050
          SUBROUTINE CA2CA3(I,B,P2,P3,N,CA1,DAM,C,A,AK,S)   MEM01060
          REAL P3(N)                                        MEM01070
C                                                           MEM01080
          COMMON/BLOCK/E1,E2,E3                             MEM01090
          E1=B/(P2-P3(I))                                   MEM01100
```

```
        E2=E1*CA1+DAM*C/A/(P2-P3(I))                          MEM01110
        E3=A*(P2-P3(I))/AK/C                                  MEM01120
        CALL FNZERO(S)                                        MEM01130
        RETURN                                                MEM01140
        END                                                   MEM01150
C                                                             MEM01160
C       ************************************************************* MEM01170
C                                                             MEM01180
        SUBROUTINE FNZERO(S)                                  MEM01190
C                                                             MEM01200
        DIMENSION X(1),XGUESS(1)                              MEM01210
        COMMON/BLOCK/E1,E2,E3                                 MEM01220
        EXTERNAL F,ZREAL                                      MEM01230
        EPS=1.E-10                                            MEM01240
        ERRABS=1.E-10                                         MEM01250
        ERRREL=1.E-6                                          MEM01260
        ETA=1.E-10                                            MEM01270
        ITMAX=200                                             MEM01280
        XGUESS(1)=0.1                                         MEM01290
        CALL ZREAL(F,ERRABS,ERRREL,EPS,ETA,1,ITMAX,XGUESS,X,INFO) MEM01300
        S=X(1)                                                MEM01310
        RETURN                                                MEM01320
        END                                                   MEM01330
C                                                             MEM01340
        REAL FUNCTION F(X)                                    MEM01350
        COMMON/BLOCK/E1,E2,E3                                 MEM01360
        F=X*(1-E1*X)/(.1-E2*X)-EXP(E3*(1.-E1*X))             MEM01370
        RETURN                                                MEM01380
        END                                                   MEM01390
C                                                             MEM01400
C       *************************************************************MEM01410
C                                                             MEM01420
        SUBROUTINE NEWVAL(I,CA3,DE,VW,DI,AJ,VISC,DX,U3,CA3BAR,P3,N) MEM01430
        REAL U3(N),CA3BAR(N),P3(N)                            MEM01440
C                                                             MEM01450
        DU3=4*DE*VW/DI**2                                     MEM01460
        IF (I.EQ.1) THEN                                      MEM01470
           U3(I)=DU3*DX                                       MEM01480
           CA3BAR(I)=CA3                                      MEM01490
           DCA3=0                                             MEM01500
        ELSE                                                  MEM01510
           U3(I)=DU3*DX+U3(I-1)                               MEM01520
           DCA3=4*DE*(AJ-VW*CA3BAR(I-1))/U3(I-1)/DI**2        MEM01530
           CA3BAR(I)=DCA3*DX+CA3BAR(I-1)                      MEM01540
        ENDIF                                                 MEM01550
                                                              MEM01560
        DP3=-32*VISC*U3(I)/DI**2                              MEM01570
        P3(I+1)=DP3*DX+P3(I)                                  MEM01580
        RETURN                                                MEM01590
        END                                                   MEM01600
C                                                             MEM01610
C       *************************************************************MEM01620
C                                                             MEM01630
        SUBROUTINE RADIAL(I,FA,R,U3,N,CA3BAR,VISC,SV,EPP,U1,CA1,P2,DR) MEM01640
        REAL U3(N), CA3BAR(N)                                 MEM01650
```

```
C
      DRU=FA*R*U3(I-1)                                  MEMO1660
      DRUCA=DRU*CA3BAR(I-1)                             MEMO1670
      DP1=-30*VISC*EPP*U1*SV**2                         MEMO1680
      DU1=(DRU-U1)/R                                    MEMO1690
      DCA1=(DRUCA-U1*CA1-R*DU1*CA1)/R/U1                MEMO1700
      U1=U1+DU1*DR                                      MEMO1710
      CA1=CA1+DCA1*DR                                   MEMO1720
      P2=P2+DP1*DR                                      MEMO1730
      RETURN                                            MEMO1740
      END                                              MEMO1750
                                                        MEMO1760
                                                        MEMO1770
                                                        MEMO1780
```

Index

reverse osmosis, see Reverse
   osmosis
ultrafiltration, see Ultrafiltration
SEPEREX System, 408
Sodium chromate, 388
Sodium nitrate, 388
Sodium silicate, 388
Solution-diffusion model, see
   Membrane transport, solu-
   tion-diffusion model
Solvent exchange method, 49
Spinodial curves, 69, 72, 83
Spiral-wound module, 305–309,
   328–334, 337
Synthetic polymeric membrane,
   169

T

Ternary system, 69–70
Tetrachloroethylene, 388
Tetrahydrofuran, 389

Thermodynamics of the polymer
   solution, 65
Toray UTC-20 Membrane, 390
Toray UTC-40 Membrane, 390
Toray-type spiral-wound module,
   317
Trichlorobenzene, 388

U

Ultrafiltration, 3, 58–61
   clear fruit juice, production of by,
   396–399
   tap water, cleaning by, 399–400
Ultrapure water, production of, 3,
   390–392
UOP TFCL-LP Module, 387
Urea, 388, 389

Z

Zinc chloride, 388